双语版 Java 程序设计
Learn Java through English and Chinese

何月顺　主编

李　祥　张　军　副主编

电子工业出版社
Publishing House of Electronics Industry
北京·BEIJING

内 容 简 介

Java 是一种良好的跨平台、可移植性好、安全性高、纯面向对象的程序设计语言，是目前软件开发的主流编程语言之一。本书由浅入深、循序渐进地介绍了 Java 语言的发展、编程环境、开发工具、基本语法、面向对象编程、异常处理、线程、输入输出流、GUI 编程、网络编程、数据库编程等内容，并结合大量的实例进行讲解。

本书是国内首次出版的中英文对照混排式双语版 Java 程序设计教材。本书内容注重理论与实践结合，参考了 Java 在线官方文档及国内外优秀的 Java 程序设计教材的知识体系。针对中国学生英文水平及实际教学状况，做了针对性的编写，并对重要的、难以理解的内容进行了中文解释，方便了读者对英文的理解。本书配有电子课件、案例分析、实验指导等教学资源，可登录电子工业出版社华信教育资源网（www.hxedu.com.cn），免费注册、下载。

通过本书的学习，可使学生了解 Java 语言的发展，理解 Java 面向对象编程的基本思想，掌握 Java 语言的基本语法、面向对象程序设计的实现思想、多线程编程、网络程序开发及数据库访问等方面的基本技术。

本书贯彻理论与实践相结合的原则，深入浅出，配以大量实例分析，同时作为中英文对照教材，非常适合高校进行"Java 程序设计"课程的双语教学。本书既可作为高等学校 Java 程序设计或专业英语课程的教材，也可供从事 Java 程序开发的从业人员学习、参考。

Brief Introduction

Java is an excellent cross-platform, portable, safe, pure object-oriented programming language. It is one of mainstream programming languages.

The book introduces Java language progressively. The main chapters of the book include development of Java, programming environment, developing tool, basic syntax, object-oriented programming, exception handling, thread, input and output stream, GUI programming, network programming and database programming. In addition, large numbers of examples are injected in each chapter in order that reader can easily understand the correspondent contents.

The content arrangement of the book highlights combination of theory and practice. Author referred to Java official document and many excellent Java programming textbooks at home and abroad.

Considering Chinese student's English level and actual teaching situation, author provides necessary comments for difficult concepts, so that reader can understand them easily. Related teaching sources of the book such as PPT, case analysis and experimental guidance could be provided to reader. So reader log on www.hxedu.com.cn and obtain these free resources.

By learning the book, students will understand basic concepts of object-oriented programming, master basic syntax of Java language and basic skill of object-oriented programming.

The outstanding features of the book such as combination of theory and practice and large number of examples makes itself suitable for bilingual textbook of "Java programming" course at college. The book as textbook is available for undergraduate students and beginning graduate students. Java programmers could also use the book for reference.

未经许可，不得以任何方式复制或抄袭本书之部分或全部内容。
版权所有，侵权必究。

图书在版编目（CIP）数据

双语版 Java 程序设计 ＝Learn Java through English and Chinese：汉英对照 / 何月顺主编．
北京：电子工业出版社，2012.6
ISBN 978-7-121-16982-3

I．①双… II．①何… III．①JAVA 语言－程序设计－高等学校－教材－汉、英 IV．①TP312

中国版本图书馆 CIP 数据核字（2012）第 092275 号

策划编辑：史鹏举
责任编辑：史鹏举
印　　刷：涿州市京南印刷厂
装　　订：涿州市京南印刷厂
出版发行：电子工业出版社
　　　　　北京市海淀区万寿路 173 信箱　邮编　100036
开　　本：787×1092　1/16　印张：22.5　字数：749 千字
版　　次：2012 年 6 月第 1 版
印　　次：2017 年 12 月第 5 次印刷
定　　价：45.00 元

凡所购买电子工业出版社图书有缺损问题，请向购买书店调换。若书店售缺，请与本社发行部联系，联系及邮购电话：(010)88254888。
质量投诉请发邮件至 zlts@phei.com.cn，盗版侵权举报请发邮件至 dbqq@phei.com.cn。
服务热线：(010)88258888。

前　言

　　Java 语言以它独特的魅力赢得了世界上大部分程序员的认可，它良好的跨平台性、可移植性、安全性等优点使之风靡全球。Sun 公司（2009 年 4 月并入甲骨文公司）最初开发 Java 语言是为了解决智能家用电器的控制和通信问题。随着 Internet 的发展，Sun 公司逐步将 Java 改造成适合计算机网络应用的程序设计语言。目前 Java 语言已经成为网络程序设计的主流编程语言之一，在全球云计算和移动互联网的产业环境下，Java 更具备了显著优势和广阔前景。

　　"Java 程序设计"是高等院校计算机及相关专业教学计划中的一门重要专业课程，主要内容包括 Java 语言基本语法、面向对象编程、字符串处理、异常处理、线程、输入输出流、GUI 编程、网络编程、数据库编程等内容。

　　本书采用中英文对照方式对内容进行编排，以英文为主，对重要的、难理解的知识用中文进行了解释，兼顾了英语基础较差的读者。全书图文并茂，通俗易懂，在介绍理论知识的同时穿插了丰富的实例进行讲解，不仅介绍了 Java 的基础语法，降低了没有编程基础读者学习的难度，又全面介绍了 Java 面向对象程序设计、多线程、异常处理机制、输入输出流、网络编程等重点内容，使读者学完后能进行初级的 Java 程序设计。

　　本书将阐释：
- Java 的发展历史及 Java 特性、Java 运行环境及编程工具
- Java 语法基础
- Java 面向对象编程，包括类、继承、接口、多态等特性
- 多线程编程
- Java 异常处理机制
- 输入输出流
- GUI 编程
- 网络程序设计
- 数据库访问

　　本教材基于作者多年来教学实践与改革的经验，以及对开展双语教学的研究，并已经在作者所在学校多次使用之后，特别是收集了学生的反馈意见，教师的教学建议并结合目前国内外 Java 程序设计优秀教材的优点且考虑到学生对双语课程学习特点而编写。主要特色包括以下几个方面：

　　（1）中英文结合，突出双语特色。本书以英文为主，中文解释为辅，一方面注重知识点的编排，另一方面也注重英语的应用技巧，可以锻炼读者的英文应用能力。

　　（2）增强理论与实践相结合，注重引导式讲解。本书对理论知识的讲解采用循序渐进的方式融入大量实例中，使得对理论知识的理解更加容易，并以国外教材常见的 step by step 的方式完成实例的分析讲解，读者在读完相应的章节后就能进行相应的程序设计。

　　（3）突出组织逻辑，增加趣味性。目前国内教材和选用的国外经典教材，用于本科教学后，学生普遍反映概念原理介绍过多，内容组织的逻辑思路不是很明显，以及介绍得比较技术性，不是很生动等。针对学生的反馈，本教材进行了改进。

（4）重点突出，本书着重论述了 Java 语言在多线程编程、异常处理机制、网络编程、数据库访问技术等常用技术，同时对于 Java 语言的特色技术 Applet 也进行了介绍，使读者能更全面了解 Java 的应用。

本书既可作为高等学校 Java 程序设计或专业英语课程的教材，也可供从事 Java 程序开发的从业人员学习、参考。

本书由何月顺主编，李祥、张军副主编。高永平、汪雪元、章伟、吴光明、王志波参加了本书的编写工作，并得到 Ashok 的大力支持，在此表示诚挚的感谢。本书是编写组成员对以上内容大量理论知识与实践经验的积累结果，因时间仓促，可能存在不妥之处，欢迎指正（Email：shipj@phei.com.cn）。

编 者

Foreword

Java is a very appealing language, specific characteristics of which such as cross-platform, portability, security have attracted attention of most programmers in the world. Originally, Java was developed by Sun (Incorporated into the Oracle Corporation in April 2009) to solve problem of control and communication among smart home-appliances. With the development of the Internet, Java was gradually reconstructed into new programming language which can be well suitable for computer network application. Currently Java language has become one of the mainstream programming languages in network programming. Combined with cloud computing technology and global mobile internet, Java's advantages and broad prospects will be more obvious.

"Java Programming" is a key professional course in teaching plan of software engineering faculty and relative faculty. The main teaching content include the basic syntax of the Java language, object-oriented programming, string manipulation, exception handling, thread, input and output streams, GUI programming, network programming and database programming.

The book is bilingual textbook in which English is first language. For some poor-English reader, some important and difficult contents are translated into Chinese. In the book structure, theoretical concepts are associated with examples and images in order that reader can easily understand these concepts. In addition, for different levels of readers, the book not only narrates Java's basic syntax, but also introduces some advanced features of object oriented programming such as multithread, exception handling, input and output streams and network programming. So reader can program after finishing this book.

The main content of the book:
- Java's history, runtime environment and programming tools
- Java syntax
- Object-oriented programming including classes, inheritance, interfaces, polymorphism and other features
- Multithread programming
- Exception handling mechanism
- Input output stream
- GUI programming
- Network programming
- Database

The author has extensive bilingual teaching experience. The draft of the book has been used as textbook at author's school. Combined with the feedback from students and teachers and other excellent textbook, the first version of the book will be published.

The main feature of the book:

(1) The combination of English and Chinese. English is the first language in the book while some

contents are translated into Chinese. The arrangement will not only contribute to understanding knowledge, but also can improve English language proficiency.

(2) The combination of theory and practice. Problem-based learning model is adopted in the book. A large number of examples are integrated with theoretical knowledge so that these knowledge can be more easily understood. The examples in the book are listed in accordance with the step-by-step procedure. So reader can complete appropriate program after finish the corresponding chapter.

(3) The combination of logicality and delight. Traditional textbook, whether home or abroad, focus much attention on concept and notion. It is difficult for the student in the undergraduate level to understand the architecture of knowledge. The abstract contents seriously impact the interest of the students on learning. So these defects are improved in our book.

(4) Emphasizing key contents. The book focuses mainly on the key features of Java language such as multithread programming, exception handling mechanism, network programming and database access technology. In addition, Applet is introduced so that reader can fully understand the Java language.

The book as textbook is available for undergraduate students and beginning graduate students. Java programmers also can refer to the book.

Prof He yue shun as editor in chief guide the whole editing process. Li Xiang and Zhang Jun as deputy editor are in charge of different chapters. At last, several people such as Gao Yong Ping, Wang Xueyuan, Zhang Wei, Ashok, Wu Guangming and Wang Zhibo need to be acknowledged. Those people make this book possible.

Completion of the book is based on theoretical knowledge and practical experience of editor group. Due to time constraint, there are some deficiencies in the book. If you find problem, please contact us!

Contents
目　录

Chapter 1　Genesis of Java ·· 1
Java 概述

1.1　Introduction ·· 1
　　　Java 简介
1.2　Java Development Today ··· 1
　　　Java 发展历史
1.3　Evolution of 'C' Based Programming Languages ······································· 1
　　　C 系列语言发展
1.4　Main Features of Java Programming Language ··· 2
　　　Java 语言的主要特性
　　　1.4.1　Portability ·· 2
　　　　　　轻量级
　　　1.4.2　Simple ·· 3
　　　　　　简单
　　　1.4.3　Robust ··· 3
　　　　　　健壮
　　　1.4.4　Multithread ·· 4
　　　　　　多线程
　　　1.4.5　Architecture-Neutral ·· 4
　　　　　　平台无关
　　　1.4.6　Interpreted and High Performance ·· 5
　　　　　　解释性和高效
　　　1.4.7　Distributed ·· 5
　　　　　　分布式
　　　1.4.8　Dynamic ··· 5
　　　　　　动态
　　　1.4.9　Security ··· 5
　　　　　　安全
1.5　Java Applet ·· 6
　　　Java Applet 小应用程序
1.6　Exercise for you ·· 7
　　　课后习题

Chapter 2　Java Overview ··· 8
Java 总览

2.1　Concepts of OOP ·· 8
　　　面向对象程序设计
　　　2.1.1　Class ··· 9
　　　　　　类
　　　2.1.2　Object ··· 9
　　　　　　对象
　　　2.1.3　Encapsulation ·· 10
　　　　　　封装
　　　2.1.4　Inheritance ··· 10
　　　　　　继承
　　　2.1.5　Polymorphism ··· 11
　　　　　　多态

	2.2	More Details on Object-Oriented Programming ·· 11
		面向对象程序设计具体实例
		2.2.1 Encapsulation of Car ·· 12
		Car 类封装
		2.2.2 Inheritance of Car ·· 12
		Car 类继承
		2.2.3 Polymorphism of Car ·· 13
		Car 类多态
		2.2.4 Conclusion on Object-Oriented Programming ································· 13
		面向对象程序设计小结
	2.3	Write the First Java Program ·· 14
		编写第一个 Java 程序
	2.4	How to Run the First Java Program ··· 15
		运行第一个 Java 程序
	2.5	Lexical Elements ··· 16
		语法规则
	2.6	White Space ·· 17
		空白符
	2.7	Comments ··· 18
		注解
		2.7.1 Single Line ·· 18
		单行注解
		2.7.2 Multi-line ·· 18
		多行注解
		2.7.3 Javadoc ··· 18
		Javadoc 注解
	2.8	Keywords ··· 19
		关键字
	2.9	Identifiers ·· 19
		标志符
	2.10	Java Class Library ·· 20
		Java 类库
	2.11	Sample Program Practice ··· 20
		程序实例
	2.12	Exercise for you ·· 21
		课后习题

Chapter 3　Data Types ·· 22
数据类型

	3.1	Data Types Overview ··· 22
		数据类型概述
	3.2	Primitive Types ·· 23
		基本数据类型
	3.3	Casting ·· 23
		类型转换
		3.3.1 Widening ·· 24
		类型扩展
		3.3.2 Narrowing ·· 24
		类型收缩
	3.4	Reference Types ··· 24
		引用类型
	3.5	Summary ·· 25
		基本类型汇总

3.6 Complex Data Types ·· 26
　　复合数据类型
　　3.6.1 Reference Data Types ·· 26
　　　　引用数据类型
　　3.6.2 Class Types ·· 26
　　　　类类型
　　3.6.3 Interface Types ·· 26
　　　　接口类型
3.7 Composite Data Types ··· 27
　　构造复合数据类型
　　3.7.1 Initializing Composite Data Types ··· 27
　　　　复合数据类型数据初始化
　　3.7.2 Predefined Composite Data Types ··· 28
　　　　预定义复合数据类型
3.8 Casting Variables to a Different Type ··· 28
　　不同数据类型转换
　　3.8.1 Automatic Casting ·· 28
　　　　自动转换
　　3.8.2 Explicit Casting ··· 28
　　　　显式转换
3.9 Java's Floating Point Types ··· 29
　　浮点数据
　　3.9.1 Primitive Floating Point Types ·· 29
　　　　基本浮点类型
　　3.9.2 Integer Operators ·· 29
　　　　整型运算符
　　3.9.3 Input and Output of Floating Point Values ······························· 30
　　　　输入输出浮点数据
　　3.9.4 Casting of Floating Point to and from Integer Values, and Floating Point Literals ············· 30
　　　　整型数据和浮点型字符转换为浮点数据
　　3.9.5 Floating Point Operations in the Standard Packages ·············· 30
　　　　系统包中的浮点运算
　　3.9.6 The Float Class ··· 31
　　　　Float 类
3.10 Variable ··· 32
　　变量
　　3.10.1 Declaring a Variable ·· 32
　　　　变量声明
　　3.10.2 Difference between Zero and '0'-Unicode Characters ··········· 33
　　　　区分数字 0 和字符 0
　　3.10.3 Initialization of the Variable ·· 33
　　　　变量初始化
　　3.10.4 Error Checking by the Compiler ·· 33
　　　　编译错误
　　3.10.5 Using the Cast Operator ··· 33
　　　　类型转换符的使用
　　3.10.6 Why Declare the Variables as Type Int? ································· 33
　　　　变量声明为整型
　　3.10.7 Shortcut Declaring Variables of the Same Type ···················· 33
　　　　同类型变量的声明
　　3.10.8 Assigning Values to Variables ·· 34
　　　　变量赋值

　　　　3.10.9　A Shortcut, Declare and Assign at the Same Time ·················· 34
　　　　　　　变量同时声明与赋值
　　3.11　Record ·· 34
　　　　　记录
　　3.12　Sample Program Practice ·· 35
　　　　　程序实例
　　3.13　Exercise for you ··· 36
　　　　　课后习题

Chapter 4　Operators ·· 37
　　　　　运算符

　　4.1　Arithmetic Operators ··· 37
　　　　　算术运算符
　　　　4.1.1　The Modulus Operators ·· 37
　　　　　　　取模运算符
　　　　4.1.2　Arithmetic Assignment Operators ··· 38
　　　　　　　算术赋值运算符
　　　　4.1.3　Increment and Decrement ·· 38
　　　　　　　自增与自减运算符
　　4.2　Relational Operators ··· 39
　　　　　关系运算符
　　4.3　Boolean Logical Operators ··· 39
　　　　　逻辑运算符
　　4.4　Bitwise and Shift Operators ··· 41
　　　　　位运算符与移位运算符
　　　　4.4.1　Bitwise Complement (~) ·· 41
　　　　　　　按位取反运算符
　　　　4.4.2　Bitwise AND (&) ·· 41
　　　　　　　按位与运算符
　　　　4.4.3　Bitwise OR (|) ·· 41
　　　　　　　按位或运算符
　　　　4.4.4　Bitwise XOR (^) ··· 42
　　　　　　　按位异或运算符
　　　　4.4.5　Left Shift (<<) ·· 42
　　　　　　　按位左移运算符
　　　　4.4.6　Signed Right Shift (>>) ··· 42
　　　　　　　带符号按位右移运算符
　　　　4.4.7　Unsigned Right Shift (>>>) ·· 43
　　　　　　　无符号按位右移运算符
　　4.5　Assignment Operators ·· 43
　　　　　赋值运算符
　　4.6　The Conditional Operator ··· 44
　　　　　条件运算符
　　4.7　The Instanceof Operator ··· 44
　　　　　instanceof 运算符
　　4.8　Special Operators ··· 45
　　　　　特殊运算符
　　　　4.8.1　Object Member Access (.) ·· 45
　　　　　　　对象成员访问符(.)
　　　　4.8.2　Array Element Access ([ ]) ·· 45
　　　　　　　数组元素访问符([ ])

	4.8.3	Method Invocation (( )) ································ 46
		方法调用操作符
	4.8.4	Object Creation (new) ··································· 46
		对象创建运算符
4.9	Type Conversion or Casting ······························ 46
	数据类型转换
4.10	Sample Program Practice ································· 46
	程序实例
4.11	Exercise for you ·· 49
	课后习题

Chapter 5 Flowing Control ································ 50
控制流

5.1	Control Statements ··· 50
	控制表达式
5.2	Selection Statements ······································· 50
	分支表达式
	5.2.1 If Statement ··· 50
	if 表达式
	5.2.2 If-else Statement ····································· 51
	if-else 表达式
	5.2.3 Switch Statement ···································· 51
	switch 表达式
5.3	Repetition Statements ······································ 52
	循环表达式
	5.3.1 While Loop Statement ······························· 52
	while 循环表达式
	5.3.2 Do-while Loop Statement ··························· 53
	do-while 循环表达式
	5.3.3 For Loop Statement ·································· 54
	for 循环表达式
5.4	Branching Statements ······································ 54
	分支跳转表达式
	5.4.1 Break Statement ····································· 54
	break 表达式
	5.4.2 Continue Statement ·································· 55
	continue 表达式
	5.4.3 Return Statement ···································· 55
	return 表达式
5.5	Sample Program Practice ································· 56
	程序实例
5.6	Exercise for you ·· 58
	课后习题

Chapter 6 Class ··· 59
类

6.1	Class Definition ··· 59
	类定义
	6.1.1 A Simple Class Definition ··························· 60
	简单类定义示例
	6.1.2 Defining a Class ····································· 60
	定义类

- 6.2 Declaring and Instantiating an Object 62
  - 对象定义与初始化
  - 6.2.1 Fields and Methods 65
    - 成员和方法
  - 6.2.2 Default Values for Primitive Members 66
    - 基本类型数据成员的缺省值
  - 6.2.3 Methods, Arguments, and Return Values 66
    - 方法、参数和返回值
  - 6.2.4 The Argument List 67
    - 参数列表
- 6.3 Constructor 68
  - 构造方法
  - 6.3.1 Calling Constructors from Constructors 70
    - 构造方法中调用构造方法
  - 6.3.2 Default Constructors 71
    - 缺省构造方法
- 6.4 Keyword "this" 72
  - this 关键字
- 6.5 Garbage Collection 74
  - 垃圾回收
  - 6.5.1 The Use of finalize( ) 75
    - finalize 方法的使用
  - 6.5.2 Cleanup 76
    - 垃圾清理
- 6.6 Static Methods and Static Variables 79
  - 静态方法和静态变量
  - 6.6.1 Static Methods 79
    - 静态方法
  - 6.6.2 Static Variables 80
    - 静态变量
- 6.7 Sample Examples 82
  - 程序实例
- 6.8 Exercise for you 84
  - 课后习题

Chapter 7 Method 85
方法

- 7.1 Method Overloading 85
  - 方法重载
  - 7.1.1 Distinguishing Overloaded Methods 87
    - 方法重载匹配
  - 7.1.2 Overloading with Primitives 87
    - 基本数据类型参数重载
  - 7.1.3 Overloading on Return Values 91
    - 基于返回值重载
  - 7.1.4 Overriding with Constructors 92
    - 构造方法重载
- 7.2 Parameter Passing in Java-By Reference or By Value 93
  - Java 参数传递：引用传递和值传递
  - 7.2.1 Passing Named Arguments to Java Programs 94
    - 给 Java 程序传递参数
  - 7.2.2 Passing Information into a Method 95
    - 方法信息传递

    7.2.3 Pass by Value ·· 97
       值传递
    7.2.4 Passing Primitive Types ··· 100
       传递基本类型参数
    7.2.5 Return Values ··· 101
       返回值
    7.2.6 Passing Object References ·· 101
       传递对象引用
    7.2.7 Passing Strings ·· 102
       传递字符串
    7.2.8 Passing Arrays ·· 103
       传递数组
  7.3 Recursion ·· 103
    递归
  7.4 Controlling Access to Members of a Class ··· 105
    类成员访问控制
    7.4.1 Class Access Level ··· 106
       类级别访问
    7.4.2 Package Access Level ··· 107
       包级别访问
  7.5 Static Import ·· 108
    静态导入
  7.6 Arrays ·· 109
    数组
    7.6.1 Array Overview ··· 109
       数组概述
    7.6.2 Java Arrays ·· 114
       Java 数组
  7.7 String ··· 116
    字符串
    7.7.1 Creating a String ··· 116
       字符串创建
    7.7.2 Strings Operation ·· 117
       字符串操作
    7.7.3 Alter Strings ·· 117
       字符串修改
  7.8 Command Line Arguments ··· 117
    命令行参数
  7.9 Sample Examples ·· 118
    程序实例
  7.10 Exercise for you ·· 122
    课后习题

Chapter 8 Inheritance ··· 123
    继承

  8.1 Derived Classes ··· 123
    派生类
  8.2 Abstract Classes ·· 128
    抽象类
  8.3 Keyword "final" ·· 130
    final 关键字
    8.3.1 Final Data ··· 130
       final 数据

· 13 ·

		8.3.2	Final Methods ·· 133
			final 方法
		8.3.3	Final Classes ·· 135
			final 类
	8.4	Sample Example ·· 136
		程序实例
	8.5	Exercise for you ·· 140
		课后习题

Chapter 9　Packages and Interfaces ·· 141
包和接口

9.1	Package ··· 141
	包
	9.1.1　Packages Overview ·· 141
	包概述
	9.1.2　Packages in Java ·· 142
	Java 包
	9.1.3　Access Specifiers ··· 142
	访问标志符
	9.1.4　How to Create a Package ··· 144
	包的创建
	9.1.5　Setting Up the CLASSPATH ··· 145
	类路径设置
	9.1.6　Subpackage (Package inside Another Package) ··············· 146
	子包(一个包在另一个包中)
	9.1.7　How to Use Package ··· 147
	使用包
9.2	Interface ·· 149
	接口
	9.2.1　Interface Overview ·· 149
	接口概述
	9.2.2　Creating and Using Interfaces ·· 150
	创建和使用接口
	9.2.3　Defining an Interface ··· 151
	接口的定义
	9.2.4　The Interface Body ··· 151
	接口体
	9.2.5　Implementing an Interface ·· 152
	接口的实现
	9.2.6　Using an Interface as a Type ··· 153
	接口类型
9.3	Sample Example ·· 153
	程序实例
9.4	Exercise for you ·· 156
	课后习题

Chapter 10　Exception Handling ·· 157
异常处理

10.1	Definition of Exception ··· 157
	异常定义
	10.1.1　What is an Exception ··· 157
	什么是异常

10.1.2 Common Exceptions ··· 157
　　　普通异常
10.1.3 The Throwable Superclass ·· 158
　　　Throwable 类
10.1.4 Effectively Using try-catch ·· 159
　　　有效使用 try-catch
10.1.5 When should You Use Exceptions ······························ 162
　　　何时使用异常
10.1.6 How do You Best Use Exceptions ······························ 163
　　　如何最大限度的使用异常处理
10.2 The Throw Statement ·· 166
　　throw 表达式
10.3 The Finally Statement ··· 166
　　finally 表达式
10.4 Runtime Exceptions ·· 167
　　运行时异常
　10.4.1 ArithmeticException ·· 167
　　　ArithmeticException 类
　10.4.2 NullPointerException ··· 167
　　　NullPointerException 类
　10.4.3 IncompatibleClassChangeException ·························· 168
　　　IncompatibleClassChangeException 类
　10.4.4 ClassCastException ·· 168
　　　ClassCastException 类
　10.4.5 NegativeArraySizeException ····································· 168
　　　NegativeArraySizeException 类
　10.4.6 OutOfMemoryException ··· 169
　　　OutOfMemoryException 类
　10.4.7 NoClassDefFoundException ····································· 169
　　　NoClassDefFoundException 类
　10.4.8 IncompatibleTypeException ······································ 169
　　　IncompatibleTypeException 类
　10.4.9 ArrayIndexOutOfBoundsException ···························· 170
　　　ArrayIndexOutOfBoundsException 类
　10.4.10 UnsatisfiedLinkException ······································ 170
　　　UnsatisfiedLinkException 类
10.5 Sample Examples ··· 170
　　程序实例
10.6 Exercise for you ··· 174
　　课后习题

Chapter 11　Multithread ·· 175
　　多线程

11.1 Multithread Overview ·· 175
　　多线程概述
11.2 Synchronization ·· 179
　　同步
　11.2.1 Inter-thread Communication ····································· 181
　　　线程间通信
　11.2.2 Java Thread Scheduling ·· 183
　　　Java 线程调度
　11.2.3 Thread Priorities ·· 184
　　　线程优先级

    11.2.4 Java Synchronization ·············· 185
      Java 线程同步
  11.3 The Life Cycle of a Thread ·············· 187
    线程生命周期
    11.3.1 Creating a Thread ·············· 187
      线程创建
    11.3.2 Starting a Thread ·············· 188
      线程启动
    11.3.3 Making a Thread Not Runnable ·············· 189
      使线程处于非运行状态
    11.3.4 Stopping a Thread ·············· 190
      线程停止
    11.3.5 Testing Thread State ·············· 191
      线程状态测试
    11.3.6 Why Pause and Resume Processes ·············· 193
      线程暂停和继续
  11.4 Sample Examples ·············· 194
    程序实例
  11.5 Exercise for you ·············· 198
    课后习题

Chapter 12 Input and Output ·············· 199
输入与输出

  12.1 Basic Java I/O ·············· 199
    Java I/O 基础
    12.1.1 Background ·············· 200
      背景
    12.1.2 Exceptions ·············· 200
      异常
    12.1.3 Applications ·············· 200
      应用
    12.1.4 File Attributes ·············· 200
      文件属性
  12.2 Streams ·············· 201
    流
    12.2.1 Byte Streams ·············· 202
      字节流
    12.2.2 Character Streams ·············· 203
      字符流
    12.2.3 Buffered Streams ·············· 206
      缓冲流
    12.2.4 Data Streams ·············· 207
      数据流
    12.2.5 Class StreamTokenizer ·············· 208
      StreamTokenizer 类
    12.2.6 Class StringTokenizer ·············· 209
      StringTokenizer 类
  12.3 Class File ·············· 210
    File 类
    12.3.1 The PrintWriter Class ·············· 211
      PrintWriter 类
    12.3.2 Constructor Method ·············· 211
      构造方法

    12.3.3 File Handling and Input/Output ·· 212
       文件处理与输入输出
    12.3.4 The Basic Input Output ··· 218
       基本输入输出流类
  12.4 Sample Examples ·· 225
    程序实例
  12.5 Exercise for you ··· 227
    课后习题

Chapter 13 String Handling ·· 228
    字符串处理

  13.1 The String Class ·· 228
    String 类
  13.2 Strings in Java ·· 229
    Java 字符串
    13.2.1 String Basics ·· 229
       字符串基础
    13.2.2 Creating a String ··· 230
       字符串创建
    13.2.3 Comparing Strings ··· 232
       字符串比较
    13.2.4 Other Operations ·· 233
       字符串其他操作
    13.2.5 StringBuffer Objects ·· 238
       StringBuffer 类
    13.2.6 String Analyzing ··· 240
       字符串分析
  13.3 Sample Examples ·· 241
    程序实例
  13.4 Exercise for you ··· 243
    课后习题

Chapter 14 Networking ··· 244
    网络

  14.1 Computer Network Basics ··· 244
    计算机网络基础
  14.2 URL Objects in Java ·· 247
    Java URL 对象
    14.2.1 Creating URL Objects ··· 247
       创建 URL 对象
    14.2.2 Query Methods on URL Objects ··· 247
       URL 对象查询方法
    14.2.3 Reading from a URL Connection ·· 248
       从 URL 连接读取数据
    14.2.4 URL Operations ·· 249
       URL 操作
  14.3 Sockets in Java ··· 259
    Java 套接字
    14.3.1 Establishing a Connection ·· 259
       建立连接
    14.3.2 The Client Side of a Socket Connection ······························· 259
       套接字连接的客户端

14.3.3 Socket Operations ··· 262
套接字操作

14.4 Sample Examples ·· 268
程序实例

14.5 Exercise for you ·· 272
课后习题

Chapter 15　Applets ··· 273
Applet 小应用程序

15.1 Applet Overview ·· 273
Applet 概述

15.2 Life Cycle, Graphics, Fonts, Colors ····················· 274
Applet 生命周期、图形、字体、颜色

15.2.1 Life Cycle of an Applet ····························· 274
Applet 生命周期

15.2.2 Parameter Passing ···································· 276
参数传递

15.2.3 Graphics Class ·· 276
Graphics 类

15.2.4 Font Class ··· 276
Font 类

15.2.5 Color Class ·· 277
Color 类

15.3 User Interface Components ································ 277
使用组件接口

15.4 Applet Fundamentals ·· 278
Applet 基础

15.4.1 The Applet Class ···································· 278
Applet 类

15.4.2 Applet Architecture ································· 279
Applet 体系结构

15.4.3 Requesting Repainting ······························ 280
请求重画

15.5 Working with URLs and Graphics ······················ 283
URL 与图形的应用

15.6 Using the instanceof Keyword in Java ················ 284
Java 中 instanceof 关键字应用

15.7 Sample Examples ··· 285
程序实例

15.8 Exercise for you ·· 288
课后习题

Chapter 16　Swing GUI Introduction ··················· 289
图形用户界面 Swing

16.1 Event-Driven Programming ································ 289
基于事件驱动编程

16.2 Event Handling ·· 291
事件处理

16.2.1 The Component Class ······························· 291
Component 类

16.2.2 The Event Class ······································· 292
Event 类

        16.2.3　Event Handling for the Mouse ······················· 293
                鼠标事件处理
        16.2.4　Keyboard Event Handling ···························· 294
                键盘事件处理
　　16.3　Buttons, Events, and Other Swing Basics ························ 295
        按钮、事件和其他 Swing 基础
        16.3.1　Buttons ··· 298
                按钮
        16.3.2　Action Listeners and Action Events ················· 298
                动作监听器和动作事件
        16.3.3　Labels ·· 300
                标签
        16.3.4　Color ··· 301
                颜色
　　16.4　Containers and Layout Managers ································ 303
        容器和布局管理器
        16.4.1　Border Layout Managers ······························· 304
                边界布局管理器
        16.4.2　Flow Layout Managers ·································· 307
                流式布局管理器
        16.4.3　Grid Layout Managers ··································· 307
                格点布局管理器
        16.4.4　Panels ·· 309
                面板
        16.4.5　The Container Class ······································ 313
                Container 类
　　16.5　Menus and Buttons ··· 314
        菜单和按钮
        16.5.1　Menu Bars, Menus, and Menu Items ················ 314
                菜单条、菜单和菜单项
        16.5.2　The AbstractButton Class ······························· 317
                AbstractButton 类
        16.5.3　The setActionCommand Method ······················ 318
                setActionCommand 方法
        16.5.4　Listeners as Inner Classes ······························· 319
                内部类监听器
　　16.6　Text Fields and Text Areas ··· 320
        文本域和文本区
        16.6.1　Text Areas and Text Fields ······························ 321
                文本域和文本区
        16.6.2　Window Listeners ·· 322
                窗口监听器
　　16.7　Sample Examples ·· 323
        程序实例
　　16.8　Exercise for you ·· 325
        课后习题

Chapter 17　Programming with JDBC ··· 326
JDBC 编程

　　17.1　JDBC Introduction ·· 326
        JDBC 介绍

17.2 Connecting to the Database ·················· 327
　　 数据库连接
　　17.2.1　A Simple Database Connection ·················· 328
　　　　　 简单的数据库连接示例
　　17.2.2　The JDBC Classes for Creating a Connection ·················· 329
　　　　　 创建连接的 JDBC 类
17.3 Basic Database Access ·················· 330
　　 数据库访问基础
　　17.3.1　Basic JDBC Database Access Classes ·················· 330
　　　　　 访问数据库的 JDBC 基础类
　　17.3.2　SQL NULL versus Java null ·················· 332
　　　　　 SQL 空值与 Java 空值
　　17.3.3　Clean Up ·················· 332
　　　　　 清理
　　17.3.4　Modifying the Database ·················· 332
　　　　　 修改数据库
17.4 SQL Data Types and Java Data Types ·················· 333
　　 SQL 数据类型和 Java 数据类型
17.5 Scrollable Result Sets ·················· 335
　　 滚动结果集
　　17.5.1　Result Set Types ·················· 335
　　　　　 结果集类型
　　17.5.2　Result Set Navigation ·················· 335
　　　　　 结果集导引
17.6 Sample Examples ·················· 336
　　 程序实例
17.7 Exercise for you ·················· 339
　　 课后习题

Chapter 1 Genesis of Java

1.1 Introduction

Java has developed from James Gosling's vision to a mission that rives the development world today. A high-level programming language developed by Sun Microsystems, Java was originally designed for handheld devices and set-top boxes. Initially titled Oak, in 1995, Sun changed the name to Java and modified the language to take advantage of the opportunities presented by the World Wide Web.

Java is a general programming language, many characteristics of which make it well adapted to the WWW application. Small Java applications known as applets, which can be downloaded from server, run easily on any Java-compatible browser.

So far, Sun Microsystems has published three versions of the Java platform—the Standard Edition (J2SE), the Enterprise Edition (J2EE), and the Micro Edition (J2ME).

> 到现在为止，Java发布了3个版本：标准版本，企业级版本，移动设备版本。

1.2 Java Development Today

More than ever before, many organizations optimize their internal processes and improve operational efficiency by using Java applications and software. These could form many competitive products such as Java databases, CRM applications, and wireless SFA applications.

As per Sun Microsystems, the Java brand is one of the hottest technology brands in the world, recognized by over 85% of tech-savvy consumers around the globe. Now 10 years old, this cup of our favorite brew truly is hot! It is estimated that Java is driving more than $100 billion of business annually, assisted by over 4.5 million software developers. That's the power of Java!

> Java作为一种比较流行的编程语言，占据了85%的市场，约450万程序员都在使用这种编程工具。

1.3 Evolution of 'C' Based Programming Languages

In 1960s structured programming language was born. This method of programming can be supported by some languages like C. By using this structured language, moderately complex programs can be completed fairly easy. However, even with such a structured programming method, once a project reaches a certain size, its complexity will be far beyond the scope of what a programmer can manage. By the early 1980s, many projects were pushing the structured approach past its limits .To solve this problem, a new method of programming came out. This approach is object-oriented programming. Object-oriented programming can solve complex program based on inheritance, encapsulation, and polymorphism.

In the final analysis, although C is one of the world's great programming languages, there are some deficiencies such as weaker capability to handle complexity. Once program codes exceed more than 25,000 lines of code, it becomes so complex that it is difficult to control as a totality. OOP programming overcomes this obstacle and manages larger programs.

> 面向过程编程的缺陷：当程序规模增大时候，复杂度也随之增加，程序员无法控制所写的代码。而面向对象编程可以克服这个缺陷。

1.4 Main Features of Java Programming Language

1.4.1 Portability

Most modern programming languages are designed to be (relatively) easy for people to write and to understand. These languages that were designed for people to read are called high-level languages. The language that the computer can directly understand is called machine language. Machine language or any language similar to machine language is called a low-level language. A program written in a high-level language, like Java, must be translated into a program in machine language before the program can be run. The program that does the translating (or at least most of the translating) is called a **compiler** and the translation process is called **compiling**. The Java compiler does not translate your program into the machine language for your particular computer. Instead, it translates your Java program into a language called **byte-code**(字节码). Byte-code is not the machine language for any particular computer. Byte-code is the machine language for a fictitious computer called the **Java Virtual Machine**(Java 虚拟机).

> Java 作为一种高级编程语言，具有很强的移植性，通过 Java 虚拟机可将字节码运行在不同的机器平台上。

The reason is: only the JVM needs to be implemented for each platform. Once the run-time package exists for a given system, any Java program can run on it. Remember, although the details of the JVM will differ from platform to platform, all interpret the same Java bytecode. If a Java program were compiled to native code, then different versions of the same program would have to exist for each type of CPU connected to the Internet. This is, of course, not a feasible solution. Thus, the interpretation of bytecode is the easiest way to create truly portable programs. The fact that a Java program is interpreted also helps to make it secure. Because the execution of every Java program is under the control of the JVM, the JVM can contain the program and prevent it from generating side effects outside of the system. As you will see, safety is also enhanced by certain restrictions that exist in the Java language.

> JVM 是 Java 跨平台的基础，Java 源程序被编译成字节码，然后交给 JVM 解释执行。这样 Java 就不需要知道自己是在何种硬件平台下运行的。

When a program is interpreted, it generally runs substantially slower than it would run if compiled to executable code. However, with Java, the differential between the two is not so great. The use of bytecode enables the Java run-time system to execute programs much faster than you might expect. Although Java was designed for interpretation, there is technically nothing about Java that prevents on-the-fly compilation of bytecode into native code. Along these lines, Sun supplies its Just In Time (JIT) compiler for bytecode, which is included in the Java 2 release. When the JIT compiler is part of the JVM, it compiles bytecode into executable code in real time, on a piece-by-piece, demand basis. It is important to understand that it is not possible to compile an entire Java program into executable code all at once, because Java performs various run-time checks that can be done only at run time. Instead, the JIT compiles code, as it is needed, during execution. However, the just-in-time approach still yields a significant performance boost. Even when dynamic compilation is applied to bytecode, the portability and safety features still apply, because the run-time system (which performs the compilation) still is in charge of the execution environment. Whether your Java program is actually interpreted in the traditional way or compiled on the fly, its functionality is the same.

> JIT 是 JVM 的一部分，它可以按要求即时地将字节码逐一地编译为可执行的代码。

1.4.2 Simple

We wanted to build a system that could be programmed easily without a lot of esoteric training and which leveraged today's standard practice. Most programmers working these days use C, and most programmers doing object-oriented programming use C++. So even though we found that C++ was unsuitable, we designed Java as closely to C++ as possible in order to make the system more comprehensible.

Java omits many rarely used, poorly understood, confusing features of C++ that in our experience bring more grief than benefit. These omitted features primarily consist of operator overloading (although the Java language does have method overloading), multiple inheritance, and extensive automatic coercions.

We added automatic garbage collection, thereby simplifying the task of Java programming but making the system somewhat more complicated. A common source of complexity in many C and C++ applications is storage management: the allocation and freeing of memory. By virtue of having automatic garbage collection (periodic freeing of memory not being referenced) the Java language not only makes the programming task easier, it also dramatically cuts down on bugs.

> Java 是一种非常简单的编程语言，优化了 C++语言中一些令人难以理解的特性，增加了诸如垃圾自动回收机制的内容，方便了初学者的学习和使用。

1.4.3 Robust

The multi-plat formed environment of the Web places extraordinary demands on a program, because the program must execute reliably in a variety of systems. Thus, the ability to create robust programs was given a high priority in the design of Java. To gain reliability, Java restricts you in a few key areas, to force you to find your mistakes early in program development. At the same time, Java frees you from having to worry about many of the most common causes of programming errors. Because Java is a strictly typed language, it checks your code at compile time. However, it also checks your code at run time. In fact, many hard-to-track-down bugs that often turn up in hard-to-reproduce run-time situations are simply impossible to create in Java. Knowing that what you have written will behave in a predictable way under diverse conditions is a key feature of Java. To better understand how Java is robust, consider two of the main reasons for program failure: memory management mistakes and mishandled exceptional conditions (that is, run-time errors).

> 在多平台环境中，健壮性是一个程序的重要指标，Java 有着严格的语法规范，它可以在编译时和运行时检查所编写的代码。Java 主要考虑了会导致程序运行失败的两个方面：内存管理和不可处理的异常。

One of the advantages of a strongly typed language (like C++) is that it allows extensive compile-time checking so bugs can be found early. Unfortunately, C++ inherits a number of loopholes in **compile-time**(编译期间) checking from C, which is relatively lax (particularly method/procedure declarations). In Java, we require declarations and do not support C-style implicit declarations. The linker understands the type system and repeats many of the type checks done by the compiler to guard against version mismatch problems. The single biggest difference between Java and C/C++ is that Java has a pointer model that eliminates the possibility of overwriting memory and corrupting data. Instead of pointer arithmetic, Java has true arrays. This allows subscript checking to be performed. In addition, it is not possible to turn an arbitrary integer into a pointer by casting.

> Java 程序在一些关键的地方做了强制性限制，这样使开发者很容易发现自己所犯的错误。同样，Java 也解除了很多开发者认为可能会引起程序错误的内容。

1.4.4 Multithread

Java was designed to meet the real-world requirement of creating interactive, networked programs. To accomplish this, Java supports multithread programming, which allows you to write programs that do many things simultaneously(同时地). The Java run-time system comes with an elegant yet sophisticated solution for multiprocessor Synchronization(多处理器同步) that enables you to construct smoothly running interactive systems. Java's easy-to-use approach to multithread allows you to think about the specific behavior of your program, not the multitasking subsystem.

> Java 提供易使用的多线程允许开发者考虑程序中所需要的特定行为。

1.4.5 Architecture-Neutral

A central issue for the Java designers was that of code longevity and portability. One of the main problems facing programmers is that no guarantee exists that if you write a program today, it will run tomorrow—even on the same machine. Operating system upgrades, processor upgrades, and changes in core system resources can all combine to make a program malfunction. The Java designers made several hard decisions in the Java language and the Java Virtual Machine in an attempt to alter this situation. Their goal was "write once, run anywhere, any time, forever." To a great extent, this goal was accomplished.

> 体系结构中性可以保证 Java 程序能在不同平台上运行，实现"一次编写，处处运行"的目标。

1.4.6 Interpreted and High Performance

As described earlier, Java enables the creation of cross-platform programs by compiling into an intermediate representation called Java byte code. This code can be interpreted on any system that provides a Java Virtual Machine. Most previous attempts at cross-platform solutions have done so at the expense of performance. Other interpreted systems, such as BASIC, TCL and PERL, suffer from almost insurmountable performance deficits. Java, however, was designed to perform well on very low-power CPUs. As explained earlier, the Java byte code was carefully designed so that it would be easy to translate directly into native machine code for very high performance by using a just-in-time compiler.

> Java 程序占有少量的 CPU 资源，其字节码能够很容易被直接翻译成机器代码。

1.4.7 Distributed

Java is designed for the distributed environment of the Internet, because it handles TCP/IP protocols. In fact, accessing a resource using a URL is not much different from accessing a file. The original version of Java included features for intra-address-space messaging. This allowed objects on two different computers to execute procedures remotely. Java revived these interfaces in a package called Remote Method Invocation. This feature brings an unparalleled level of abstraction to client/server programming.

> Java 是一种能够在 Internet 的分布式环境下运行的语言，通过远程方法调用的接口包来实现远程访问。

1.4.8 Dynamic

Java programs carry with them substantial amounts of run-time type information that is used to verify and resolve accesses to objects at run time. This makes it possible to dynamically link code in a safe and expedient manner. This is crucial to the robustness of the applet environment, in which small fragments of byte code may be dynamically updated on a running system.

> Java 程序可以在运行时动态地加载所需要的对象，这对于 Applet 小程序的运行非常重要。

1.4.9 Security

The fundamental problem in the internet associated with Java is security. Java's security model is one of the language's key architectural features that make it an appropriate technology for networked environments. Security is important because networks provide a potential avenue of attack to any computer hooked to them. This concern becomes especially strong in an environment in which software is downloaded across the network and executed locally, as is done with Java applets, for example. Because the class files for an applet are automatically downloaded when a user goes to the containing Web page in a browser, it is likely that a user will encounter applets from untrusted sources.

Without any security, this would be a convenient way to spread viruses. Thus, Java's security mechanisms help make Java suitable for networks because they establish a needed trust in the safety of network-mobile code.

> Java安全模型是Java语言体系结构的关键性特性。它使得Java语言能够适应网络环境。因为这种安全措施在网络代码上建立了一种信任感。

1.5 Java Applet

A Java applet is an applet delivered to users in the form of Java bytecode. Java applets can run in a Web browser using a Java Virtual Machine (JVM), or in Sun's AppletViewer, a stand-alone tool for testing applets. Java applets were introduced in the first version of the Java language in 1995, and are written in programming languages that compile to Java bytecode, usually in Java, but also in other languages such as Jython, JRuby, or Eiffel (via SmartEiffel).

Java applets run at speeds comparable to, but generally slower than, other compiled languages such as C++, but until approximately 2011 many times faster than JavaScript. In addition they can use 3D hardware acceleration that is available from Java. This makes applets well suited for non trivial, computation intensive visualizations. When browsers have gained support for native hardware accelerated graphics in the form of Canvas and WebGL, as well as Just In Time compiled JavaScript, the speed difference has become less noticeable.

> Applet或Java小应用程序是一种在Web环境下，运行于客户端的Java程序组件。它也是20世纪90年代中期，Java在诞生后得以一炮走红的功臣之一。通常，每个Applet的功能都比较单一（如仅用于显示一个舞动的Logo），因此它被称做"小应用程序"。

Since Java's bytecode is cross-platform or platform independent, Java applets can be executed by browsers for many platforms, including Microsoft Windows, Unix, Mac OS and Linux. It is also trivial to run a Java applet as an application with very little extra code. This has the advantage of running a Java applet in offline mode without the need for any Internet browser software and also directly from the integrated development environment (IDE).

The applet can be displayed on the web page by making use of the deprecated applet HTML element, or the recommended object element. Embed element can be used with Mozilla family browsers (embed is no longer deprecated in since HTML 5). This specifies the applet's source and location. Object and embed tags can also download and install Java virtual machine (if required) or at least lead to the plug-in page. Applet and object tags also support loading of the serialized applets that start in some particular (rather than initial) state. Tags also specify the message that shows up in place of the applet if the browser cannot run it due any reason.

> Applet必须运行于某个特定的"容器"，这个容器可以是浏览器本身，也可以是通过各种插件，或者包括支持Applet的移动设备在内的其他各种程序来运行。

However, despite object being officially a recommended tag, as of 2010, the support of the object tag was not yet consistent among browsers and Sun kept recommending the older applet tag for deploying in multibrowser environments, as it remained the only tag consistently supported by the most popular browsers. To support multiple browsers, the object tag currently requires JavaScript (that recognizes the browser and adjusts the tag), usage of additional browser-specific tags or delivering adapted output from the server side. Deprecating applet tag has been criticized. Oracle now provides a maintained JavaScript code to launch applets with cross platform workarounds

> 与一般的Java应用程序不同，Applet不是通过main方法来运行的（参见Java的Hello World程序和Java Applet的Hello World程序）。在运行时Applet通常会与用户进行互动，显示动态的画面，并且会遵循严格的安全检查，阻止潜在的不安全因素（如根据安全策略，限制Applet对客户端文件系统的访问）。

1.6 Exercise for you

1. Why do you think that Java is stronger, easier and safer?
2. Compare Java with other programming languages and find out the differences.

Chapter 2　Java Overview

This chapter will explain the concept of OOP and several basis of Java language. In almost all the programming languages when we work, we need some prior knowledge to work as the programming language itself is a whole bunch of ideas together. Here are some. The points given here would give you a better assistance to have a hold on Java language and help you create the basic programs using Java.

2.1　Concepts of OOP

As you know, all computer programs include two elements: code and data. Conceptually, a program can be organized based on code and data. That is, some programs are written around "what is happening" and others are written around "who is being affected". These are the two examples that show how a program is constructed. The first way is called the process-oriented model(面向过程模式).

> 所有程序一般包括两个元素：代码和数据。其中，一些程序侧重处理的过程，一些程序侧重处理的对象。

This approach characterizes a program as a series of linear steps (that is, code). The process-oriented model can be viewed of as series of linear steps (that is, code). In the process-oriented model code is acted on data. C is paradigm for using this model. However problems appear as program grows larger and more complex.

To manage increasing complexity(复杂性), the second approach was invented. We called it called object-oriented programming(面向对象编程). This model is described as data controlling access to code. As you will know, by switching the controlling entity to data, you can achieve several organization benefits.

> 面向对象编程能够管理比较复杂的项目，这种编程模型使用数据来控制访问代码。

The following problem is that how to obtain object. The same types of entity have the same properties. So we can extract those same features from real-entities. This approach is called abstraction.

Abstraction: Abstraction is the presentation of simple concept (or object) to the external world. Abstraction concept is to describe a set of similar concrete entities. It focuses on the essential, inherent aspects of an entity and ignoring its accidental properties. The concept is usually used during analysis: deciding only with application-domain concepts, not making design and implementation decisions.

> 抽象是对外部世界简单概念(对象)的表达，是对同类实体集合的描述。它侧重于实体的本质、内在的方面，而忽略其偶然性。抽象概念通常在分析阶段使用，在设计阶段和实施阶段不使用。

Two popular abstractions: procedure abstraction and data abstraction. Procedure abstraction is to decompose problem to many simple sub-works. Data abstraction is to collect essential elements composing to a compound data. These two abstractions have the same objective: reusable and replaceable. Figure 2.1 shows an example of data abstraction to abstract doors as a data structure with essential properties.

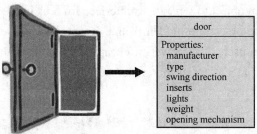

Figure 2.1   Abstract of Door

Hierarchical classification is used to manage abstraction. This allows you to layer the semantics of complex systems, breaking them into more manageable pieces. From the outside, the car is a single object. Once inside, you see that the car consists of several subsystems: steering, brakes, sound system, seat belts, heating, cellular phone, and so on. In turn, each of these subsystems is made up of more specialized units. For instance, the sound system consists of a radio, a CD player, and /or tape player. The point is that you manage the complexity of the car (or any other complex system) through the use of hierarchical abstraction.

> 层次化分类用来管理抽象，允许将一个复杂的系统分层，将一个系统分解成一些可以管理的部分。

2.1.1   Class

A class is a construct that is used as a blueprint to create instances of itself. For example: There may be thousands of other bicycles in existence, all of the same make and model. Each bicycle was built from the same set of blueprints and therefore contains the same components. In object-oriented terms, we say that your bicycle is an instance of the class of objects known as bicycles. A class also is the blueprint from which individual objects are created.

> 类是一种被用来创建实例的模板。

2.1.2   Object

In everyday life, an object is anything that is identifiably a single material item. An object can be a car, a house, a book, a document, or a paycheck. For our purposes, we're going to extend that concept a bit and think of an object as anything that is a single item that you might want to represent in a program. We'll therefore also include living "objects", such as a person, an employee, or a customer, as well as more abstract "objects", such as a company, a database, or a country.

> 任何事物都可以看作对象，对象可以代表整个程序，也可以包括一些有生命的"对象"，如一个人、雇员、顾客，还包括一些比较抽象的"对象"，如一个企业、一个数据库或一个国家。

2.1.3 Encapsulation

Encapsulation is the mechanism that binds together code and the data it manipulates, and keeps both safe from outside interference and misuse.

Encapsulation allows an object to separate its interface from its implementation. The data and the implementation code for the object are hidden behind its interface. So encapsulation hides internal implementation details from users. The power of encapsulated code is that everyone knows how to access it and thus can use it regardless of the implementation details and without fear of unexpected side effects.

> 封装将对象的接口与实现分开,实现对数据和代码的隐藏,提高数据的安全性。

In Java the basis of encapsulation is the class. A class defines the structure and behavior (data and code) that will be shared by a set of objects. Each object of a given class contains the structure and behavior defined by the class, as if it were stamped out by a mold in the shape of the class. For this reason, objects are sometimes referred to as instance of class. Thus, a class is a logical construct; an object has physical reality.

When you create a class, you will specify the code and data that constitute that class. Collectively, these elements are called members of the class. Specifically the code that operates on that data is referred to as member variables or instance variables. The code that operates on that data is referred to as member methods or just methods. (If you are familiar with C/C++, it may help to know that what a Java programmer calls a method a C/C++ programmer calls a function.) In properly written Java programs, the methods define how the member variables can be used. This means that the behavior and interface of a class are defined by the methods that operate its instance data.

> 类由代码和数据组成,代码和数据是类的成员。类本身体现了封装特性。

2.1.4 Inheritance

In object-oriented programming (OOP), inheritance is a way to compartmentalize and reuse code by creating collections of attributes and behaviors called objects which can be based on previously created objects. In classical inheritance where objects are defined by classes, classes can inherit other classes. The new classes, known as subclasses (or derived classes), inherit attributes and behavior (i.e. previously coded algorithms) of the pre-existing classes, which are referred to as superclasses, ancestor classes or base classes. The inheritance relationships of classes give rise to a hierarchy. In prototype-based programming, objects can be defined directly from other objects without the need to define any classes, in which case this feature is called differential inheritance. The inheritance concept was invented in 1967 for Simula.

> 在面向对象编程(OOP)中,继承可以通过创造一系列属性和行为的方式来划分和重用代码。这些属性和方法依赖先前已经创建的对象,从而提高代码的重用率。

Without the use of hierarchies, each object would need to define all of its characteristics explicitly. However, by use of inheritance, an object need only define those qualities that make it unique within its class. It can inherit its general attributes from its parent. Thus, it is the inheritance mechanism that makes it possible for one object to be a specific instance of a more general case. Let's take a closer look at this process.

Most people naturally view the world as made up of objects that are related to each other in hierarchical way, such as animals, mammals, and dogs. If you wanted to describe animals in an abstract way, you would say they have some attributes, such as size intelligence, and type of skeletal system. Animals also have certain behavioral aspects; they eat, breathe, and sleep. This description of attributes and behavior is the class definition for animals.

2.1.5 Polymorphism

Polymorphism (from the Greek, meaning "many forms") is the ability to create a variable, a function, or an object that has more than one form. In Java, polymorphism is the ability of objects to react differently when presented with different information, known as parameters. In a functional programming language, it is the only way to complete two different tasks is to have two functions with different names.

> 多态性：发送消息给某个对象，让该对象自行决定响应何种行为。通过将子类对象引用赋值给超类对象引用变量，动态实现方法调用。

2.2 More Details on Object-Oriented Programming

The example lets use is that of a car. Object-oriented programming thinks of real world things as objects, so in the case of the example the car is an object. Objects have two parts to them, data and operations, which can be carried out on this data. So in the car example, the data might contain the speed of the car, the height of the seat and whether the horn is currently being rung or not. There are several different operations, which can be carried out on the car, the rider may drive faster, and they might want to change the gear. So in this simple object we have the following:

> 作为对象应该具备两个部分：数据和操作，操作是基于数据来完成相应动作。以车为例来说明，车的数据包括车速、座位的高度、喇叭是否能响。车可有几种不同的操作。当司机加速的时候，改变车速这个数据就可以实现。

Data
- Speed
- Height of seat
- Status of the horn

Operations
- Drive Faster
- Adjust gear
- Blow horn

2.2.1 Encapsulation of Car

Encapsulation is an important part of OO programming, but it's not difficult. In Java Encapsulation is implemented by a **class**(类), a class is the generic form of an object.

So in our example the class is Car while the object (an instance of the class) is, perhaps MyCar. Within the object there are so-called instance variables, in our example MyCar will have instance variables for speed, height of seat and status of the horn. So when an operation (in Java a method) operates on an object it changes the instance variables for that object. If the Faster Speed method were called for the object MyCar then the instance variable of speed, within MyCar, would be altered accordingly.

> Car 是类，MyCar 是 Car 创建的对象(类的实例)。

2.2.2 Inheritance of Car

When a class is extended, to create a **sub-class**(子类), all of the properties (variables and methods) of the original class still exist within the new class along with others, which have been added.

The Car in the example is a very simple one; if we wanted to extend it to have gears we would simply create a new class based on Car but which had a new variable called gear and two new methods, one to change up a gear and one to change down.

> 当一个类创建了子类，父类的所有属性(变量和方法)仍将与新创建的子类的属性共存。

Note that Java does not directly support multiple inheritances.

> 注：Java 不支持类的多重继承。

Figure 2.2 shows the Inheritance properties, which can constitute a Car finally.

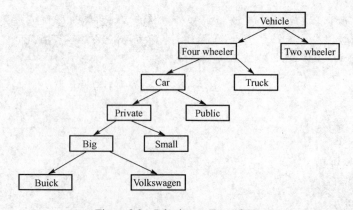

Figure 2.2　Inheritance Tree of Car

Figure 2.3 shows the Buick car inherits the encapsulation of all its super classes.

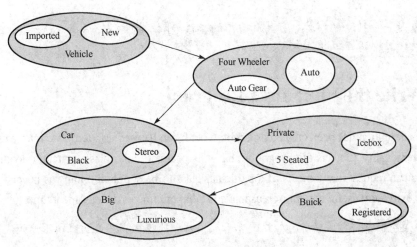

Figure 2.3　Inheritance of Buick Car

2.2.3　Polymorphism of Car

Object-oriented languages, which include Java, allow different methods to be run depending on what type of parameters is specified. So if our car were to be involved in a collision with a object, the object would be passed as a parameter to a collision method, obviously collision with a fly will have very different affects to collision with a Bus. The fly may cause the car no damage while collision with a Bus may cause a sudden loss of speed and major damage.

This ability of a method to react to different parameters is achieved by overloading, a number of methods are written (with the same name) but each one has a different set of input parameters.

So for our example there will be two methods called collision, one, which accepts a parameter of type fly while the other, accepts a parameter of type Bus.

> 多态性是指同一操作对不同的类可以呈现不同的行为，如参数的个数，类型不同。重载是通过不同的参数列表实现方法的不同功能，一系列的方法可以用相同的名字来编写，但是每个方法却有不同的输入参数。

2.2.4　Conclusion on Object-Oriented Programming

Object-oriented programming is fundamentally different to traditional functional programming, used correctly it can lead to the development of very robust, easily expandable and maintainable code. Objects are everywhere around us, while the class they belong to is not always obvious. We use encapsulation all the time, when we change gear in a car we do not need to know what happens with the clutch or gearbox we simply need to know that moving the gear lever will change gear. If this were likened to an object, the gear lever would be the method while the internal workings of the clutch and gearbox would be the instance variables. Expanding this simple car to be a racing car with wings and spoilers is examples of inheritance while the car's ability to cope with steering movements to the left and right is an example of polymorphism.

Java is an object-oriented programming language. But what are objects? An object is a self-contained entity which has its own private collection of properties (i.e. data) and methods (i.e. operations) that encapsulate functionality into a reusable and dynamically loaded structure.

> 对象是一个独立的实体，它具有自己的属性(即数据)和方法(即操作)，这些属性和方法封装在一个可以被重复使用的结构中。

2.3　Write the First Java Program

A class definition has been created as a prototype. It can be used as a template for creating new classes that add functionality. Objects are programming units of a particular class. Dynamic loading implies that applications can request new objects of a particular class to be supplied on an "as needed" basis. Objects provide the extremely useful benefit of reusable code that minimizes development time.

> 对象是类的具体实现。动态加载意味着应用程序能够在需要的时候请求一个特殊类的实例。

A Java application resembles programs in most compiled languages. Code in object format resides on the user's machine and is executed by a run-time interpreter that normally has to be user installed.

You wrote a simple 'hello world!' application to test the development environment. Now comes the explanation of the basic structure of Java applications using it as an example. Applications are stand-alone and are invoked (or executed) by using a Java interpreter.

```java
/**
 * The HelloWorldApp 类实现一个应用
 * 显示"Hello World!".
 */
public class HelloWorldApp {
    public static void main(String[] args) {
            System.out.println("Hello World!");
    }
}
```

The first four lines is a comment on the application's function. Comments are not required but are extremely useful as documentation within the source. Other notes and doc files may get lost but these stay right where they are most useful. A long comment starts with a /* or /** and ends with a */ .Short one line comments begin with // and end with the <return>.

The fifth line starts with the reserved word *public*. This indicates that objects outside the object can invoke (or use) it. The reserved word *class* indicates that we are building a new object with the name that follows. HelloWorldApp is the object name (your choice) and is case sensitive. Java 'style' is to capitalize the first letter of each word only. The line concludes with an opening curly bracket that marks the start of the object definition.

> 注释对于理解代码有很重要的作用，有单行注释和多行注释。public 修饰符表示其他对象可以调用该程序。HelloWorldApp 是一个类名字，而该类的类体包含在花括号内。

Line six is an invocation of an initialization method (or procedure in older languages). Static indicates that it calls the class and not an 'instance' of the class. The concept of 'instance' will be discussed later in the tutorials. The method's name is main and the reserved word void indicates that no result is returned back. Main has arguments (aka parameters) in round brackets. String[] indicates the variable type is an array and args is a reserved word indicating that it is the command line values that are being passed over to main. The line finishes with an opening bracket for the main method.

> main 函数是该程序的主函数，也是程序执行的入口点。main 函数没有返回类型，但是具有一个参数 String 数组，该数组可以接收来自控制台的输入参数。

Line eight invokes the println method of the system.out object. What is to be printed is passed in the argument as a string parameter. Note that each Java statement concludes with a semicolon.

Finally closing curly brackets are used for the main and for the object definition.

2.4 How to Run the First Java Program

Now we finish the first program. But we do not know how tor runs this application.

Steps to follow:

(1) Install the JDK (Java Development Kit) any version above 1.2 which is freely downloadable from the http://java.sun.com/j2se/1.5.0/download.jsp in to your computer and run it.//Sun 公司官网上下载 JDK 程序包

(2) Check and set the path where the jdk/bin directory is and check if the Java compiler (javac) is running well. //检查和配置 jdk/bin 路径

(3) Open the command and type "javac" to check the Java compiler is running, if not, check the path again..//打开命令行，输入 javac 检测 Java 编译器是否能够运行

(4) Application or applet program must be saved on the main class name and extension .java ("dot"java). //Java 应用程序必须以.java 为后缀保存

(5) To compile the program type javac and the program name (javac HelloWorldApp as per the case here) and enter. It would compile the Java file and create the .class file.//通过 javac 来编译 Java 源程序，生成.class 字节码程序

(6) The result or out put could be viewed by writing "java" and the file name (java Hello World App in this case) and enter. You can see the result.//通过 java 来运行.class 文件得到相应结果

Compiling the program

To compile the first program, execute the compiler, javac, specifying the name of the source file on the command line, as shown here:

C:\>javac first.java

The javac compiler creates a file called first.class that contains the bytecode version of the program. As discussed earlier, the Java bytecode is the intermediate representation of your program that contains instructions the Java interpreter will execute. Thus, the output of javac is not code that can be directly executed. //将源文件编译成字节码程序

To actually run the program, you must use the Java Interpreter, called java. To do so, pass the class name first as a command-line argument, as shown here:

C:\>java first //运行这个字节码程序

When the program is run the following output is displayed:

Hello World

source code→Compiler→byte-code file→Interpreter→output screen
first.java   javac   first.class   java   Hello World

When Java source code is compiled, each individual class is put into its own output file named after the class and using the **.class** extension.

> 当 Java 源代码被编译后，每个独立的类被编译后的输出文件名称由类名和.class 扩展名组成。

This is why it is a good idea to give your Java source files the same name as the class they contain-the name of the source file will match the name of the **.class** file.

> 给出的 Java 源代码文件名与它包容的类名相同，这样 Java 源代码文件名将与编译后的.class 名匹配。

When you execute the Java interpreter as just shown, you are actually specifying the name of the class that you want the interpreter to execute. It will automatically search for a file by that name that has the .class extension. If it finds the file, it will execute the code contained in the specified class. The Figure 2.4 shows life cycle of Java Program:

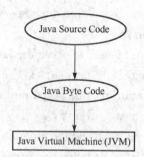

Figure 2.4   Presentation of Java Programming Life Cycle

2.5 Lexical Elements

Java is a programming language and like any other language there are certain rules about how "words" in the language are formed and what constitutes a legal "sentence" in the language. In addition, Java has rules about how the sentences can be put together into a complete program, somewhat like the less formal rules about how to format a formal letter (e.g. with the address of the recipient, and opening salutation etc.) or how to structure a book (e.g. with chapters, a table of contents etc.). In this section we will begin by looking at the various "words" and symbols, called lexical elements, that are used to construct Java programs.

> Java 和其他程序语言一样，有一些特定的语句规则，这些规则约束了哪些单词可以使用，以及如何构成程序语言，这有点类似于写一封规范的信（例如，收信人的称呼、开场白等），或者是如何构建一本书（例如，篇章、目录等）。

The most fundamental element in the structure of a program is a single character that can be displayed on a computer screen or typed at a computer keyboard. Prior to Java, most programming languages in common use during the last 10-20 years have used what is known as the ASCII character set. This character set provides for 256 different characters. This may seem like a lot, but when you consider all of the human languages in the world and the various symbols they use, it turns out that 256 is not enough. Therefore, Java has adopted what is called the Unicode character set. This set provides for more than 64,000 different characters.

When a Java compiler first begins to analyze a Java program it groups the individual characters together into larger lexical elements usually called tokens(符号). Some of the tokens will only be one character long such as the plus sign "+" which is the Java symbol used to add two numbers together (among other things). Other tokens will be many characters long such as the keywords like class and public. These basic tokens are then combined into larger language structures such as expressions, method calls, class definitions etc.

> 当 Java 编译器开始分析 Java 程序时，它将各个字符集合在一起成为更大的元素，通常称为符号。有些符号只有一个字符长，例如"+"在 Java 语言中将两个数字加在一起。其他符号可以为多个字符，如关键词 class 和 public。

2.6　White Space

The first rule we present is a rule that does not result in any tokens or larger language structures. White space in Java refers to the space character (what you get when you type the space bar on the keyboard), the tab character (this is actually one character although it may appear as several spaces on your screen) and the new line character (what you get when you type the return or enter key on the keyboard). White space is used primarily to make the program look nice and also serves to separate adjacent, multi-character tokens that are not separated by any other punctuation(标点符号) and would otherwise be considered a single, longer token.

For example, white space is used to separate the words (tokens) public static void that we saw in example. In such situations, where one white space character is required, any number of white space characters can be used. We could have, for example, put each of the words public, static and void on separate lines or put lots of spaces between them like this:

public　　　static　　　void　　　main(...

Except with String literals discussed below, any number of one or more adjacent white space characters (even mixing tab, space and newline) is the same as just one white space character as far as the structure and meaning of the program are concerned. Stated another way, if you can legally put in one space, you can put in as many spaces, tabs and newlines as you want. You cannot put white space in the middle of a keyword or identifier such as a variable or class name.

> 除了即将讨论的 String 字符外，就程序的结构和意义而言，一个或多个相邻的空格字符（即使混合的 tab、空格和换行字符）等同于一个空格字符。

2.7 Comments

Comments are very important (and often neglected). The primary purpose of a comment is to provide additional information to a person that is reading a program. As far as the computer is concerned, a comment is just like more white space. The comment does not result in a token, it only separates other tokens or gets ignored completely when it is not needed to separate tokens. Java has 3 ways to specify comments.

2.7.1 Single Line

Typing // causes the rest of the line (all characters up to the next newline) to be treated as a comment and ignored by the compiler. This is called a single line comment because the comment cannot be longer than a single line. By definition, the comment ends at the end of the line containing the //.

单行的注释用//。

2.7.2 Multi-line

As the names suggests, this style of comment can extend across several lines in a program. The comment beginning is marked with /* and the comment ending is marked with */. Everything between the marks and the marks themselves is a comment and is ignored. Here is the multi-line comment from our first Java program.

多行的注释用/* */。

```
/* HelloWorld.java
 * Purpose:
 *This program is the classic "Hello World" program for Java. It simply prints
a message to the screen.
 */
```

The single asterisks on the 2nd through 6th lines are not required and are just used to accent the extent of the comment. These comments are also called block comments.

2.7.3 Javadoc

The third style of comment is a minor variation on the previous multi-line comment. The comment beginning marker has an additional asterisk. These comments are identical to the multi-line comment except that they are recognized by a special program called "javadoc" that will automatically extract such comments and produce documentation for the program organized as an HTML document. HTML stands for Hyper Text Markup Language and it is the formatting language used for World Wide Web documents.

javadoc注释是一种多行注释，以/**开头，以*/结束，注释可以包含一些HTML标志符和专门的关键词。

```
/** starts with slash DOUBLE asterisk and
    continues
```

```
    until asterisk slash
*/
```

2.8 Keywords

Keywords cannot be used for anything but their predefined purpose. There are 49 in all, These words each have a special meaning to the Java compiler. A keyword must be separated from other keywords or identifiers by white space, a comment or some other punctuation symbol.

The keywords const and goto have no meaning in Java but are included as keywords to facilitate error reporting. These are keywords in C++, a language that many Java programmers will have learned before learning Java.

> 关键字只能使用在预定义部分，关键字和保留字之间需要用一个空格分隔开。

In addition the symbols true and false may appear to be keywords because they cannot be used as identifiers, they are in fact boolean literals. Table 2.1 lists main keywords in Java:

Table 2.1   keywords in Java

Abstract	Boolean	break
cast	catch	char
default	do	double
final	finally	float
goto	if	implements
int	interface	long
operator	outer	package
rest	return	short
synchronized	this	throw
try	Var	Void

2.9 Identifiers

These are the names you get to make up for classes, methods and variables like args and HelloWorld in the HelloWorld example. An identifier is any sequence of uppercase letters, lowercase letters, digits, $, and _ (underscore), except that it cannot start with a digit to avoid confusion with numbers. The set of "letters" in Java is much broader than in most other programming languages. Java letters include the symbols from most written languages used in the world today.

> 必须给类、方法和变量加上名字，例如args and HelloWorld in the HelloWorld example。标志符可以为任何大写字母、小写字母、阿拉伯数字、$和_(下画线)，但是不能以阿拉伯数字开头，以防与数字混淆。

Here are some legal identifiers:
    helloWorld $$$$$$ you_can_almost_make_a_sentence
    _something   x12345 $__$123

Here are some illegal identifiers:
    123   x+y   some***name   no space   1floor

Some naming conventions (not required rules) used by many Java programmers are

(1) **class** names start with uppercase and capitalize embedded words - e.g. HelloWorld

(2) **public** methods and variables start with lowercase and capitalize embedded words-e.g. drawOval

(3) **private** and local variables are all lowercase, separating words with underscore-e.g. loop_index, input_value

(4) Although legal, the dollar sign "$" should not be used except in machine generated Java programs.

2.10 Java Class Library

Java environment consists of several built in class libraries that contain many built in methods that support the I/O, String Handling, Networking and graphics. So Java is combination of Java language it self and all its classes together.

2.11 Sample Program Practice

Example 2.1 Static Demo

```
//对静态修饰符修饰的变量，可以直接用类名来调用
class StaticDemo{
    static int a = 23;
    static int b = 23;
    static void callme(){
        System.out.println("a = " + a);
    }
}
class Static{
    public static void main(String args[]){
        staticDemo.callme();
        System.out.println("b = " + staticDemo.b);
    }
}
```

Example 2.2 Inheritance

```
//实现继承机制
class Box{
    double width;
    double height;
    double depth;
    Box(){//构造函数1
        width = height = depth = -1;
    }
    Box(double len) {//构造函数2
        width = height = depth = len;
    }
    Box(double w, double h, double d) {//构造函数3
        width = w;
        height = h;
        depth = d;
    }
    double volume(){
        return width*height*depth;
```

```
        }
    }
    class boxWeight extends Box {     //继承了父类
        double weight;
        boxWeight(double w, double h, double d, double wh){
                width = w;
                height = h;
                depth = d;
                weight = wh;
        }
        double volume(){
                return width*height*depth*weight;
        }
    }
    class Sample{
        public static void main(String args[]){
                boxWeight mybox1 = new boxWeight(10,11,12,13);
                double vol;
                vol = mybox1.volume();
                System.out.println(vol);
        }
    }
```

2.12 Exercise for you

1. What is the difference between Encapsulation, Inheritance and Polymorphism? How the Polymorphic character of Java helps to create program flow?

2. Write a simple application in Java that can multiply two numbers and print.

3. Write an application program that can print numbers from 1 to 10 and then from 10 to 1.

Chapter 3　Data Types

This chapter describes the most common features of all the languages. That is Variables, Data types and Arrays. As a developed and strong language Java supports various types of data. These types could be used to declare variables, create arrays and changes the types. Java's approach to these variables, data types and arrays are simple, clean and strong.

> 本章将讲述程序语言中的共同特性，即变量、数据类型和数组。作为一种先进而功能强大的语言，Java 支持各种数据类型。这些类型可用于变量声明、数组创建和类型转换。

3.1　Data Types Overview

Java is known as a strongly typed language. It means that Java is a language that will only accept specific values within specific variables or parameters. Some languages, such as JavaScript, are weakly typed languages. This means that you can readily store whatever you want into a variable. Here is an example of the difference between strongly typed and weakly typed languages:

JavaScript (weakly typed)（Java 脚本）

```
1: var x;            // Declare a variable
2: x = 1;            // Legal
3: x = "Test";       // Legal
4: x = true;         // Legal
```

Java (strongly typed)

```
1: int x;            // Declare a variable of type int
2: x = 1;            // Legal
3: x = "Test"        // Compiler Error
4: x = true;         // Compiler Error
```

In a weakly typed language, such as JavaScript, you simply declare a variable without assigning it a type. In a strongly typed language, such as Java, you must give a variable a type when you declare it. Once you've declared a variable to be that type, it will remain of that type definitely and will only accept values that are within that types range. You should note that this is one of the many differences between Java and JavaScript. Despite their names, they have very little to do with one another.

> Java 是一种强类型编程语言。当声明一个变量的时候，必须给出该变量的类型。一旦给定了类型，所赋予的值必须在相应的精度内。

Now that we know that Java is a strongly typed language, you can probably see how important it is to know what data types there are in Java. There are 9 data types in Java, 8 primitive types and a reference type. First, let's look at primitive types and then we'll move along to reference types.

> Java 是强制类型检查的语言。Java 中有 9 种数据类型，其中 8 种是基本类型，一种是引用类型。

3.2 Primitive Types

Let us make a couple quick notes about a couple of the data types as shown in table 3.1 before we continue. The char type is a Unicode character in Java, rather than an ASCII character, as most people are used to seeing. That means that any character (regardless of language) can be represented with a char variable. That's why a char variable requires 2 bytes of storage rather than the 1 byte that is required for all ASCII characters. Also, you'll notice in the sample initialization of the float variable the extra 'f' character. This is required because, by default, the Java compiler views all decimal literals as doubles. In order to place a double into a float, you need to do a cast. I'll look at the reason for this next.

> Java 中字符类型属于 unicode 编码，每个字符占据 2 字节的存储空间。

Table 3.1 Main primitive types in Java

Type Name	Description	Size	Range	Sample Declaration & Initialization
boolean	true or false	1 bit	{true, false}	boolean myBool = true;
Char	Unicode Character	2 bytes	\u0000 to \uFFFF	char myChar = 'a';
Byte	Signed Integer	1 byte	−128 to 127	int myInt = 100;
Short	Signed Integer	2 bytes	−32768 to 32767	short myShort = 1000;
Int	Signed Integer	4 bytes	−2147483648 to 2147483647	int myInt = 100000;
Long	Signed Integer	8 bytes	−9223372036854775808 to 9223372036854775807	long myLong = 0;
Float	IEEE 754 floating point	4 bytes	±1.4E−45 to ±3.4028235E+38	float myFloat = 10.0f;
double	IEEE 754 floating point	8 bytes	±4.9E−324 to ±1.7976931348623157E+308	double myDouble = 20.0;

3.3 Casting

Because Java is a strongly typed language, it is sometimes necessary to perform a cast of a variable. Casting is the explicit(显式) or implicit(隐式) modification(修改) of a variable's type. Remember before when I mentioned that once you declare a variable as a specific type, it has that type indefinitely? Well, casting allows us to view the data within a given variable as a different type than it was given. Casting is a rather large topic (especially when dealing with reference types). Instead of trying to tackle(解决) all of it, I'll just hit on a few aspects relating to Java's primitive data types.

> 类型转换允许将一个变量的类型改变为不同的类型。

In order to perform a cast, you place the type that you want to view the data as in parenthesis prior to the variable you want to cast, like this:

```
1: short x = 10;
2: int y = (int)x;
```

This code snippet(片段) causes the data in x, which is of type short, to be turned into an int and that value is placed into y. Please note, the variable x is not changed! The value within x is copied into a new variable (y), but as an int rather than a short.

There are two types of conversions that can be done to data, widening conversions and narrowing conversions. Widening conversions can be done implicitly (by the compiler, that is) and narrowing conversions must be done explicitly, by you, the programmer. The difference is simple. A widening conversion is when you take a variable and assign it to a variable of larger size. A narrowing conversion is when you take a variable and assign it to a variable of smaller size. Let's look at some examples:

> 对数据来讲有两种转换方式：扩展转换和收缩转换。扩展转换是隐式的(通过编译器)，而收缩转换必须显式声明。

3.3.1 Widening

```
1: byte x = 100;       // Size = 1 byte
2:                     // In memory: x = 01100100 = 100
3: short y;            // Size = 2 bytes
4:                     // In memory: y = 0000000000000000 = 0
5: y = x;              // Legal
6:                     // In memory: y = 0000000001100100 = 100
```

3.3.2 Narrowing

```
1: short x = 1000;     // Size = 2 bytes
2:                     // In memory: x = 0000001111101000 = 1000
3: byte y;             // Size = 1 byte
4:                     // In memory: y =00000000 = 0
5: y = x;              // Illegal
6:                     // Not enough space to copy all data
7: y = (byte)x;        // Legal
8:                     // In memory: y =11101000 = 232
```

You can see from the first example that it is easy to copy the contents of a byte into a short because a short has enough space to accommodate whatever data was originally stored in the byte. In the second example, however, there isn't enough space in a byte to accommodate what may be in the short. In order to get around that, you are required to explicitly cast, as seen in the seventh line. Notice, however, even though you are allowed to perform this cast, some data may be lost, as was the case in this example. This potential data loss is the reason for the required explicit cast.

> 字节型转换为短整型只是简单地将内容复制，因为短整型提供了足够大的空间来存储原存储在字节型中的任何数。如果没有提供足够的空间来存储短整型，为了实现这种转换就必须显式地进行转换。

3.4 Reference Types

We have now looked at 8 of the 9 data types in Java. The final type is what is known as a reference type. Basically, anything that is not a primitive (an int, a float, etc.) is a reference. That means that arrays are references, as are instances of classes. The notion of a reference is similar to that of a pointer in C++. Let's say you have created a class called MyClass. When you want to create an instance of that class, you'll probably want to name it so that you can work with it like any variable. In doing so, the variable that you create does not have the object you've created in it. Rather, it has a reference to that object in it. Let's look at an example and see if we can make this make sense:

```
1:  // 创建一个新的对象
2:  MyClass anInstanceOfMyClass = new MyClass();
```

The variable anInstanceOfMyClass does not really contain an instance of MyClass. Rather, it contains the memory address where that instance really exists. The notion of a reference is very important in Java when it comes to multiple accesses to the same object and parameter passing. For now, it's sufficient enough to know that when you declare an array or the instance of a class, the variable you store it in is a reference, rather than the object itself.

> anInstanceOfMyClass 引用包含着被创建对象的内存首地址。

3.5 Summary

That's it for data types in Java. Just remember that Java is a strongly typed language. So much so, in fact, that you'll learn these data types in a big hurry if you use Java very much at all. For those of you coming from a C++ type background, you'll probably notice right away that Java is even more strongly typed than C++. For example, no more assigning 1 to a boolean variable. In Java, booleans are true or false - nothing else. Most of the information here is probably review/common sense if you've done any sort of programming at all. The most important thing for you to get out of this is to remember that Java is strongly typed.

As you will already know a variable is a value in a program which can be changed, this value is then used in some way by the program. For example, if a program were to ask the user for their age then print out how old they would be in 10 years time the program would store the age as a variable and then add 10 to it.

In Java variables are *declared* with the following syntax(语法）:

```
data_type variable_name;
```

The data_type can be from the **simple types** or from user defined types (which will be covered in later tutorials).

The simple types are:

byte, **short**, **int** and **long**. All these types hold integer numbers of varying minimum and maximum sizes.

float and **double** which hold floating point numbers, again with different minimum and maximum values.

char is used to hold the value of characters.

boolean variables are either true or false.

> byte, short, int 和 long，这些类型都存储具有不同范围的最大值和最小值的整数。
> float 和 double 存储浮点数字，同样具有不同的最小值和最大值。
> char 用来存储字符的值。
> boolean 变量为真值或假值。

3.6　Complex Data Types

3.6.1　Reference Data Types

Along with primitive data types, the Java Virtual Machine (JVM) defines data types known as reference data types. Reference types "refer" to variables and, even though Java doesn't have an explicit pointer type, can be thought of as pointers to the actual value or set of values represented by variables. An object can be "pointed to" by more than one reference. The JVM never addresses objects directly; instead, the JVM operates on the objects' references.

There are three distinct reference types: class, interface, and array. Reference types can contain references to dynamically created instances of classes, instances, and arrays. References may also contain a special value known as the null reference. The null reference has no type at runtime but may be cast to any reference type. The default value for reference types is null.

> 引用类型被认为是实际值的指针或变量的指针。一个对象可以拥有多个引用。有三种不同的引用类型：类、接口和数组。引用类型能够指向动态创建的类实例、接口和数组。

3.6.2　Class Types

Classes refer to data types that define methods and data. Internally, the JVM typically implements a class type object as a set of pointers to methods and data. Variables defining class types can refer only to instances of the class types or to the *null* reference, as the following snippet demonstrates:

```
MyObject anObject = new MyObject();      // 合法
MyObject anotherObject = null;           // 合法
MyObject stillAnotherObject = 0;         // 不合法
```

3.6.3　Interface Types

An interface acts as a template defining methods that an object must implement in order to be referred to as an instance of the interface. An interface cannot be instantiated. A class can implement multiple interfaces and be referred to by any of the interfaces that it implements. An interface variable can refer only to instances of classes that implement that interface.

> 接口扮演着方法模板的角色，对象通过引用接口来实现接口中的方法。

For example, assuming we have defined an interface called Comparable and a class called SortItem, which implements Comparable, we can define the following code:

```
Comparable c = new SortItem();//对象SortItem实现接口Comparable
```

If the Comparable interface defines one method, public void compare(Comparable item), then the SortItem class must provide an implementation of the compare method, as follows:

```
public class SortItem implements Comparable {
    public void compare(Comparable item) {
```

```
       ...method implementation here
    }
}
```

> 接口是定义方法的模板，对象必须执行该方法的目的是引用接口的实例。接口不能实例化。一个类可以实现多个接口，并且可以引用任何被实现的接口。一个接口变量仅引用执行该接口的类的实例。

3.7 Composite Data Types

The Java language does not support structs or unions. Instead, classes or interfaces are used to build composite types. In defense of this limitation, classes provide the means to bundle data and methods together, as well as a way to restrict(限制) access to the private data of the class. For example, using a struct in C, we might define a car this way:

```
struct Car
{
    char*model;
    char*make;
    int year;
    Engine*engine;
    Body*body;
    Tire**tires;
}
```

The example above implies that we have previously defined Engine, Body, and Tire structs. In Java, the example above assumes that we have previously defined the CarModels and CarMakes interfaces and that we have previously defined the classes Engine, Body, Tire, DurangoEngine, DurangoBody, and GoodyearTire.

3.7.1 Initializing Composite Data Types

Composite data types are initialized by the code defined in the constructor called when a creation technique, such as "new," is applied to a variable of a class type. Fields of a class are initialized with their default values or with explicit value assignments before the constructor is called.

> 复合数据类型的初始化由构造方法的代码实现，构造方法是在 new 创建对象。

In Listing B, when the instance of *myClass* is created using the *new* operator, the constructor, *public MyClass*( ), is called to allow the class to initialize itself. The initialization process occurs as follows:

(1) The statement "int myInt;" is called and assigns the default value, 0, to myInt.

(2) The statement "AnotherClass anotherClass;" is called and assigns the default value, null, to anotherClass.

(3) The statement within the constructor "myint = 2;" is called and assigns the value 2 to myInt.

3.7.2 Predefined Composite Data Types

Java supplies a vast array of predefined composite data types(预定义复合数据类型). One of these is the String class belonging to the java.lang package. The String class provides methods to perform a number of commonly used string operations, such as *length*( ), *substring*(*int beginIndex*), *toUpperCase*( ), *toLowerCase*( ), *equals*( ) and others. Another commonly used composite data type Java provides is the Vector class belonging to the java.util package. The Vector class defines methods to perform commonly used operations on an expandable array of objects. Some of these methods are *add*(*int index, Object element*), *elementAt*(*int index*), *isEmpty*( ), and *remove*(*int index*). These are a small sample of the predefined composite data types that Java supplies. In future articles, we'll discuss these and other predefined types in detail.

3.8 Casting Variables to a Different Type

There will be occasions when you will have a variable which you wish to store in a variable of another type.

3.8.1 Automatic Casting

When the Java compiler knows that the new data type has enough room to store the new value the casting is automatic and there is no need to specify it explicitly. For example, if you had a value stored in a byte typed variables and wanted to store it in a integer typed variable the compiler will let you because a byte value will always fit in an integer.

```
int a;          // 定义一个变量a
byte b=10;      // 定义一个字节b并分配10
a = b;          // 将b赋值给a
```

This kind of casting is commonly known as widening since the value has been made wider because it is now stored in a type which is larger than it actually needs, it is useful to view the widened value as padded by zeros.

> 这种扩展是一种普通的转换,它的值会扩展,因为它存储在一个比它实际还大的类型中,通过填补0来完成这种扩展。

3.8.2 Explicit Casting

When an int value is to be stored in a byte it is possible that the byte will not be able to hold the full value. In this case as explicit casting must be made, the syntax for this is as follows:

```
variable_of_smaller_type = (smaller_type) value_of_larger_type;
```

> 该赋值语句将右边较大的类型变量转换为左边类型较小的变量。

At first this appears complicated but it isn't really, what the code is really saying is that the variable on the left is made equal to the smaller_type version of the value on the right. For the example above, this is:

```
class CastTest{
    public static void main (String args[]){
        byte a;                   //声明一个字节变量a
        int b=10000;              //声明一个整型变量并赋值10000
        a = (byte) b;             //显式地将b转换成a
        System.out.println (a);   //将a打印出来
    }
}
```

In other words the variable a is assigned the byte value of the int variable b. When the full value will not fit in the new type, as in this example, then the value is reduced modulo（模）the byte's range. In the above example the result stored in a is 16, that is 10000 % 128 = 16, where 128 is the range of a byte (0 to 127). If you want to try this for yourself, the just copy and paste it, save as CastTest.java and compile with the javac command and run with java - as you did for the first program in the last lesson.

> 将整型变量b的值赋值给字节变量。在这个例子中，变量的值不能完整地装入新的类型中，这个值被缩小为字节型的范围内。

3.9 Java's Floating Point Types

3.9.1 Primitive Floating Point Types

Java supplies two different primitive floating point data types.Table 3.2 lists characteristics of two floating point types.

> Java 提供了两种浮点数据类型，其特征如表3.2所示。

Table 3.2 Java's primitive floating point types.

type	size(bits)	default value	minimum value	maximum value
float	32	0.0	1.40129846432481707e-45	3.40282346638528860e+38
double	64	0.0	4.94065645841246544e-324	1.79769313486231570e+308

The minimum and maximum values of the types are declared as manifest constant values in the corresponding Number class.For example the minimum and maximum double values are declared as java.lang.Double.MIN_VALUE and java.lang.Double.MAX_VALUE.

3.9.2 Integer Operators

The operators for the double data types are listed with examples. The operators for the float type are essentially identical.Table 3.3 lists relative operators of float data types.

Table3.3 The operators for **float** data type

name	example	other
Addition（加）	other = dis + that	17.0
Subtraction（减）	other = dis - that	13.0
Multiplication（乘）	other = dis * that	30.0
Division（除）	other = dis / that	7.5

3.9.3 Input and Output of Floating Point Values

The considerations(事项)for the input and output of floating point values are essentially identical to those for integer data types.

3.9.4 Casting of Floating Point to and from Integer Values, and Floating Point Literals

Casting a floating point value from a float to a double representation(表示法)is never dangerous, as it can be guaranteed that a double can always store the value accurately.

For example the following can never result in a dangerous operation and upon completion both variables will have the same value.

```
float floatVariable = someFloatValue;
double doubleVariable;
doubleVariable = floatVariable;
```

Although this is acceptable by Java it is regarded as good style to always indicate that the developer is aware that a type conversion is taking place by explicitly indicating a cast in the assignment.

```
doubleVariable = (double) floatVariable;
```

Likewise any integer value can be automatically(自动地) or explicitly(显式地) cast to either floating point type without any danger, as the floating point variable will always be able to represent it. However the converse(相反的) conversion, from any floating point type to any integer type, can be accomplished by an explicit cast but will result in an unpredictable(不可预知的) value unless the floating point value is within the limits of the integer type's range. This can be guarded against as shown in the following fragment.

```
if (((long) doubleVariable) >= java.lang.Long.MIN_VALUE &&
    ((long) doubleVariable) <= java.lang.Long.MAX_VALUE )
  {
  longVariable = (long) doubleVariable;
  }
else {
  // Value cannot be converted
  } // End if
```

Floating point literals can be expressed in conventional (e.g. 345.678) or exponent notation (e.g. 3.45678e+02), but both mantissa and exponent must always be expressed as denary (base 10) values. By default such literals are always assumed to be of type double. Should it ever be required a double value can be indicated by appending a D and a float by appending a F, e.g. 345.678D or 345.678F.

> 浮点型数字可以用传统的符号和指数符号来表示。但它们的尾数和指数必须用十进制数表示。这些数字默认为 double 类型。为了明确表示 double 值要附加一个 D，而 float 值要附加一个 F，如 345.678D，345.678F。

3.9.5 Floating Point Operations in the Standard Packages

A random float or double value between 0.0 and 1.0 can be obtained from the class-wide nextFloat() or nextDouble() actions in the java.util.Random class. The random number package is automatically seeded from the system clock which will ensure that a different sequence of random numbers is obtained every time an instance is constructed.

The java.lang.Math class supplies a class-wide random( ) action which can sometimes be more conveniently used to obtain a double value between 0.0 and 1.0. Other actions and class wide constants in java.lang.Math. Table 3.4 lists functions in java.lang.Math.

Table 3.4   Java.lang.Math

manifest values(常数)	
E	E is implemented as 2.7162816284590452354F
PI	Pi is implemented as 3.14159265358979323846F
class wide general actions	
public static double exp(double *number*)	Exponential, e raised to the power *number*
public static double log(double *number*) throws ArithmeticException	Natural logarithm of *number*. Throws exception if number is less than 0.0
public static double sqrt(double *number*) throws ArithmeticException	Square root of *number*. Throws exception if number is less than 0.0
public static double ceil(double *number*)	*Ceil*ing, the smallest whole number >= *number*
class wide general actions	
public static double floor(double *number*)	*Floor*, the smallest whole number <= *number*
public static double pow(double *this*,double *that*) throws ArithmeticException	Power, *this* raised to the power *that*. Throws exception under various conditions
public static int round(float *number*)	Rounds by adding 0.5 and then returning the largest int value less or equal to it
public static long round(double *number*)	As for int round( ) action, but using long
public static synchronized double random( )	Generates a random number between 0.0 and 1.0
public static float max(float *this*, float *that*)	Returns greater of *this* and *that*, (double version also supplied)
public static float min(float *this*, float *that*)	Returns least of *this* and *that*, (double version also supplied)
public static double sin(double *angle*)	Sine value of *angle* (in radians)
public static double cos(double *angle*)	Cosine value of *angle* (in radians)
public static double tan(double *angle*)	Tangent value of *angle* (in radians)
public static double asin(double *angle*)	Arc sine value of *angle* (in radians)
public static double acos(double *angle*)	Arc cosine value of *angle* (in radians)
public static double atan(double *angle*)	Arc tangent value of *angle* (in radians)
For the Arc variants the value of angle has to be between -1.0 and +1.0 and the value returned is between -/2 and /2	

3.9.6   The Float Class

In addition to the primitive floating point types Java also supplies two classes which implement the double and float types as objects, the major reason for this is to allow them to be used with Java's utility classes. For example if you wanted to store double values in a hash table then only Double instances could be used, not primitive double variables.

The two packages are java.lang.Number.Double and java.lang.Number.Float., the Double class is essentially identical. Table 3.5 lists functions in class java.lang.Number.Float.

Table 3.5 java.lang.Number.Float

Manifest values	
MIN_VALUE	The minimum Float value
MAX_VALUE	The minimum Float value
NEGATIVE_INFINITY	Value returned by some Math actions
POSITIVE_INFINITY	Value returned by some Math actions
NaN ( )	Not a Number, Value returned by some Math actions
Constructors (构造器)	
public Float(float *aFloat*)	Constructs a Float with the value *aFloat*
public Float(double *aDouble*)	Constructs a Float with the value *aDouble*
public Float(String *aString*) throws NumberFormatException	Constructs a Float with the value represented in *aString*. Throws exception if *aString* is invalid
instance actions	
public int intValue( )	Obtains value as primitive int, by casting
public long longValue( )	Obtains value as primitive long, by casting
public float floatValue( )	Obtains value as primitive float
public double doubleValue( )	Obtains value as primitive double
public String toString( )	Formats value in a String
public boolean isNaN( )	True if value Not a Number
public boolean isInfinite( )	True if value is, positive or negative, infinity
public int hashCode( )	Supplies hash code
public boolean equals(Object *object*)	True if *object* contains same Float value
class-wide actions	
public static String toString(float *aFloat*)	Formats *aFloat* in a String
public static Float valueOf(String *aString*) throws NumberFormatException	Creates new Float initialized to the value in *aString*. Exception thrown if *aString* cannot be so interpreted
public static boolean isNaN(float *aFloat*)	True if *aFloat* is Not a Number
public static boolean isInfinite(float *aFloat*)	True if *aFloat* is, positive or negative, infinity

3.10 Variable

Variables are used in a Java program to contain data that changes during the execution of the program.

3.10.1 Declaring a Variable

To use variables, you must notify the compiler of the name and the type of the variable (declare the variable).

The syntax for declaring a variable in Java is to precede the name of the variable with the name of the type of the variable as shown below. It is also possible (but not required) to initialize a variable in Java when it is declared as shown.

> Java 中声明变量：在变量名称前加变量的类型名称。Java 在声明变量时也可初始化(但不是必需的)变量。

```
int ch1, ch2 = '0';
```

This statement declares two variables of type **int**, initializing one of them to the value of the zero character (**0**).

3.10.2 Difference between Zero and '0'-Unicode Characters

The value of the zero character is not the same as the numeric value of zero, but hopefully you already knew that. As an aside, characters in Java are 16-bit entities called Unicode characters instead of 8-bit entities as is the case with many programming languages. The purpose is to provide many more possible characters including characters used in alphabets(字母表) other than the one used in the United States..

3.10.3 Initialization of the Variable

Initialization of the variable named **ch2** in this case was necessary to prevent a compiler error. Without initialization of this variable, the compiler recognized and balked at the possibility that an attempt might be made to execute the following statement with a variable named **ch2** that had not been initialized.

3.10.4 Error Checking by the Compiler

The strong error-checking capability of the Java compiler refused to compile this program until that possibility was eliminated by initializing the variable.

```
System.out.println("The char before the # was" + (char)ch2);
```

Java编译器错误检测功能会拒绝编译一个没有初始化变量的程序。

3.10.5 Using the Cast Operator

You should also note that the variable **ch2** is being cast as a **char** in the above statement. Recall that **ch2** is a variable of type **int**, but we want to display the character that the numeric value represents. Therefore, we must cast it (purposely change its type for the evaluation of the expression). Otherwise, we would not see a character on the screen. Rather, we would see the numeric value that represents that character.

3.10.6 Why Declare the Variables as Type Int?

It was necessary to declare these variables as type **int** because the highlighted portion of the following statement returns an **int**.

```
while((ch1 = System.in.read()) != '#') ch2 = ch1;
```

Java provides very strict type checking and generally refuses to compile statements with type mismatches(错配).

3.10.7 Shortcut Declaring Variables of the Same Type

There is a short cut which can be taken when declaring many variables of the same type, you can put them all on the same line, after the data type, separating the names with commas.

> 声明同一类型的多个变量，可以将它们放置在数据类型后的同一行，并用逗号分开。

```
//declaring variables of the same type on the same line
int counter, age, number;
```

3.10.8 Assigning Values to Variables

Once a variable has been declared then assigning a value to it takes the following syntax:

```
variable_name = value;
```

But the value must be of the same type, see the examples below:

```
//声明一个变量
int counter, age, number;
char letter;
float pi;
boolean flag;
//给以上变量赋值
counter = 3;
number = counter;
letter = 'z';
pi = 3.1415927
flag = true
counter = 'z'
number = 7.6
flag = 1;
```

3.10.9 A Shortcut, Declare and Assign at the Same Time

Often you will want to declare a variable and assign a value to it at the same time, you can do this as follows:

```
int counter = 0;
boolean flag = false;
```

3.11 Record

When a number of data items are chunked together into a unit, the result is a data structure. Data structures can have very complex structure, but in many applications, the appropriate data structure consists simply of a sequence of data items. Data structures of this simple variety can be either arrays or records.

The term "record" is not used in Java. A record is essentially the same as a Java object that has instance variables only, but no instance methods. Some other languages, which do not support objects in general, nevertheless do support records. The C programming language, for example, is not object-oriented, but it has records, which in C go by the name "struct." The data items in a record — in Java, an object's instance variables—are called the fields of the record. Each item is referred to using a field name. In Java, field names are just the names of the instance variables. The distinguishing characteristics of a record are that the data items in the record are referred to by name and that different fields in a record are allowed to be of different types. For example, if the class Person is defined as:

> record 在 Java 中没有被使用。本质上讲，一个记录与 Java 中的对象实例是一样的，但是它没有实例的方法。在 Java 中，字段名正好就是实例变量的名字。区分一个记录中的字段是通过记录中的数据项名字来实现的，并且记录中不同的字段允许有不同的类型。

```
class Person {
     String name;
     int id_number;
     Date birthday;
     int age;
}
```

then an object of class Person could be considered to be a record with four fields. The field names are name, id_number, birthday, and age. Note that the fields are of various types: String, int, and Date.

Because records are just a special type of object, I will not discuss them further.

3.12 Sample Program Practice

Example 3.1

```
//扩展精度
class typetest {
   public static void main (String args[]) {
    long bigInteger = 98765432112345678L;
    float realNo = bigInteger;
    System.out.println(bigInteger);
    System.out.println(realNo);
   }
}
```

Example 3.2

```
//显式转换
class typetest2 {
   public static void main(String args[]) {
       byte   b = 0;
       char   c = 'A';
       short  s = 0;
       int    i = 0;
       long   l = 0L;
       float  f = 0.0F;
       double d = 0.0;
       f = l;
       l = (long)f;
       d = b;
       b = (byte)s;
       s = (short)c;
       c = (char)s;
       b = (byte)c;
       b = (byte)s;
       c = (char)b;
```

```
            i = b;
            System.out.print (......what ever you want to print here);
        }
    }
```

Example 3.3

```
//十六进制的转换
public class Tester {
    public static void main(String[] args) {
        short s1 = 0x1234;
        int   i1 = 0x0000abcd;
        byte b1;
        b1 = (byte)s1;
        System.out.println("s1=" + s1 + ", b1=" + b1);
        s1 = (short)i1;
        b1 = (byte)s1;
        System.out.println("i1=" + i1 + ", s1=" + s1 + ", b1=" + b1);
    }
}
```

3.13 Exercise for you

1. Write one program to print one byte, one short, one int, one long, one float, one double, one char and one boolean value.

2. Write a program that can initialize a value dynamically.

3. Write a program which cast one integer to byte, one double to int and one double to byte.

Chapter 4   Operators

Java provides rich operator environment. Most of its operators can be divided into the following groups: arithmetic, relational, logical.

Java 提供了丰富的运算符。运算符可分为算术运算符，关系运算符，逻辑运算符。

4.1 Arithmetic Operators

Arithmetic operators are used in mathematical expression(数学表达式)in the same way that they are used in algebra(代数学). The Table 4.1 lists the arithmetic operators:

Table 4.1 Arithmetic Operators

Operator	Result	Operator	Result
+	Addition	+=	Addition assignment
-	Subtraction	-=	Subtraction assignment
*	Multiplication	*=	Multiplication assignment
/	Division	/=	Division assignment
%	Modulus	%=	Modulus assignment
++	Increment	--	Decrement

The operands of the arithmetic operators must be of numeric type. You cannot use them on boolean types, but you can use them on char types, since the char type in Java is, essentially, a subset of int.

```
class example1 {
   public static void main(String args[]) {
      System.out.println("Integer Arithmetic");
      int a = 1 + 1;
      System.out.println("a="+a);
   }
}
```

when you run this program you will see the following output：
Integer Arithmetic
a = 2

4.1.1 The Modulus Operators

The modulus operator, %, return the remainder of a division operation. It can be applied to floating-point types as well as integer types. The following program demonstrates.

求余运算符%，返回除法操作的余数。它可以应用于浮点型，也可以应用于整型。下面将给出示例。

```
class example2 {
   public static void main(String args[]) {
```

```
            int x = 42;
            double y = 42.3;
            System.out.println("x mod 10 = " + x % 10);
            System.out.println("y mod 10 = " + y % 10);
        }
    }
```

when you run this program you will get following output:

x mod 10 = 2

y mod 10 = 2.299999999999997

4.1.2 Arithmetic Assignment Operators

Java provides special operators that can be used to combine(结合) an arithmetic operation with an assignment. As you probably know, statements like the following are quite common in programming.

```
    a=a+4;
```

In Java, you can write this statement as shown here :

```
    a+=4;
```

This version uses the += assignment operator. Both statements performs the same action: they increase the value of a by 4;here are some more examples.

```
a=a%2;  can be written as a%=2;
b=b*3;  can be written as b*=3;
c=c/5;  can be written as c/=5;
d=d-7;  can be written as d-=7;
```

4.1.3 Increment and Decrement

The ++ and the – are Java's increment and decrement operators. The increment operator increases its operand by one. The decrement operator decreases its operand by one.

For example, this statement x=x+1; can be rewritten like this by use of increment operator x++;

Similarly, this statement x=x-1; is equivalent to x--;

These operators are unique in that they can appear both in postfix form, where they follow the operand as just shown, and prefix form, where they precede the operand.

> 当这些运算符跟在操作数的后面时,以后缀形式出现;当这些运算符放在操作数的前面时,以前缀形式出现。

```
    class example3 {
        public static void main(String args[]) {
            int a = 23;
            int b = 5;
            System.out.println("a & b : " + a + " " + b);
            a += 30;
            b *= 5;
            System.out.println("after arithmetic assignment a & b: "+a+" "+b);
            a++;
```

```
            b--;
            System.out.println("after increment & decrement a & b: "+a+" "+b);
    }
}
```

when you run this program you will get following output:

a & b : 23 5

after arithmetic assignment a & b : 53 25

after increment & decrement a & b : 54 24

4.2 Relational Operators

The relational operator determine the relationship that one operand has to the other. Specifically, they determine equality and ordering. Table 4.2 shows relational operators.

Table 4.2   Relational Operators

Operator	Result
==	Equal to
!=	Not equal to
>	Greater than
<	Less than
>=	Greater than or equal to
<=	Less than or equal to

The outcome of these operations is a boolean value. The relational operators are most frequently used in the expressions that control the if statement and the various loop statements.

Any type in Java, including integers, floating-point numbers, characters, and Boolean can be compared using the equality test, = =, and the inequality test, !=. Notice that in Java equality is denoted with two equal signs, not one. (single equal sign is the assignment operator.)

eg:
```
    int a = 4;
    int b = 1;
    boolean c = a < b;
```
In this case, the result of a < b (which is false) is stored in c.

4.3 Boolean Logical Operators

The boolean logical operators shown here operate only on boolean operands. All of the binary logical operators combine two boolean values to form a resultant boolean value. Table 4.3 lists logical operators.

Table 4.3   Logical Operators

Operator	Result
&&	AND
||	OR
!	Logical unary NOT
==	Equals to
!=	Not Equals to
?:	Ternary if-then-else

```
class example4 {
    public static void main(String args[]) {
        boolean b;
        b = (2 > 3) && (3 < 2);
        System.out.println("b = "+b);
        b = false || true ;
        System.out.println("b = "+b);
    }
}
```

The output of the program is shown here:

b = false

b = true

Ternary Operator

Java includes a special ternary (three-way) operator that can replace certain types of if-then-else statement.

> Java 包含一个特殊的运算符(三元运算符)，可以替代语句 if-then-else 类型。

```
expression1 ? expression2 : expression3
```

Here, expression1 can be any expression that evaluates to a boolean value. If expression1 is true, then expression2 is evaluated; otherwise, expression3 is evaluated. The result of the ? operation is that of the expression evaluated. Both expression2 and expression3 are required to return the same type, which can't be void.

> 先求解表达式 1 的布尔值，若为真则求解表达式 2 的值，否则求解表达式 3 的值。?运算的结果是被求解的表达式的值。要求表达式 2 和表达式 3 返回相同类型的值，这个值不能为空。

```
eg: int i=20;
    int j=30;
    int max = ( i > j ) ? x : j;
```

this section checks for the condition, if i is greater than j then value of i is assigned to max else value of j is assigned to max.

```
class example5 {
    public static void main(String args[]) {
        int i,k;
        i = 10;
        k = i < 0 ? -i : i;
        System.out.print("Absolute value of ");
        System.out.println(i + " is " + k);
        i = -10;
        k = i < 0 ? -i : i;    //get absolute value of i
        System.out.print("Absolute value of ");
        System.out.println(i + " is " + k);
    }
}
```

when you run this program you will get following output:
Absolute value of 10 is 10
Absolute value of -10 is 10

4.4 Bitwise and Shift Operators

The bitwise and shift operators are low-level operators that manipulate the individual bits that make up an integer value. The bitwise operators are most commonly used for testing and setting individual flag bits in a value. In order to understand their behavior, you must understand binary (base-2) numbers and the twos-complement format used to represent negative integers. You cannot use these operators with floating-point, boolean, array, or object operands. When used with boolean operands, the &, |, and ^ operators perform a different operation, as described in the previous section.

If either of the arguments to a bitwise operator is a long, the result is a long. Otherwise, the result is an int. If the left operand of a shift operator is a long, the result is a long; otherwise, the result is an int.

> 位运算和移位运算是底层的运算，可以实现将单个的二进制位组装成整型。位运算符通常被用来测试和设置单个二进制位的值。如果位运算操作数是 long 型，其结果也是 long 型；否则，结果就是 int 型。

4.4.1 Bitwise Complement (~)

The unary ~ operator is known as the bitwise complement, or bitwise NOT, operator. It inverts each bit of its single operand, converting ones to zeros and zeros to ones. For example:

```
byte b = ~12;          // ~00000110 ==> 11111001 or -13 decimal
flags = flags & ~f;
```

4.4.2 Bitwise AND (&)

This operator combines its two integer operands by performing a Boolean AND operation on their individual bits. The result has a bit set only if the corresponding bit is set in both operands. For example:

> &运算符将参加运算的两个整数操作数按二进制位进行布尔 AND 运算。如果相应的二进制位都为 1，则该位的结果为 1。

```
10 & 7                 // 00001010 & 00000111 ==> 00000010 or 2
if ((flags & f) != 0)
```

When used with boolean operands, & is the infrequently used Boolean AND operator described earlier.

4.4.3 Bitwise OR (|)

This operator combines its two integer operands by performing a Boolean OR operation on their individual bits. The result has a bit set if the corresponding bit is set in either or both of the operands. It has a zero bit only where both corresponding operand bits are zero. For example:

> |运算符将参加运算的两个整数操作数按二进制位进行布尔 OR 运算。如果相应的二进制位为 1 或者全都为 1，则该位的结果为 1。相应的二进制位全都为 0，则该位的结果为 0。

```
10 | 7              // 00001010 | 00000111 ==> 00001111 or 15
flags = flags | f;
```

When used with boolean operands, | is the infrequently used Boolean OR operator described earlier.

4.4.4 Bitwise XOR (^)

This operator combines its two integer operands by performing a Boolean XOR (Exclusive OR) operation on their individual bits. The result has a bit set if the corresponding bits in the two operands are different. If the corresponding operand bits are both ones or both zeros, the result bit is a zero. For example:

```
10 & 7              // 00001010 ^ 00000111 ==> 00001101 or 13
```

When used with boolean operands, ^ is the infrequently used Boolean XOR operator.

4.4.5 Left Shift (<<)

The << operator shifts the bits of the left operand left by the number of places specified by the right operand. High-order bits of the left operand are lost, and zero bits are shifted in from the right. Shifting an integer left by n places is equivalent to multiplying that number by 2^n. For example:

> <<运算符将它左边的操作数的二进制数左移该运算符右边操作数指定的位数。

```
10 << 1    // 00001010 << 1 = 00010100 = 20 = 10*2
7 << 3     // 00000111 << 3 = 00111000 = 56 = 7*8
-1 << 2    // 0xFFFFFFFF << 2 = 0xFFFFFFFC = -4 = -1*4
```

If the left operand is a long, the right operand should be between 0 and 63. Otherwise, the left operand is taken to be an int, and the right operand should be between 0 and 31.

4.4.6 Signed Right Shift (>>)

The >> operator shifts the bits of the left operand to the right by the number of places specified by the right operand. The low-order bits of the left operand are shifted away and are lost. The high-order bits shifted in are the same as the original high-order bit of the left operand. In other words, if the left operand is positive, zeros are shifted into the high-order bits. If the left operand is negative, ones are shifted in instead. This technique is known as sign extension; it is used to preserve the sign of the left operand. For example:

> >>运算符将它左边的操作数的二进制数右移该运算符右边操作数指定的位数。如果左边的这个操作数是正数，移位后高位补 0。如果左边的这个操作数是负数，移位后高位补 1。这被称为符号扩展，常用来保存左边操作数的符号。

```
10 >> 1    // 00001010 >> 1 = 00000101 = 5 = 10/2
27 >> 3    // 00011011 >> 3 = 00000011 = 3 = 27/8
-50 >> 2   // 11001110 >> 2 = 11110011 = -13 != -50/4
```

If the left operand is positive and the right operand is n, the >> operator is the same as integer division by 2^n.

4.4.7 Unsigned Right Shift (>>>)

This operator is like the >> operator, except that it always shifts zeros into the high-order bits of the result, regardless of the sign of the left-hand operand. This technique is called *zero extension*; it is appropriate when the left operand is being treated as an unsigned value (despite the fact that Java integer types are all signed). Examples:

```
-50 >>> 2     // 11001110 >>> 2 = 00110011 = 51
0xff >>> 4    // 11111111 >>> 4 = 00001111 = 15 = 255/16
```

4.5 Assignment Operators

The assignment operators store, or assign, a value into some kind of variable. The left operand must evaluate to an appropriate local variable, array element, or object field. The right side can be any value of a type compatible with the variable. An assignment expression evaluates to the value that is assigned to the variable. More importantly, however, the expression has the side effect of actually performing the assignment. Unlike all other binary operators, the assignment operators are right-associative, which means that the assignments in a=b=c are performed right-to-left, as follows: a=(b=c).

The basic assignment operator is =. Do not confuse it with the equality operator, = =. In order to keep these two operators distinct, I recommend that you read = as "is assigned the value".

> 不要混淆赋值运算符 "=" 与等号 "= =" 运算符。

In addition to this simple assignment operator, Java also defines 11 other operators that combine assignment with the 5 arithmetic operators and the 6 bitwise and shift operators. For example, the += operator reads the value of the left variable, adds the value of the right operand to it, stores the sum back into the left variable as a side effect, and returns the sum as the value of the expression. Thus, the expression x+=2 is almost the same as x=x+2. The difference between these two expressions is that when you use the += operator, the left operand is evaluated only once. This makes a difference when that operand has a side effect. Consider the following two expressions, which are not equivalent:

> 除了简单的赋值符号以外，Java 还定义其他 11 种运算符与赋值运算符共同使用。

```
a[i++] += 2;
a[i++] = a[i++] + 2;
```

The general form of these combination assignment operators is:

```
var op = value
```

This is equivalent (unless there are side effects in var) to:

```
var = var op value
```

The available operators are:

```
+=   -=   *=   /=   %=     // Arithmetic operators plus assignment
&=   |=   ^=               // Bitwise operators plus assignment
<<=  >>=  >>>=             // Shift operators plus assignment
```

The most commonly used operators are += and −=, although &= and |= can also be useful when working with boolean flags. For example:

```
i += 2;              // Increment a loop counter by 2 增加2
c -= 5;              // Decrement a counter by 5 减少5
flags |= f;
flags &= ~f;
```

4.6 The Conditional Operator

The conditional operator ?: is a somewhat obscure ternary (three-operand) operator inherited from C. It allows you to embed a conditional within an expression. You can think of it as the operator version of the if/else statement. The first and second operands of the conditional operator are separated by a question mark (?), while the second and third operands are separated by a colon (:). The first operand must evaluate to a boolean value. The second and third operands can be of any type, but they must both be of the same type.

> 条件运算符 ?:是一个三元运算符(三个操作数)，条件运算符的前面两个运算符是由一个问号分开的，而第二个和第三个运算符是由冒号分开的。第一个运算符的计算结果必须为boolean型的值。第二个和第三个操作数可以为任何类型，但两者必须是同一类型。

The conditional operator starts by evaluating its first operand. If it is true, the operator evaluates its second operand and uses that as the value of the expression. On the other hand, if the first operand is false, the conditional operator evaluates and returns its third operand. The conditional operator never evaluates both its second and third operand, so be careful when using expressions with side effects with this operator. Examples of this operator are:

```
int max = (x > y) ? x : y;
String name = (name != null) ? name : "unknown";
```

Note that the ?: operator has lower precedence than all other operators except the assignment operators, so parentheses are not usually necessary around the operands of this operator. Many programmers find conditional expressions easier to read if the first operand is placed within parentheses, however. This is especially true because the conditional if statement always has its conditional expression written within parentheses.

4.7 The Instanceof Operator

The instanceof operator requires an object or array value as its left operand and the name of a reference type as its right operand. It evaluates to true if the object or array is an *instance* of the specified type; it returns false otherwise. If the left operand is null, instanceof always evaluates to false. If an instanceof expression evaluates to true, it means that you can safely cast and assign the left operand to a variable of the type of the right operand.

> 实例运算符要求一个对象或数组值作为其左边的操作数，引用类型的名称作为其右边的操作数。

The instanceof operator can be used only with array and object types and values, not primitive types and values. Object and array types are discussed in detail later in this chapter. Examples of instanceof are:

```
"string" instanceof String
//True: all strings are instances of String
"" instanceof Object
//True: strings are also instances of Object
new int[] {1} instanceof int[]
//True: the array value is an int array
new int[] {1} instanceof byte[]
//False: the array value is not a byte array
new int[] {1} instanceof Object
//True: all arrays are instances of Object
null instanceof String
//False: null is never instanceof anything
//Use instanceof to make sure that it is safe to cast an object
if (object instanceof Point) {
  Point p = (Point) object;
}
```

4.8 Special Operators

There are five language constructs in Java that are sometimes considered operators and sometimes considered simply part of the basic language syntax. These "operators" are in order to show their precedence relative to the other true operators. The use of these language constructs is detailed elsewhere in this chapter, but is described briefly here, so that you can recognize these constructs when you encounter them in code examples:

4.8.1 Object Member Access (.)

An *object* is a collection of data and methods that operate on that data; the data fields and methods of an object are called its members. The dot (.) operator accesses these members.

If o is an expression that evaluates to an object reference, and f is the name of a field of the object, o.f evaluates to the value contained in that field. If m is the name of a method, o.m refers to that method and allows it to be invoked(调用)using the ( ) operator shown later.

> 对象是数据和操作数据方法的集合；对象的数据域和方法是其成员，可以通过.操作访问其成员。

4.8.2 Array Element Access ([ ])

An array is a numbered list of values. Each element of an array can be referred to by its number, or index. The [] operator allows you to refer to the individual elements of an array.

If a is an array, and i is an expression that evaluates to an int, a[i] refers to one of the elements of a. Unlike other operators that work with integer values, this operator restricts array index values to be of type int or narrower.

> 假如a是一个数组，i是一个整型的表达式，a[i]就是a的一个元素。和其他整型操作数不同，这个运算符限制了数组的下标为整型或更小的类型。

4.8.3 Method Invocation (())

A method is a named collection of Java code that can be run, or invoked, by following the name of the method with zero or more comma-separated expressions contained within parentheses. The values of these expressions are the arguments to the method. The method processes the arguments and optionally returns a value that becomes the value of the method invocation expression.

If o.m is a method that expects no arguments, the method can be invoked with o.m( ). If the method expects three arguments, for example, it can be invoked with an expression such as o.m(x,y,z). Before the Java interpreter invokes a method, it evaluates each of the arguments to be passed to the method. These expressions are guaranteed to be evaluated in order from left to right (which matters if any of the arguments have side effects).

4.8.4 Object Creation (new)

In Java, objects are created with the new operator, which is followed by the type of the object to be created and a parenthesized list of arguments to be passed to the object *constructor*. A constructor is a special method that initializes a newly created object, so the object creation syntax is similar to the Java method invocation syntax. For example:

```
new ArrayList();
new Point(1,2)
```

> new 运算符被用来创建一个对象，对象后面的括号用来对应相应的构造方法。

4.9 Type Conversion or Casting

As we've already seen, parentheses can also be used as an operator to perform narrowing type conversions, or casts. The first operand of this operator is the type to be converted to; it is placed between the parentheses. The second operand is the value to be converted; it follows the parentheses. For example:

```
(byte) 28
(int) (x + 3.14f)
(String)h.get(k)
```

4.10 Sample Program Practice

Example 4.1

```
//右移
class Rightshift {
    public static void main(String args[]) {
        char hex[]={'0','1','2','3','4','5','6','7','8','9','a','b','c','d','e','f'};
        byte b = (byte)0xff;
        c = (byte)(b >> 4);
        d = (byte)(b >>> 4);
        e = (byte)(b & 0xff) >> 4;
        System.out.println("b = 0x" + hex[(b>>4) & 0x0f] + hex[b & 0x0f]);
        System.out.println("b >>4 = 0x" + hex[(c>>4) & 0x0f] + hex[c & 0x0f]);
    }
}
```

Exapmple 4.2
```
//左移
class L_shift{
    public static void main(String args[]){
        int i,j;
        int num = 0xfffe;//???
        for(i=0; i<4; ++i){
            num <<= 1;
            System.out.print(num);
        }
    }
}
```

Example 4.3
```
//递归
class Factorial{
  int Fact(int n){
    if(n == 1){
      return 1;
    }
    else{
      return Fact(n-1)*n;
    }
  }
}
class Recursion {
  public static void main(String args[]){
    Fact f = new Fact();
    System.out.println(f.Fact(3));
    System.out.println(f.Fact(4));
    System.out.println(f.Fact(5));
  }
}
```

Example 4.4
```
//位运算
class Bitlogical{
  public static void main(String args[]){
        String binary[] = {
                    "0000","0001","0010","0011",
                    "0100","0101","0110","0111",
                    "1000","1001","1010","1011",
                    "1100","1101","1110","1111"
                };
        int a = 3,
            b = 6,
            c = a | b,
            d = a & b,
            e = a ^ b,
            f = (~a&b) | (a&~b),
            g = ~a & 0x0f;
        System.out.println("a = " + binary[a]);
```

```java
            System.out.println("b = " + binary[b]);
            System.out.println("c = " + binary[c]);
            System.out.println("d = " + binary[d]);
            System.out.println("e = " + binary[e]);
            System.out.println("f = " + binary[f]);
            System.out.println("g = " + binary[g]);
        }
    }
```

Example 4.5
```java
    //二维数组实例
    class DouDimArray  {
      public static void main(String args[]){
            double D[][] = {{0*0,1*0,2*0,3*0},
                            {0*1,1*1,2*1,3*1},
                            {0*2,1*2,2*2,3*2},
                            {0*3,1*3,2*3,3*3}};
            int i,j;
            for(i=0; i<4; ++i){
                for(j=0; j<4; ++j){
                    System.out.println(D[i][j]+" ");
                }
                System.out.println();
            }
        }
    }
```

Example 4.6
```java
    //三维数组实例
    class ThrDimArray  {
        public static void main(String args[]){
            double D[][][] = new double[3][4][5];
            int i,j,k;
            for(i=0; i<3; ++i){
                for(j=0; j<4; ++j){
                    for(k=0; k<5; ++k){
                        D[i][j][k] = i*j*k;
                    }
                }
            }
            for(i=0; i<3; ++i){
                for(j=0; j<4; ++j){
                    for(k=0; k<5; ++k){
                        System.out.println(D[i][j][k] + " ");
                    }
                    System.out.println("    ");
                }
                System.out.println();
            }
        }
    }
```

4.11 Exercise for you

1. Write one program to show the absolute value(绝对值) of 10 and –10.
2. What is assignment operator and how you can assign the value of x=y=z=20?
3. Write one program to show the Boolean logical operator.

Chapter 5　Flowing Control

5.1　Control Statements

Whenever the computer runs a Java program, it goes straight from the first line of code to the last. Control statements allow you to change the computer's control from automatically reading the next line of code to reading a different one. Let's say you only want to run some code on condition. For example, let's say that you are making a program for a bank. If someone wants to see his records, and gives his password, you don't always want to let him see the records. First, you need to see if his password is correct. You then create a control statement saying "if" the password is correct, then run the code to let him see the records. "else" run the code for people who enter the wrong password. You can even put one control statement inside another. In our example, the banking system has to worry about people trying to get someone else's records. So, as the programmer, you give the user three shots. If he misses them, then the records are locked for a day. You can have code with control statements that look like this. (not real code, only meant to help you understand. also, Java doesn't have go back to earlier lines- again just trying to focus on the control part.) There are basically three types of controlling statement in java. There are Selection, Repetition and Branching.

(1) Selection Statements(选择条件语句)
(2) Repetition Statements(循环语句)
(3) Branching Statements(分支语句)

> 计算机执行 Java 程序是严格地一条代码一条代码的顺序执行。而流控制语句允许改变程序执行次序，并不一定按照计算机的顺序执行次序执行代码。

5.2　Selection Statements

5.2.1　If Statement

This is a control statement to execute a single statement or a block of code, when the given condition is true and if it is false then it skips if block and rest code of program is executed.

Syntax:
```
if(conditional_expression){
<statements>;
...;
...;
}
```

Example: If n%2 evaluates to 0 then the "if" block is executed. Here it evaluates to 0 so if block is executed. Hence "This is even number" is printed on the screen.

> 如果 n 模 2 的值为 0，if 中的程序块被执行。"This is even number" 将被打印输出到屏幕上。

```
int n = 10;
if(n%2 = = 0){
   System.out.println("This is even number");
}
```

5.2.2　If-else Statement

The "if-else" statement is an extension of if statement that provides another option when 'if' statement evaluates to "false" i.e. else block is executed if "if" statement is false.

> if-else 语句是 if 语句的扩展，当 if 语句中的条件为假时，提供另一种选择。

Syntax:
```
if(conditional_expression){
<statements>;
...;
...;
}
else{
<statements>;
....;
....;
}
```

If n%2 doesn't evaluate to 0 then else block is executed. Here n%2 evaluates to 1 that is not equal to 0 so else block is executed. So "This is not even number" is printed on the screen.

```
int n = 11;
if(n%2 = = 0){
   System.out.println("This is even number");
}
else{
   System.out.println("This is not even number");
}
```

5.2.3　Switch Statement

This is an easier implementation to the if-else statements. The keyword "switch" is followed by an expression that should evaluates to byte, short, char or int primitive data types, only. In a switch block there can be one or more labeled cases. The expression that creates labels for the case must be unique. The switch expression is matched with each case label. Only the matched case is executed, if no case matches then the default statement (if present) is executed.

> if-else 语句能够比较容易地实现。关键字 switch 中包含一个表达式，这个表达式只能为 byte、short、char 或 int 数据类型。

Syntax:
```
switch(control_expression){
case expression 1:
   <statement>;
case expression 2:
   <statement>;
```

```
    ...
    ...
    case expression n:
        <statement>;
    default:
        <statement>;
}//end switch
```

Here expression "day" in switch statement evaluates to 5 which matches with a case labeled "5" so code in case 5 is executed that results to output "Friday" on the screen.

```
int day = 5;
switch (day) {
    case 1:
        System.out.println("Monday");
        break;
    case 2:
        System.out.println("Tuesday");
        break;
    case 3:
        System.out.println("Wednesday");
        break;
    case 4:
        System.out.println("Thursday");
        break;
    case 5:
        System.out.println("Friday");
        break;
    case 6:
        System.out.println("Saturday");
        break;
    case 7:
        System.out.println("Sunday");
        break;
    default:
        System.out.println("Invalid entry");
        break;
}
```

5.3　Repetition Statements

5.3.1　While Loop Statement

This is a looping or repeating statement. It executes a block of code or a statement till the given condition is true. The expression must be evaluated to a Boolean value. It continues testing the condition and executes the block of code. When the expression results to false control comes out of loop.

> 循环语句是指只要给定的条件为真，代码块或语句将被执行。Expression 表达式的值必须为 Boolean 类型。程序将一直测试条件的真假以判断是否执行代码块。当表达式的值为 false 则循环结束。

Syntax:
```
while(expression){
<statement>;
...;
...;
}
```

Example: Here expression i<=10 is the condition which is checked before entering into the loop statements. When "i" is greater than value 10 control comes out of loop and next statement is executed. So here "i" contains value "1" which is less than number "10" so control goes inside of the loop and prints current value of i and increments value of i. Now again control comes back to the loop and condition is checked. This procedure continues until "i" becomes greater than value "10". So this loop prints values 1 to 10 on the screen.

```
int i = 1;
//print 1 to 10
while (i <= 10){
  System.out.println("Num " + i);
  i++;
}
```

5.3.2 Do-while Loop Statement

This is another looping statement that tests the given condition past so you can say that the do-while looping statement is a past-test loop statement. First the **do** block statements are executed then the condition given in **while** statement is checked. So in this case, even the condition is false in the first attempt, do block of code is executed at least once.

> 这种循环将后检测给定的条件。因此可以说 do-while 循环是一种后检测循环。首先 do 语句块被执行，然后 while 语句中的给定条件被检查。因此在这个例子中，尽管给定的条件是 false，但程序还是运行一次。

Syntax:
```
do{
<statement>;
...;
...;
}while (expression);
```

Example: Here first do block of code is executed and current value "1" is printed then the condition i<=10 is checked. Here "1" is less than number "10" so the control comes back to do block. This process continues till value of i becomes greater than 10.

```
int i = 1;
do{
  System.out.println("Num: " + i);
   i++;
}while(i <= 10);
```

5.3.3　For Loop Statement

This is also a loop statement that provides a compact way to iterate over a range of values. From a user point of view, this is reliable because it executes the statements within this block repeatedly till the specified conditions are true.

Syntax:

```
for (initialization; condition; increment or decrement){
   <statement>;
   ...;
   ...;
}
```

initialization: The loop is started with the value specified.(初始化)

condition: It evaluates to either 'true' or 'false'. If it is false then the loop is terminated.(计算条件，如果计算的结果为假，循环将被结束)

increment or decrement: After each iteration, value increments or decrements.(每次执行后，值被增加或减少)

Example: Here num is initialized to value "1", condition is checked whether num<=10. If it is so then control goes into the loop and current value of num is printed. Now num is incremented and checked again whether num<=10.If it is so then again it enters into the loop. This process continues till num>10. It prints values 1 to10 on the screen.

```
for (int num = 1; num <= 10; num++){
   System.out.println("Num: " + num);
}
```

5.4　Branching Statements

5.4.1　Break Statement

The break statement is a branching statement that contains two forms: labeled and unlabeled. The break statement is used for breaking the execution of a loop (while, do-while and for). It also terminates the switch statements.

> break语句属于分支语句，它包含两种形式：有标签的和没有标签的。break语句被用来终止循环的执行(while, do-while, for)，也可以用来终止switch语句。

Syntax:

```
break;            // breaks the innermost loop or switch statement.
break label;      // breaks the outermost loop in a series of nested loops.
```

Example: When if statement evaluates to true it prints "data is found" and comes out of the loop and executes the statements just following the loop. The code shows that program will jump loop when break statement is executed.

> 当if语句是真的时候，data is found被打印出来，并调出循环，执行循环后面的语句。

```
int num[]={2,9,1,4,25,50};
int search=4;
for(int i=1;i<num.length;i++)
{
   if (num[i]==search)
   {
      System.out.println("data is found");
      break;
   }
}
```

5.4.2 Continue Statement

This is a branching statement that are used in the looping statements (while, do-while and for) to skip the current iteration of the loop and resume the next iteration. The code shows that program will end the current iteration when num[i]! =search.

> continue 语句被用来跳出当前本次循环，但是会继续执行下次循环。

Syntax:
```
continue;
```
Example:
```
int num[]={2,9,1,4,25,50};
int search=4;
for(int i=1;i<num.length;i++)
{
   if (num[i]==search)
   {
      System.out.println("data is found");
      continue;
   }
   if(found==search)
   {
      System.out.println("data is found");
      break;
   }
}
```

5.4.3 Return Statement

It is a special branching statement that transfers the control to the caller of the method. This statement is used to return a value to the caller method and terminates execution of method. This has two forms: one that returns a value and the other that can not return. The returned value type must match the return type of method.

Syntax:
```
return;
return values;
```
return;   //This returns nothing. So this can be used when method is declared with void return type.
return expression;   //It returns the value evaluated from the expression.

Example: Here Welcome( ) function is called within println( ) function which returns a String value "Welcome to roseIndia.net". This is printed to the screen. The code shows the value will be return from Welcome( ).

> 在 println( )方法中 welcome( )被调用，返回一个 string 值 "Welcome to roseindia.net"，并将其打印输出到屏幕上。

```
public static void hello()
{
    System.out.println("Hello"+welcome());
}
static string welcome()
{
    Return("welcome to roseindia.net");
}
```

5.5 Sample Program Practice

Example 5.1

```
//闰年的条件是符合下面二者之一：能被 4 整除，但不能被 100 整除；能被 4 整除，同时又能被 100 整除
public class KY2_6 {
    public static void main(String args[]) {
        boolean leap;
        int year=2005;
        if((year%4==0 && year%100!=0) || (year%400==0)) // 方法 1
            System.out.println(year+" 年是闰年");
        else
            System.out.println(year+" 年不是闰年");
        year=2008; // 方法 2
        if(year%4!=0)
            leap=false;
        else if (year%100!=0)
            leap=true;
        else if (year%400!=0)
            leap=false;
        else
            leap=true;
        if (leap==true)
            System.out.println(year+" 年是闰年");
        else
            System.out.println(year+" 年不是闰年");
        year=2050; // 方法 3
        if(year%4==0) {
            if(year%100==0) {
                if(year%400==0)
                    leap=true;
                else
                    leap=false;
            }
            else
                leap=false;
```

```
            }
            else
                leap=false;
            if(leap==true)
                System.out.println(year+" 年是闰年");
            else
                System.out.println(year+" 年不是闰年");
        }
```

Example 5.2
```
//使用 switch 语句
    class KY2_7{
        public static void main(String args[]) {
            int c=38;
            switch (c<10?1:c<25?2:c<35?3:4) {
                case 1:
                    System.out.println(" "+c+"℃ 有点冷。要多穿衣服。");
                case 2:
                    System.out.println(" "+c+"℃ 正合适。出去玩吧。");
                case 3:
                    System.out.println(" "+c+"℃ 有点热。");
                default:
                    System.out.println(" "+c+"℃ 太热了!开空调。");
            }
        }
    }
```

Example 5.3
```
//for 循环
    class KY2_8{
        public static void main (String args[]) {
            int h,c;
            System.out.println("摄氏温度 华氏温度");
            for(c=0; c<=40; c+=5) {
                h=c*9/5+32;
                System.out.println(" "+c+"          "+h);
            }
        }
    }
```

Example 5.4
```
//while 循环
    import java.io.*;
    class KY2_9 {
        public static void main(String args[]) throws IOException {
            char ch;
            System.out.println("按 1/2/3 数字键可得大奖!");
            System.out.println("按空格键后回车可退出循环操作。");
            while ((ch=(char)System.in.read())!=' '){
                System.in.skip(2);       //跳过回车键
                switch (ch) {
                    case '1':
                        System.out.println("恭喜你得大奖,一辆汽车!");
```

```
                    break;
                case '2':
                    System.out.println("不错呀，你得到一台笔记本电脑!");
                    break;
                case '3':
                    System.out.println("没有白来，你得到一台冰箱!");
                    break;
                default:
                    System.out.println("真遗憾，你没有奖品！下次再来吧。");
            }
        }
    }
```

Example 5.5
```
//do-while 循环
class KY2_10 {
    public static void main(String args[]) {
        int n=1, sum=0;
        do{
            sum+=n++;
        }while (n<=100);
        System.out.println("1+2+...+100 ="+sum);
    }
}
```

5.6 Exercise for you

1. Write a program in which all the 12 months of the year would be declared and divide in to 4 parts like spring, summer, autumn and winter. Your month name should show what is the state of the season (for example April is spring, June is summer and so on). Use if..else If.

2. Write a program to print 10 lines one billow another where line one prints 10 stars ("*") line two prints 9 stars…..like that till line 10 prints one.

3. Write a program to print the loop in side another loop.

Chapter 6　Class

It is important to realize that classes have two related meanings in software development. First, a class is a modeling construction, that is, a way of representing entities in the real world. In the 'analysis' stage of software development, the developer may build a model of the system to be developed based on classes. Classes are also programming constructs, that is, units of organization of a program. Ideally there should be a strong correspondence between these two views: classes in the model should correspond to classes in the program. In practice the correspondence is rarely one-to-one; the programmer will usually have to introduce additional classes to support the operation of the program at the technical level. The Java 'String' class is an example of this use of a class.

> 软件开发中类有两层含义。一方面，类是一种建模方法，用来表示现实世界中的实体集。在软件开发的分析阶段，开发人员可以建立系统的基于类的模型。另一方面，类又是程序的构成部件。这两层含义在某种意义上又有很强的一致性，即建模中设计的类和构成程序的类是一致的。实际上这种一致性是不完全的，主要是有的时候从技术的角度，程序员不得不为了实现某些功能而设计一些额外的类。

Classes are the single most important language feature that facilitates object-oriented programming (OOP), the dominant programming methodology in use today. An object is a value of a class type and is referred to as an instance of the class. An object differs from a value of a primitive type in that it has methods (actions) as well as data. For example, "Hello" is an object of the class String. It has the characters in the string as its data and also has a number of methods, such as length.

> 类是目前程序开发的主流程序设计语言的最为重要的特征，它使得面向对象程序设计更加方便快捷。对象是类类型的值，是类实例的引用。对象和基本类型数据的值有区别，它除了有数据还包括方法。

6.1　Class Definition

A Java program consists of objects from various classes interacting with one another. Before we go into the details of how you define classes, let's review some of the terminology used with classes. Among other things, a class is a type and you can declare variables of a class type. A value of a class type is called an object. An object has both data and actions. The actions are called methods. Each object can have different data, but all objects of a class have the same types of data and all objects in a class have the same methods. An object is usually referred to as an object of the class or as an instance of the class rather than as a value of the class, but it is a value of the class type. To make this abstract discussion come alive, we need a sample definition.

For example, if A is a class, then the phrases "bla is of type A", "bla is an instance of the class A", and "bla is an object of the class A" all mean the same thing.

> Java 程序由多个交互的不同类对象组成。和其他数据一样，类也是一种数据类型，可以声明该类型的变量。类中包含的行为称为方法。每个对象都包含不同的数据，但是类定义的所有对象包含数据的类型是一致的，并拥有相同的方法。对象通常都是类实例的引用，而不是类的值，但是它是类类型的值。

6.1.1　A Simple Class Definition

```
class Count {
    public static void main(String args[])      //定义一个main方法
        throws Java.io.IOException              //抛出输入输出异常
    {
        int count = 0;
        while (System.in.read() != -1)
            count++;
        System.out.println("Input has " + count + " chars.");
    }
}
```

In the Java language, all methods and variables must exist within a class. So, the first line of the character-counting application defines a class, Count, that defines the methods, variables, and any other classes needed to implement the character-counting application. Since this program is such a simple one, the Count class just defines one method named main.

> 在 Java 语言中，所有的方法和变量都必须存在于一个类中。在字符统计应用程序的第一行定义了一个 Count 类，在类中定义变量和方法。

6.1.2　Defining a Class

The implementation of a class in Java always looks similar to the following:

```
class Customer
  {
    // variables
    <variable1>
    <variable2>
    <variable...>
    // methods
    <method1>
    <method2>
    <method...>
  }
```

A class defines a set of variables (technical instance variables, or fields, or attributes) and a set of methods. A class definition acts as a template for each instance of the class created when a Java program runs. When a Java program runs none, one or many instances (objects) may be created from a single class.

> 类定义了一组变量(技术上又称实例变量、实例成员或实例属性)和一组方法。Java 程序运行时，类充当了实例化对象创建的模板。Java 程序运行时，一个类可能产生一个或多个实例化对象，或者一个都不产生。

Classes are central to a Java program. Every program must have at least one. The program may use classes to represent real world (concrete) categories of object, or the program classes may be entirely internal (such as user interface classes like Button, MenuItem etc.). Program instructions can only exist inside methods, which are inside classes.

> 类是 Java 的核心，每个 Java 程序都必须至少有一个类。Java 程序用类表现现实世界中不同类型的对象，也可以表现对象的内部特征(用户界面相关类中例如 Button 类和 MenuItem 类等)。Java 程序的指令只能定义在类内部的方法中。

The program designer decides what variables and methods to define as part of a class after analysis and modeling activities, prior to coding at a computer.

6.1.2.1 Variables

Variables store values. Variables have a name, a type, and a value. For example, in a program concerned with company accounts, we may define a class called Customer, representing and managing the details of the company's customers. Variables of Customer may include name, address, credit limit, etc. Each object of class customer may have different values of these variables.

Taking the 'Customer' example, let's assume that the important features of a customer are name, address, account balance and credit limit. We might define this as below:

```
class Customer
{
    // Variables of customer
    String name;
    String address;
    int accountBalance;
    int creditLimit;
    // Methods of Customer go here
}
```

Defining a more complex Class.

Considering a library records system, we shall define a class for text items as follows:

```
class TextItem
{
    // variables
    String title;
    int numPages;
    String shelfMark;
    boolean onLoan;
    int daysLate;
    // methods
    void TextItem( String itemTitle, int itemNumPages, String itemShelfMark)
    { // method implementation }
    String getTitle() { // method implementation }
    int getNumPages() { // method implementation }
    String getShelfMark() { // method implementation }
    boolean getLoanStatus() {// method implementation }
    void setLoanStatus(boolean newLoanStatus)
    { // method implementation }
    int getDaysLate() { // method implementation }
    void setDaysLate(int newDaysLate)
```

```
            {//method implementation  }
            double getFine() { // method implementation    }
            void informBorrowerOfFine( <args> )
            {// method implementation }
    }
```

For now the implementation of the methods has been replaced with comments.

As can be seen from the class definition, the class defines 5 variables:

```
title
numPages
shelfMark
onLoan
daysLate
```

These correspond to attributes identified during analysis and modeling. However, another attribute was also identified, which has not been implemented as a variable in this class. That other attribute was the fine due on a TextItem object.

A general rule of thumb, when performance is not a major issue, is never to store a value that can be calculated when it is needed. If we assume that the fine for a TextItem is always 0.05 pounds for each day late, since we have the variable daysLate already stored as a variable, we can always calculate the fine when needed.

> 一般来说，当不必考虑性能的时候，可以计算得到的值一般是不需要存储的。可以假设如果每天的罚款都是0.5英镑，并且定义变量daysLate存储超期的天数，则可以随时计算所需要的罚款。

A method that needs to find out the fine is the getFine() method, which we might implement as follows:

```
double getFine()
{
    return (0.05 * daysLate);
}
```

This method calculates the fine and returns the result of the calculation as a reply.

Not every attribute identified in analysis and modeling is implemented as a variable.

If a value can be calculated form other variables, then a get( ) method is all that is needed — the implemented variable will be set via changes to the variables that make up its calculation.

> 并不是分析建模阶段确立的每个属性都必须在实现的时候定义成变量，如果某个值可以计算后得到其他的值，则只需要定义一个get方法，在该方法中对其他的变量进行计算生成最终的值。

6.2 Declaring and Instantiating an Object

```
public class DateFirstTry
{
    public String month;
    public int day;
```

```
            public int year; //a four digit number.
            public void writeOutput()
            {
                System.out.println(month + " " + day + ", " + year);
            }
    }
    public class DateFirstTryDemo
    {
            public static void main(String[] args)
            {
                DateFirstTry date1, date2;
                date1 = new DateFirstTry();
                date2 = new DateFirstTry();
                date1.month = "December";
                date1.day = 31;
                date1.year = 2007;
                System.out.println("date1:");
                date1.writeOutput();
                date2.month = "July";
                date2.day = 4;
                date2.year = 1776;
                System.out.println("date2:");
                date2.writeOutput();
            }
    }
```

The example contains a definition for a class named DateFirstTry and a program that demonstrates using the class. Objects of this class represent dates such as December 31, 2007 and July 4, 1776. This class is unrealistically simple, but it will serve to introduce you to the syntax for a class definition. Each object of this class has three pieces of data: a string for the month name, an integer for the day of the month, and another integer for the year. The objects have only one method, which is named writeOutput. Both the data items and the methods are sometimes called members of the object, because they belong to the object. The data items are also sometimes called fields. We will call the data items instance variables and call the methods methods.

> 这个例子包含了 DateFirstTry 类的定义及测试该类的一个测试程序。该类的每个对象都包含三个数据：表示月份的字符串，表示该月第几日的整数，以及表示年份的整数。这些对象都只有一个 writeObject 方法。对象中的所有数据项和方法称为该对象的成员，它们都属于该对象。数据项有时也称为数据域，本书中称数据项为实例变量，称方法为方法。

The following three lines from the start of the class definition define three instance variables (three data members):

```
public String month;
public int day;
public int year;      //a four digit number
```

The word public simply means that there are no restrictions on how these instance variables are used. Each of these lines declares one instance variable name. You can think of an object of the class as a complex item with instance variables inside of it. So, an instance variable can be thought of as a smaller variable inside each object of the class. In this case, the instance variables are called month, day, and year.

An object of a class is typically named by a variable of the class type. For example, the program DateFirstTryDemo declares the two variables date1 and date2 to be of type DateFirstTry, as follows:

```
DateFirstTry date1, date2;
```

This gives us variables of the class DateFirstTry, but so far there are no objects of the class. Objects are class values that are named by the variables. To obtain an object, you must use the new operator to create a "new" object. For example, the following creates an object of the class DateFirstTry and names it with the variable date1:

> 可以将类的对象看做内部实例变量的组合体,一个实例变量是对象内部的变量。类的对象一般称为类类型的变量。对象是类的值,对象名就是定义的变量名。必须通过 new 操作符才能生成类的对象。

```
date1 = new DateFirstTry();
```

We will discuss this kind of statement in more detail later in this chapter when we discuss something called a constructor. For now simply note that:

```
Class_Variable= newClass_Name();
```

creates a new object of the specified class and associates it with the class variable. Because the class variable now names an object of the class, we will often refer to the class variable as an object of the class. (This is really the same usage as when we refer to an int variable n as "the integer n", even though the integer is, strictly speaking, not n but the value of n.)

Unlike what we did in above example, the declaration of a class variable and the creation of the object are more typically combined into one statement, as follows:

```
DateFirstTry date1 = new DateFirstTry();
import Java.util.Date;      //导入Java.util包
class DateApp {
    public static void main (String args[]) {
        Date today = new Date();
        //定义一个today对象,并将其初始化,为其分配内存空间
        System.out.println(today);
    }
}
```

The main( ) method of the DateApp application creates a Date object named today. This single statement performs three actions: declaration, instantiation, and initialization. Date today declares to the compiler that the name today will be used to refer to an object whose type is Date, the new operator instantiates a new Date object, and Date( ) initializes the object.

> 在 DateApp 应用程序的 main 方法中创建了 Date 类的对象 today,这条简单的语句执行了三个操作:对象声明、实例化、初始化。Date today 语句告诉编译器 today 是指向 Date 类型的对象的引用,new 操作符实例化一个新的 Date 对象,调用构造方法 Date( )初始化该对象。

6.2.1 Fields and Methods

When you define a class, you can put two types of elements in your class: data members (sometimes called fields), and member methods (typically called functions). A data member is an object of any type that you can communicate with via its reference. It can also be one of the primitive types (which is not a reference). If it is a reference to an object, you must initialize that reference to connect it to an actual object (using new) in a special function called a constructor (described fully later). If it is a primitive type you can initialize it directly at the point of definition in the class. (As you'll see later, references can also be initialized at the point of definition.)

> 定义一个类时，在类中定义了两种类型的成员：数据成员(有时称为数据域)和成员方法(有时称为成员函数)。通过数据成员的引用可以和对象进行数据交互，数据成员可以是基本数据类型(基本数据类型不是引用类型)。如果数据成员是关于一个对象的引用，则必须在构造方法中将其初始化并关联到一个真正的对象(用 new 操作符)。如果数据成员是基本数据类型，则可以在类定义该数据成员时直接初始化。

Each object keeps its own storage for its data members; the data members are not shared among objects. Here is an example of a class with some data members:

> 类的每个对象都有各自数据成员的存储空间，这些对象不共用数据成员。

```
class DataOnly {
  int i;
  float f;
  boolean b;
}
```

This class doesn't do anything, but you can create an object:

```
DataOnly d = new DataOnly();
```

You can assign values to the data members, but you must first know how to refer to a member of an object. This is accomplished by stating the name of the object reference, followed by a period (dot), followed by the name of the member inside the object:

```
objectReference.member
```

> 可以给数据成员赋值，但首先要知道如何引用该对象的数据成员。可以通过对象的引用后面加上"."运算符来引用对象内部的数据成员。

For example:
```
d.i = 47;
d.f = 1.1f;
d.b = false;
```

It is also possible that your object might contain other objects that contain data you'd like to modify. For this, you just keep "connecting the dots". For example:

```
myPlane.leftTank.capacity = 100;
```

The DataOnly class cannot do much of anything except hold data, because it has no member functions (methods). To understand how those work, you must first understand arguments and return values, which will be described shortly.

6.2.2 Default Values for Primitive Members

When a primitive data type is a member of a class, it is guaranteed to get a default value if you do not initialize it, table 6.1 shows the default value:

> 如果基本数据类型的数据成员没有初始化，系统会确保该数据成员有缺省值，具体的缺省值如下：

Table 6.1   Default Value of Primitive Data

Primitive type	Default
boolean	false
char	'\u0000' (null)
byte	(byte)0
short	(short)0
int	0
long	0L
float	0.0f
double	0.0d

Note carefully that the default values are what Java guarantees when the variable is used as a member of a class. This ensures that member variables of primitive types will always be initialized (something C++ doesn't do), reducing a source of bugs. However, this initial value may not be correct or even legal for the program you are writing. It's best to always explicitly initialize your variables.

> 缺省值使得 Java 能保证成员变量在任何时候都能被使用，这样确保基本数据类型的成员变量总能被初始化(有些语言如 C++没有这样处理)，减少了出错的可能。但是有时这些初始值对程序本身来说可能不正确甚至不合理，最好是显式地对成员变量进行初始化。

This guarantee doesn't apply to "local" variables—those that are not fields of a class. Thus, if within a function definition you have:

> 但这并不适合局部变量，局部变量不是类的成员，Java 并不对局部变量进行自动初始化。

```
int x;
```

Then x will get some arbitrary value (as in C and C++); it will not automatically be initialized to zero. You are responsible for assigning an appropriate value before you use x. If you forget, Java definitely improves on C++: you get a compile-time error telling you the variable might not have been initialized. (Many C++ compilers will warn you about uninitialized variables, but in Java these are errors.)

6.2.3 Methods, Arguments, and Return Values

Up until now, the term function has been used to describe a named subroutine. The term that is more commonly used in Java is method, as in "a way to do something." If you want, you can continue thinking in terms of functions. It's really only a syntactic difference, but from now on "method" will be used in this book rather than "function."

Methods in Java determine the messages an object can receive. In this section you will learn how simple it is to define a method.

The fundamental parts of a method are the name, the arguments, the return type, and the body. Here is the basic form:

```
returnType methodName( /* Argument list */ ) {
  /* Method body */
}
```

The return type is the type of the value that pops out of the method after you call it. The argument list gives the types and names for the information you want to pass into the method. The method name and argument list together uniquely identify the method.

> 返回类型是方法调用后返回值的类型，参数列表表示了需要传递给方法的信息的类型和名称；方法名和参数列表共同来标识某个方法。

Methods in Java can be created only as part of a class. A method can be called only for an object, and that object must be able to perform that method call. If you try to call the wrong method for an object, you'll get an error message at compile-time. You call a method for an object by naming the object followed by a period (dot), followed by the name of the method and its argument list, like this: objectName.methodName(arg1, arg2, arg3). For example, suppose you have a method f( ) that takes no arguments and returns a value of type int. Then, if you have an object called a for which f( ) can be called, you can say this:

```
int x = a.f();
```

The type of the return value must be compatible with the type of x.

This act of calling a method is commonly referred to as sending a message to an object. In the above example, the message is f( ) and the object is a. Object-oriented programming is often summarized as simply "sending messages to objects."

> Java 中的方法是类的一部分。方法只能被对象所调用，且该对象必须能够调用此方法。如果对象试图调用不能访问的方法，将出现编译时错误。通过在对象名后面用 "." 运算符并加上适当的方法名和参数列表调用方法，即类似 objectName.methodName(arg1, arg2, arg3) 的形式。方法调用通常是指给一个对象发送消息，面向对象程序设计通常称为"给对象发送消息"。

6.2.4 The Argument List

The method argument list specifies what information you pass into the method. As you might guess, this information—like everything else in Java—takes the form of objects. So, what you must specify in the argument list are the types of the objects to pass in and the name to use for each one. As in any situation in Java where you seem to be handing objects around, you are actually passing references. The type of the reference must be correct, however. If the argument is supposed to be a String, what you pass in must be a string.

> 方法的参数列表定义了要传递给方法的信息，这些信息在 Java 中也是封装在对象中的。因此方法中的参数列表需要指明传递的每个参数的对象类型和参数名。任何情况下传递对象参数时，实际上传递的是指向该对象的引用。而且，传递的引用类型必须是正确的。例如，如果传递的是字符串，该引用必须是字符串类型的。

Consider a method that takes a String as its argument. Here is the definition, which must be placed within a class definition for it to be compiled:

```
int storage(String s) {
    return s.length() * 2;
}
```

This method tells you how many bytes are required to hold the information in a particular String. (Each char in a String is 16 bits, or two bytes, long, to support Unicode characters.) The argument is of type String and is called s. Once s is passed into the method, you can treat it just like any other object. (You can send messages to it.) Here, the length( ) method is called, which is one of the methods for Strings; it returns the number of characters in a string.

> 该方法计算了保存一个特定字符串需要的字节数，参数 s 是 string 类型的。传递给该方法的参数和其他对象一样进行处理。这里调用了字符串的一个方法 length( )，返回该字符串包含的字符个数。

You can also see the use of the return keyword, which does two things. First, it means "leave the method, I'm done." Second, if the method produces a value, that value is placed right after the return statement. In this case, the return value is produced by evaluating the expression s.length( ) * 2.

> 关键字 return 完成了两件事件：首先表示完成了该方法的处理，结束方法的执行并返回；其次，如果该方法产生了一个值，该值将放置在 return 表达式的后面。

You can return any type you want, but if you don't want to return anything at all, you do so by indicating that the method returns void. Here are some examples:

```
boolean flag() { return true; }
float naturalLogBase() { return 2.716f; }
void nothing() { return; }
void nothing2() {}
```

When the return type is void, then the return keyword is used only to exit the method, and is therefore unnecessary when you reach the end of the method. You can return from a method at any point, but if you've given a non-void return type then the compiler will force you (with error messages) to return the appropriate type of value regardless of where you return.

> 如果返回类型是 void，则 return 关键字仅表示结束该方法并返回，如果到了方法的最后可以不使用该关键字。也可以从方法的任何地方返回，但是在非 void 类型的方法中不论从何处返回，编译器都要求必须返回对应类型的值。

6.3 Constructor

You can imagine creating a method called **initialize( )** for every class you write. The name is a hint that it should be called before using the object. Unfortunately, this means the user must remember to call the method. In Java, the class designer can guarantee initialization of every object by providing a special method called a **constructor**. If a class has a constructor, Java automatically calls that constructor when an object is created, before users can even get their hands on it. So initialization is guaranteed.

The next challenge is what to name this method. There are two issues. The first is that any name you use could clash with a name you might like to use as a member in the class. The second is that because the compiler is responsible for calling the constructor, it must always know which method to call. The C++ solution seems the easiest and most logical, so it's also used in Java: The name of the constructor is the same as the name of the class. It makes sense that such a method will be called automatically on initialization.

> Java 提供了一个特定的方法即构造方法以确保每个对象在创建时被初始化。如果类定义了构造方法，则当用户使用该对象之前、创建对象时 Java 会自动调用构造方法以确保对象的初始化。Java 中构造方法的名字和类名相同，给人感觉构造方法在初始化时会被自动调用。

Here's a simple class with a constructor:

```java
//: SimpleConstructor.Java
// Demonstration of a simple constructor
package c04;
class Rock {
    Rock() {        //定义一个构造方法
        System.out.println("Creating Rock");
    }
}
public class SimpleConstructor {
    public static void main(String[] args){
        for(int i = 0; i < 10; i++)
            new Rock();//调用构造方法初始化一个对象
    }
}
```

Now, when an object is created:

```java
new Rock();
```

Storage is allocated and the constructor is called. It is guaranteed that the object will be properly initialized before you can get your hands on it.

Note that the coding style of making the first letter of all methods lower case does not apply to constructors, since the name of the constructor must match the name of the class exactly.

Like any method, the constructor can have arguments to allow you to specify how an object is created. The above example can easily be changed so the constructor takes an argument:

> 按照一般的代码风格，方法的第一个字母小写，但这不适合构造方法命名，因为构造方法的名字必须与类名严格匹配。和其他的方法一样，构造方法也可以带参数，允许用户定制对象的创建。

```java
class Rock {
    Rock(int i) {      //定义入口参数 i 为整型的构造方法
        System.out.println(
        "Creating Rock number " + i);
    }
}
public class SimpleConstructor {
    public static void main(String[] args) {
```

```
        for(int i = 0; i < 10; i++)
            new Rock(i);
    }
}
```

Constructor arguments provide you with a way to provide parameters for the initialization of an object. For example, if the class **Tree** has a constructor that takes a single integer argument denoting the height of the tree, you would create a **Tree** object like this:

```
Tree t = new Tree(12); // 12-foot tree
```

If **Tree(int)** is your only constructor, then the compiler won't let you create a **Tree** object any other way.

Constructors eliminate a large class of problems and make the code easier to read. In the preceding code fragment, for example, you don't see an explicit call to some **initialize( )** method that is conceptually separate from definition. In Java, definition and initialization are unified concepts – you can't have one without the other.

The constructor is an unusual type of method because it has no return value. This is distinctly different from a **void** return value, in which the method returns nothing but you still have the option to make it return something else. Constructors return nothing and you don't have an option. If there were a return value, and if you could select your own, the compiler would somehow need to know what to do with that return value.

> 构造方法是一种类型很特殊的方法，它没有返回值。这与返回值类型为 void 有严格的区别，如果返回值类型为 void 时一般不返回值，但是也可以选择返回其他值；而构造方法不返回任何值。

6.3.1 Calling Constructors from Constructors

When you write several constructors for a class, there are times when you'd like to call one constructor from another to avoid duplicating code. You can do this using the **this** keyword.

Normally, when you say **this**, it is in the sense of "this object" or "the current object," and by itself it produces the handle to the current object. In a constructor, the **this** keyword takes on a different meaning when you give it an argument list: it makes an explicit call to the constructor that matches that argument list. Thus you have a straightforward way to call other constructors:

> 当一个类定义了几个构造方法时，为了避免代码重复，很多时候可以让一个构造方法调用另外的构造方法，这时需要使用 this 关键字。一般来说，this 是指 "这个对象" 或 "当前对象"。当传递不同的参数时，this 引用的是不同的构造方法，可以显式地调用参数匹配的构造方法。

```
//: Flower.java
// Calling constructors with "this"             //通过this调用
public class Flower {
    private int petalCount = 0;
    private String s = new String("null");  //定义一字符串s
    Flower(int petals) {                         //参数为整型的构造方法
        petalCount = petals;
        System.out.println("Constructor w/ int arg only, petalCount= "
```

```
                + petalCount);
        }
        Flower(String ss) {
                System.out.println("Constructor w/ String arg only, s=" + ss);
                s = ss;
        }
        Flower(String s, int petals) {            //定义带两个参数的构造方法
                this(petals);
                //!   this(s); // Can't call two!
                this.s = s; // Another use of "this"
                System.out.println("String & int args");
        }
        Flower(){
                this("hi",47);//调用该类中相匹配的构造方法Flower(String s,int petals)
                System.out.println( "default constructor (no args)");
        }
        void print() {
                //!this(11);
                // Not inside non-constructor!//非构造方法中不可以调用构造方法
                System.out.println( "petalCount = " + petalCount + " s = "+ s);
        }
        public static void main(String[] args){
                Flower x = new Flower();
                x.print();
        }
}
```

The constructor **Flower(String s, int petals)** shows that, while you can call one constructor using **this**, you cannot call two. In addition, the constructor call must be the first thing you do or you'll get a compiler error message.

This example also shows another way you'll see **this** used. Since the name of the argument **s** and the name of the member data **s** are the same, there's an ambiguity. You can resolve it by saying **this.s** to refer to the member data. You'll often see this form used in Java code, and it's used in numerous places in this book.

In **print( )** you can see that the compiler won't let you call a constructor from inside any method other than a constructor.

6.3.2 Default Constructors

As mentioned previously, a default constructor is one without arguments, used to create a "vanilla object." If you create a class that has no constructors, the compiler will automatically create a default constructor for you. For example:

> 缺省构造方法不带参数。如果定义的类中没有定义构造方法，编译器会自动创建一个缺省的构造方法，如下例：

```
//: DefaultConstructor.Java
class Bird {
    int i;
}

public class DefaultConstructor {
```

```
        public static void main(String[] args) {
            Bird nc = new Bird(); // default!
        }
    }
```

The line

```
new Bird();
```

creates a new object and calls the default constructor, even though one was not explicitly defined. Without it we would have no method to call to build our object. However, if you define any constructors (with or without arguments), the compiler will not synthesize one for you:

> 其中 `new Bird( );` 创建了一个新的对象，并调用其缺省构造方法。没有构造方法，则无法创建对象。若用户在类中自行定义了带参数或不带参数的构造方法，编译器将不会创建缺省的构造方法。

```
class Bush {
    Bush(int i) {}
    Bush(double d) {}
}
```

Now if you say:

```
new Bush();
```

The compiler will complain that it cannot find a constructor that matches. It's as if when you don't put in any constructors, the compiler says "You are bound to need some constructor, so let me make one for you." But if you write a constructor, the compiler says "You've written a constructor so you know what you're doing; if you didn't put in a default it's because you meant to leave it out."

> 创建对象时，若无法找到匹配的构造方法，编译时会报错。

6.4 Keyword "this"

As we noted earlier, if today is of type DateFirstTry , then

```
today.writeOutput();
```

is equivalent to

```
System.out.println(today.month + " " + today.day+ ", " + today.year);
```

This is because, although the definition of writeOutput reads

```
public void writeOutput()
{
    System.out.println(month + " " + day + ", " + year);
}
```

It really means

```
public void writeOutput()
{
    System.out.println(<the calling object>.month + " "+ <the calling object>.day + ", " + <the calling object>.year);
}
```

The instance variables are understood to have **<the calling object>**. in front of them. Sometimes it is handy, and on rare occasions even necessary, to have an explicit name for the calling object. Inside a Java method definition, you can use the keyword this as a name for the calling object. So, the following is a valid Java method definition that is equivalent to the one we are discussing:

```
public void writeOutput()
{
    System.out.println(this.month + " " + this.day+ ", " + this.year);
}
```

The definition of writeOutput in class DateFirstTry defined earlier could be replaced by this completely equivalent version. Moreover, this version is in some sense the true version. The version without the this and a dot in front of each instance variable is just an abbreviation for this version. However, the abbreviation of omitting the this is used frequently.

The keyword this is known as the this parameter. The this parameter is a kind of hidden parameter. It does not appear on the parameter list of a method, but is still a parameter. When a method is invoked, the calling object is automatically plugged in for this.

> 实例变量可以理解成在它们的前面加上了<the calling object>.，但是很少在实例变量前显式加上对象名。在Java的方法定义内部，可以使用this关键字来代表访问的对象名。this关键字也被称为this参数，它是一种隐含的参数，它不会出现在方法的参数列表中。当方法被调用的时候，访问对象会自动用this替换并传递到该方法中。

Note: Within a method definition, you can use the keyword this as a name for the calling object. If an instance variable or another method in the class is used without any calling object, then this is understood to be the calling object.

> 注意：方法定义的内部，可以用this关键字替换当前访问对象。如果类的实例变量和成员方法没有指定调用的对象，则可以用this来替换访问的对象。

There is one common situation that requires the use of the this parameter. You often want to have the parameters in a method such as setDate(you can add the method to the class DateFirstTry) be the same as the instance variables. A first, although incorrect, try at doing this is the following writing of the method setDate:

```
public void setDate(String month, int day, int year) //Not correct
{
    month = month;
    day = day;
    year = year;
}
```

This written version does not do what we want. When you declare a local variable in a method definition, then within the method definition, that name always refers to the local variable. A parameter is a local variable, so this rule applies to parameters. Consider the following assignment statement in our written method definition:

```
day = day;
```

Both the identifiers day refer to the parameter named day. The identifier day does not refer to the instance variable day. All occurrences of the identifier day refer to the parameter day. This is often described by saying the parameter day masks or hides the instance variable day. Similar remarks apply to the parameters month and year. This written method definition of the method setDate will produce a compiler error message.

> 两个标识符 day 都指向参数 day，没有一个标识符 day 指向实例变量 day。所有的标识符 day 都指向参数 day，这也是经常所说的参数 day 隐藏了同名的实例变量 day。这里对参数 month 和 year 进行同样的处理。setDate 方法的定义将产生编译错误信息。

The correct rewriting of the method setDate is as follows:

```
public void setDate(String month, int day, int year)
{
    this.month =month;
    this.day = day;
    this.year = year;
}
```

6.5　Garbage Collection

Programmers know about the importance of initialization, but often forget the importance of cleanup. After all, who needs to clean up an **int**? But with libraries, simply "letting go" of an object once you're done with it is not always safe. Of course, Java has the garbage collector to reclaim the memory of objects that are no longer used. Now consider a very special and unusual case. Suppose your object allocates "special" memory without using **new**. The garbage collector knows only how to release memory allocated with **new**, so it won't know how to release the object's "special" memory. To handle this case, Java provides a method called **finalize( )** that you can define for your class. Here's how it's supposed to work. When the garbage collector is ready to release the storage used for your object, it will first call **finalize( )**, and only on the next garbage-collection pass will it reclaim the object's memory. So if you choose to use **finalize( )**, it gives you the ability to perform some important cleanup at the time of garbage collection .

> 某对象运用完毕后就弃之不顾的做法并非总是安全的。Java 的垃圾回收器用来回收不再使用的对象所占用的内存空间。垃圾回收器只知道释放那些经由 new 分配的内存空间，不知道如何释放对象分配所得的特殊内存资源。不是由 new 方式创建的对象该如何处理呢？针对这种情况，Java 允许在类中定义一个名为 finalize( )的方法。当垃圾回收器准备释放对象所占用的内存空间时，将首先调用 finalize( )方法，并在下一次垃圾回收器动作发生时，才回收对象所占用的内存空间。因此在使用 finalize( )方法时，可在垃圾回收时执行一些重要的清理工作。

This is a potential programming pitfall because some programmers, especially C++ programmers, might initially mistake **finalize( )** for the destructor in C++, which is a function that is always called when an object is destroyed. But it is important to distinguish between C++ and Java here, because in C++ objects always get destroyed (in a bug-free program), whereas in Java objects do not always get garbage-collected. Or, put another way:

Garbage collection is not destruction.

If you remember this, you will stay out of trouble. What it means is that if there is some activity that must be performed before you no longer need an object, you must perform that activity yourself. Java has no destructor or similar concept, so you must create an ordinary method to perform this cleanup. For example, suppose in the process of creating your object it draws itself on the screen. If you don't explicitly erase its image from the screen, it might never get cleaned up. If you put some kind of erasing functionality inside **finalize( )**, then if an object is garbage-collected, the image will first be removed from the screen, but if it isn't, the image will remain. So a second point to remember is:

Your objects might not get garbage collected.

> Java 没有析构函数或相似的概念，必须手动创建一个普通的方法去执行回收功能。比如假设创建某个对象并将该对象绘制在屏幕上。如果不显式地从屏幕上将其擦除，它可能永远不会被释放。如果在 finalize 方法中添加一些擦除功能将此对象回收，则此图像将从屏幕上擦除；否则，如果垃圾回收未发生，此图像将一直保留。

You might find that the storage for an object never gets released because your program never nears the point of running out of storage. If your program completes and the garbage collector never gets around to releasing the storage for any of your objects, that storage will be returned to the operating system en masse as the program exits. This is a good thing, because garbage collection has some overhead, and if you never do it you never incur that expense.

> 当程序退出时，其占用的存储空间将一并返回给操作系统。

6.5.1　The Use of finalize( )

You might believe at this point that you should not use **finalize( )** as a general-purpose cleanup method. What good is it?

A third point to remember is: Garbage collection is only about memory.

That is, the sole reason for the existence of the garbage collector is to recover memory that your program is no longer using. So any activity that is associated with garbage collection, most notably your **finalize( )** method, must also be only about memory and its deallocation.

Does this mean that if your object contains other objects **finalize( )** should explicitly release those objects? Well, no-the garbage collector takes care of the release of all object memory regardless of how the object is created. It turns out that the need for **finalize( )** is limited to special cases, in which your object can allocate some storage in some way other than creating an object. But, you might observe, everything in Java is an object so how can this be?

> 使用垃圾回收器的唯一原因就是为了回收程序不再使用的内存。无论对象如何创建，垃圾回收器都会负责释放对象占据的所有内存空间。

It would seem that **finalize( )** is in place because of the possibility that you'll do something C-like by allocating memory using a mechanism other than the normal one in Java. This can happen primarily through native methods, which are a way to call non-Java code from Java. (Native methods are discussed in Appendix A.) C and C++ are the only languages currently supported by native methods, but since they can call subprograms in other languages, you can effectively call anything.

Inside the non-Java code, C's **malloc( )** family of functions might be called to allocate storage, and unless you call **free( )** that storage will not be released, causing a memory leak. Of course, **free( )** is a C and C++ function, so you'd need call it in a native method inside your **finalize( )**.

After reading this, you probably get the idea that you won't use **finalize( )** much. You're correct; it is not the appropriate place for normal cleanup to occur. So where should normal cleanup be performed?

6.5.2 Cleanup

To clean up an object, the user of that object must call a cleanup method at the point the cleanup is desired. This sounds pretty straightforward, but it collides a bit with the C++ concept of the destructor. In C++, all objects are destroyed. Or rather, all objects should be destroyed. If the C++ object is created as a local, i.e. on the stack (not possible in Java), then the destruction happens at the closing curly brace of the scope in which the object was created. If the object was created using **new** (like in Java) the destructor is called when the programmer calls the C++ operator **delete** (which doesn't exist in Java). If the programmer forgets, the destructor is never called and you have a memory leak, plus the other parts of the object never get cleaned up.

> 要回收一个对象,用户需要在垃圾回收时调用执行回收操作的方法,但与C++中的析构函数的概念有些不一致。在C++中,在堆栈中可创建一个局部对象,在对象创建的作用域结束处对该对象进行析构,用 new 创建对象,用 delete 操作符调用相应的析构函数进行对象占用空间的释放。但如果编程人员忘了运用 delete,就不会调用相应的析构函数,则会出现内存泄露现象,而对象的其他部分也不会得到释放。

In contrast, Java doesn't allow you to create local objects-you must always use **new**. But in Java, there's no "delete" to call for releasing the object since the garbage collector releases the storage for you. So from a simplistic standpoint you could say that because of garbage collection, Java has no destructor. You'll see as this book progresses, however, that the presence of a garbage collector does not remove the need or utility of destructors. (And you should never call **finalize( )** directly, so that's not an appropriate avenue for a solution.) If you want some kind of cleanup performed other than storage releases you must still call a method in Java, which is the equivalent of a C++ destructor without the convenience.

> 相反,Java必须使用 new 操作符创建局部对象。Java中没有 delete 命令回收对象,而是采用垃圾回收器来回收对象所占内存。Java没有析构函数。Java中的垃圾回收器不能完全代替析构函数(其不能直接调用 finalize( ))。在Java中除释放存储空间以外若还想执行其他的清理工作,必须调用某个方法,其等同于C++中的析构函数,只是没有析构函数方便而已。

One of the things **finalize( )** can be useful for is observing the process of garbage collection. The following example shows you what's going on and summarizes the previous descriptions of garbage collection:

```
//: Garbage.Java
// Demonstration of the garbage
// collector and finalization
class Chair {
```

```java
        static boolean gcrun = false;//静态成员变量，可以用类本身，实现成员变量的调用
        static boolean f = false;
        static int created = 0;
        static int finalized = 0;
        int i;
        Chair(){
            i = ++created;
            if(created == 47)
                System.out.println("Created 47");
        }
        protected void finalize() {
            if(!gcrun) {
                gcrun = true;
                System.out.println("Beginning to finalize after " +
                    created + " Chairs have been created");
            }
            if(i == 47) {
                System.out.println("Finalizing Chair #47, " +
                    "Setting flag to stop Chair creation");
                f = true;
            }
            finalized++;
            if(finalized >= created)
                System.out.println("All " + finalized + " finalized");
        }
    }
    public class Garbage {
       public static void main(String[] args) {
            if(args.length == 0) {        //程序未带任何命令行参数
                System.err.println("Usage: \n" +
                    "Java Garbage before\n  or:\n" +
                    "Java Garbage after");
                return;
            }
            while(!Chair.f){//程序带命令行参数，Chair.f为false时，进行对象的创建
                new Chair();
                new String("To take up space");
            }
            System.out.println("After all Chairs have been created:\n" +
                "total created = " + Chair.created +
                ", total finalized = " + Chair.finalized);
            if(args[0].equals("before")) {//程序所带命令行参数为"before"
                System.out.println("gc():");
                System.gc();          //强制进行终结动作
                System.out.println("runFinalization():");
                System.runFinalization();
            }
            System.out.println("bye!");
            if(args[0].equals("after"))//程序所带命令行参数为"after"
                System.runFinalizersOnExit(true);
        }
    }
```

The above program creates many **Chair** objects, and at some point after the garbage collector begins running, the program stops creating **Chair**s. Since the garbage collector can run at any time, you don't know exactly when it will start up, so there's a flag called **gcrun** to indicate whether the garbage collector has started running yet. A second flag **f** is a way for **Chair** to tell the **main( )** loop that it should stop making objects. Both of these flags are set within **finalize( )**, which is called during garbage collection.

Two other **static** variables, **created** and **finalized**, keep track of the number of **obj**s created versus the number that get finalized by the garbage collector. Finally, each **Chair** has its own (non-**static**) **int i** so it can keep track of what number it is. When **Chair** number 47 is finalized, the flag is set to **true** to bring the process of **Chair** creation to a stop.

All this happens in **main( )**, in the loop

```
while(!Chair.f) {
    new Chair();
    new String("To take up space");
}
```

You might wonder how this loop could ever finish, since there's nothing inside that changes the value of **Chair.f**. However, the **finalize( )** process will, eventually, when it finalizes number 47.

The creation of a **String** object during each iteration is simply extra garbage being created to encourage the garbage collector to kick in, which it will do when it starts to get nervous about the amount of memory available.

When you run the program, you provide a command-line argument of "before" or "after." The "before" argument will call the **System.gc( )** method (to force execution of the garbage collector) along with the **System.runFinalization( )** method to run the finalizers. These methods were available in Java 1.0, but the **runFinalizersOnExit( )** method that is invoked by using the "after" argument is available only in Java 1.1 and beyond. (Note you can call this method any time during program execution, and the execution of the finalizers is independent of whether the garbage collector runs).

> 程序运行时,所提供的命令行参数若为 "before",将运行 System.gc( )方法和 System.runFinalization( )方法,其中 System.gc( )方法表示要强制进行终结动作;所提供的命令行参数若为 "after",将运行 System.runFinalizersOnExit( )方法。

The preceding program shows that, in Java 1.1, the promise that finalizers will always be run holds true, but only if you explicitly force it to happen yourself. If you use an argument that isn't "before" or "after" (such as "none"), then neither finalization process will occur, and you'll get an output like this:

Created 47
Beginning to finalize after 8694 Chairs have been created
Finalizing Chair #47, Setting flag to stop Chair creation
After all Chairs have been created:
total created = 9834, total finalized = 108
bye!

Thus, not all finalizers get called by the time the program completes. [20] To force finalization to happen, you can call **System.gc( )** followed by **System.runFinalization( )**. This will destroy all the objects that are no longer in use up to that point. The odd thing about this is that you call **gc( )** before you call **runFinalization( )**, which seems to contradict the Sun documentation, which claims that finalizers are run first, and then the storage is released. However, if you call **runFinalization( )** first, and then **gc( )**, the finalizers will not be executed.

> 并非所有的回收器都在程序运行结束时被调用。为了强制垃圾回收，可以在运行 System.gc( )方法后执行 System.runFinalization( )方法。这样将销毁所有的对象，使对象不再有用。如果先运行 runFinalization( )，后运行 gc( )，回收操作不会被执行。

One reason that Java 1.1 might default to skipping finalization for all objects is because it seems to be expensive. When you use either of the approaches that force garbage collection you might notice longer delays than you would without the extra finalization.

6.6 Static Methods and Static Variables

6.6.1 Static Methods

Some methods do not require a calling object. Methods to perform simple numeric calculations are good examples. For example, a method to compute the maximum of two integers has no obvious candidate for a calling object. In Java, you can define a method so that it requires no calling object. Such methods are known as static methods. You define a static method in the same way as any other method, but you add the keyword static to the method definition heading, as in the following example:

```
public static int maximum(int n1, int n2)
{
  if (n1 > n2)
    return n1;
  else
    return n2;
}
```

Although a static method requires no calling object, it still belongs to some class, and its definition is given inside the class definition. When you invoke a static method, you normally use the class name in place of a calling object. So if the above definition of the method maximum were in a class named SomeClass, then the following is a sample invocation of maximum:

> Java 中，不需要通过对象调用的方法称为静态方法，静态方法的定义和其他方法的定义基本一致，但需要在方法定义的头部加上关键字 static。尽管静态方法不需要访问对象，它仍然隶属于某个类，必须将其定义在类的内部。通常情况下可以使用类名而不是访问对象名来调用静态的方法。

```
int budget = SomeClass.maximum(yourMoney, myMoney);
```

Here yourMoney and myMoney are variables of type int that contain some values. We have already been using one static method. The method exit in the class System is a static method. To end a program immediately, we have used the following invocation of the static method exit:

```
System.exit(0);
```

Note that with a static method, the class name serves the same purpose as a calling object.

> 注意，静态方法调用需要使用类的名称。

Within the definition of a static method, you cannot do anything that refers to a calling object, such as accessing an instance variable. This makes perfectly good sense, because a static method can be invoked without using any calling object and so can be invoked when there are no instance variables. (Remember instance variables belong to the calling object.) The best way to think about this restriction is in terms of the this parameter. In a static method, you cannot use the this parameter, either explicitly or implicitly. So you cannot use an instance variable in the definition of a static method.

Invoking a Nonstatic Method Within a Static Method

If myMethod( ) is a nonstatic (that is, ordinary) method in a class, then within the definition of any method of this class, an invocation of the form:

```
myMethod();
```

means:

```
this.myMethod();
```

and so it is illegal within the definition of a static method. (A static method has no this.)

However, it is legal to invoke a static method within the definition of another static method.

There is one way that you can invoke a nonstatic method within a static method: if you create an object of the class and use that object (rather than this) as the calling object. For example, suppose myMethod( ) is a nonstatic method in the class MyClass. Then, as we already discussed, the following is illegal within the definition of a static method in the class MyClass:

```
myMethod();
```

> 在静态方法的定义中可以合法地访问静态方法。有一种方式可以在静态方法中访问非静态方法：在该静态方法中，创建类的对象并使用该对象来访问非静态方法。

However, the following is perfectly legal in a static method or any method definition:

```
MyClass anObject = new MyClass();
anObject.myMethod();
```

6.6.2 Static Variables

A class can have static variables as well as static methods. A static variable is a variable that belongs to the class as a whole and not just to one object. Each object has its own copies of the instance variables. However, with a static variable there is only one copy of the variable, and all the objects can use this one variable. Thus, a static variable can be used by objects to communicate between the objects. One object can change the static variable, and another object can read that change. To make a variable static, you declare it like an instance variable but add the modifier static as follows:

> 类中也可以定义静态成员变量。静态成员变量属于类的全局变量，而不仅属于某个对象。每个对象都有实例成员变量的一份副本。但是静态成员变量对所有的对象来说只有一份副本。因此，静态成员变量可以用来在对象之间进行通信。一个对象改变静态成员变量的值，另一个对象可以读取被修改的值。定义静态变量和定义实例成员变量方法一样，只需要添加一个修饰符 static，例如：

```
private static int turn;
```

Or if you wish to initialize the static variable, which is typical, you might declare it as follows instead:

```
private static int turn = 0;
```

If you do not initialize a static variable, it is automatically initialized to a default value: Static variables of type boolean are automatically initialized to false. Static variables of other primitive types are automatically initialized to the zero of their type. Static variables of a class type are automatically initialized to null. However, we prefer to explicitly initialize static variables, either as just shown or in a constructor.

> 如果静态成员变量没有被初始化，系统会给其赋值一个默认的初始值：静态布尔类型的成员变量被初始化为 false；其他基本数据类型的静态成员变量被自动初始化为相应类型的零值。类类型的静态成员变量被自动初始化为 null。但是，一般都显式地对其进行初始化，或像刚才讨论过的方式或在构造方法中进行。

The following example shows a class with a static variable along with a demonstration program. Notice that the two objects, lover1 and lover2, access the same static variable turn.

```
public class TurnTaker
{
    private static int turn = 0;
    private int myTurn;
    private String name;
    public TurnTaker(String theName, int theTurn){
        name = theName;
        if (theTurn >= 0)
            myTurn = theTurn;
        else{
            System.out.println("Fatal Error.");
            System.exit(0);
        }
    }
    public TurnTaker(){
        name = "No name yet";
        myTurn = 0;   //Indicating no turn.
    }
    public String getName(){
        return name;
    }
    public static int getTurn(){
        turn++;
        return turn;
    }
    public boolean isMyTurn(){
        return (turn == myTurn);
    }
}
public class StaticDemo{
    public static void main(String[] args){
        TurnTaker lover1 = new TurnTaker("Dadiv", 1);
        TurnTaker lover2 = new TurnTaker("James", 3);
```

```
        for (int i = 1; i < 5; i++){
            System.out.println("Turn = " + TurnTaker.getTurn());
            if (lover1.isMyTurn())
                System.out.println("Love from " + lover1.getName());
            if (lover2.isMyTurn())
                System.out.println("Love from " + lover2.getName());
        }
    }
}
```

As we already noted, you cannot directly access an instance variable within the definition of a static method. However, it is perfectly legal to access a static variable within a static method, because a static variable belongs to the class as a whole. This is illustrated by the method getTurn in the above example. When we write turn in the definition of the static method getTurn, it does not mean this.turn; it means TurnTaker.turn. If the static variable turn were marked public instead of private, it would even be legal to use TurnTaker.turn outside of the definition of the class TurnTaker. Defined constants(we will discuss in other chapter), such as the following, are a special kind of static variable:

```
public static final double PI = 3.14159;
```

The modifier final in the above means that the static variable PI cannot be changed. Such defined constants are normally public and can be used outside the class. Good programming style dictates that static variables should normally be marked private unless they are marked final, that is, unless they are defined constants.

> 在静态方法内部不能直接访问实例成员变量。然而，在静态方法中可以访问静态的成员变量，因为静态成员变量属于类的全局变量。静态常量定义一般都定义为公有的，因此这些常量可以在类的外部被使用。好的编程风格强调静态变量一般都被定义为私有，静态常量常定义为公有。

6.7 Sample Examples

Example 6.1
```
class simpleClass_Demo1{
    public static void main(String args[]){
        simpleClassDemo1 obj = new simpleClassDemo1();   //生成对象obj
        int n = obj.simpleMethod(10);    //调用simpleMethod方法
        System.out.println("The result is: " + n);       //输出结果应为11
    }
    public int simpleMethod(int n){
        int temp = n+1;
        return temp;                     //入口参数为n，返回n+1
    }
}
```

Example 6.2
```
class Add{
    int num1;
    int num2;
    public int addNumber(){              //定义addNumber()方法
        int sum = 0;
```

```
            sum = num1 + num2;
            return sum;                    //返回两数的和
        }
    }
    class Sub extends Add{                 //定义一个Addition类的子类Sub
        public int subNumber(){
            int sub = 0;
            sub = num1 - num2;
            return sub;                    //返回两数的差
        }
    }
    public class simpleClassDemo3{
        public static void main(String args[]){
            Add obj1 = new Add();          //定义一个Add对象obj1
            Sub obj2 = new Sub();
            obj2.num1 = 1;
            obj2.num2 = 2;
            System.out.println("The sum is: " + obj2.addNumber());
            //调用父类的addNumber()方法
            System.out.println("The sub is: " + obj2.subNumber());
        }
    }
```

Example 6.3

```
    class Student{
        public void print(){               //定义print()方法
            System.out.println("I'm a student");
        }
    }
    class Freshman extends Student{        //定义一student子类Freshman
        public void print(){               //重写父类的print()方法
            System.out.println("I'm a first year student");
        }
    }
    class Sophomore extends Student{       //定义子类Sophomore
        public void print(){               //重写父类的print()方法
            System.out.println("I'm a second year student");
        }
    }
    public class simpleClassDemo10{
        public void identify(Student s){
            if(s instanceof Freshman){     //如果s为类fresnman的对象
                s.print();                 //调用类Freshman中的print()方法
                System.out.println("Instance of the Freshman");
            }
            if(s instanceof Sophomore)
            {}
        }
    }
```

Example 6.4
```
class A{
    int i;                              //定义一变量 i
}
class B extends A{
    int i;
    B(int a, int b){                    //定义有两个参数的构造方法
        super.i = a;                    //将 a 的值赋给父类中变量 i
        i = b;                          //将 b 赋值给 B 类中变量 i
    }
    void show(){                        //定义一个 show 方法
        System.out.println("In super class A:" + super.i);
        System.out.println("In sub class B:" + i);
    }
}
class useSuper{
    public static void main(String args[])   {
        B subob = new B(1,2);           //生成一个 subob 对象并初始化对象
        subob.show();                   //调用 show 方法输出父类及子类中 i 的值
    }
}
```

Example 6.5
```
class Base{
    Base(){                             //定义一个无参的构造方法
        System.out.println("Calling the base class constructor");
    }
}
class simpleClassDemo6 extends Base{//定义一个 Base 子类 simpleClassDemo6
    simpleClassDemo6(){                 //定义子类的构造方法
        System.out.println("Calling the derived class constructor");
    }
    public static void main(String args[])   {
        simpleClassDemo6 obj = new simpleClassDemo6();
        //利用子类的构造方法初始化一个对象 obj
    }
}
```

6.8 Exercise for you

1. Write a program to define a method that can take a parameter and use it.
2. Write a program to print a stack from 1 to 10 and then print 10 to 1.
3. Write a program to show the "this" keyword using the present class, method properties.

Chapter 7 Method

This chapter discusses some practical oriented fact of methods and classes. It relates several topics relating to methods, including overloading, parameter passing, recursion and overriding. It also discuss the classes, access control in the class, use of some keywords like "static" and one very important Java built in class "string".

> 本章的主题是方法和类，将介绍方法的重载、参数的传递、递归及方法的覆盖，并讨论类、类中权限的控制，一些关键词如 static，以及 Java 中 String 类的使用。

7.1 Method Overloading

One of the important features in any programming language is the use of names. When you create an object, you give a name to a region of storage. A method is a name for an action. By using names to describe your system, you create a program that is easier for people to understand and change. It's a lot like writing prose-the goal is to communicate with your readers.

> 编程语言中的一个重要特点就是名称的使用。当创建一个对象时，就为一个存储区域命名。方法是一个命名的操作。通过使用名称来描述系统，使人们更容易理解和修改所创建的程序。这样做更方便与读者交流。

You refer to all objects and methods by using names. Well-chosen names make it easier for you and others to understand your code.

A problem arises when mapping the concept of nuance in human language onto a programming language. Often, the same word expresses a number of different meanings-it's overloaded. This is useful, especially when it comes to trivial differences. You say "wash the shirt," "wash the car," and "wash the dog." It would be silly to be forced to say, "shirtWash the shirt," "carWash the car," and "dogWash the dog" just so the listener doesn't need to make any distinction about the action performed. Most human languages are redundant, so even if you miss a few words, you can still determine the meaning. We don't need unique identifiers-we can deduce meaning from context.

Most programming languages (C in particular) require you to have a unique identifier for each function. So you could not have one function called **print( )** for printing integers and another called **print( )** for printing floats-each function requires a unique name.

In Java, another factor forces the overloading of method names: the constructor. Because the constructor's name is predetermined by the name of the class, there can be only one constructor name. But what if you want to create an object in more than one way? For example, suppose you build a class that can initialize itself in a standard way and by reading information from a file. You need two constructors, one that takes no arguments (the default constructor), and one that takes a **String** as an argument, which is the name of the file from which to initialize the object. Both are constructors, so they must have the same name – the name of the class. Thus method overloading is essential to allow the same method name to be used with different argument types. And although method overloading is a must for constructors, it's a general convenience and can be used with any method.

> 在 Java 中，构造方法强制重载方法名是因为构造方法名和类名一致，故构造方法的名称是唯一的。如何要用多种方式来创建对象？例如，假设需要创建一个类，该类需要标准方式或通过读取文件信息来初始化。这时需要两个不同的构造方法，一个不带参数，另一个构造方法需要一个 String 类型的参数。由于都是构造函数，它们的名字都必须和类名相同。这样就必须使用方法的重载允许同名的方法使用不同的参数类型区分。构造方法一定需要重载，重载其他的方法也非常方便。

Here's an example that shows both overloaded constructors and overloaded ordinary methods:

```
//: Overloading.Java
// Demonstration of both constructor
// and ordinary method overloading.
import Java.util.*;
class Tree {
    int height;
    Tree() {                                        //构造方法
        prt("Planting a seedling");
        height = 0;
    }
    Tree(int i) {                                   //构造方法的重载
        prt("Creating new Tree that is " + i + " feet tall");
        height = i;
    }
    void info() {                                   //普通成员方法的定义
        prt("Tree is " + height + " feet tall");
    }
    void info(String s) {                           //普通成员方法的重载
        prt(s + ": Tree is " + height + " feet tall");
    }
    static void prt(String s) {
        System.out.println(s);
    }
}
public class Overloading {
    public static void main(String[] args) {
        for(int i = 0; i < 5; i++) {
            Tree t = new Tree(i);                   //调用带参数的构造方法实例化对象
            t.info();             //调用 t 对象的成员方法 info()，此方法是不带参数的
            t.info("overloaded method");
                                  //调用 t 对象的成员方法 info()，此方法是带参数的
        }
        // Overloaded constructor:
        new Tree();                                 //调用不带参数的构造方法实例化对象
    }
}
```

A **Tree** object can be created either as a seedling, with no argument, or as a plant grown in a nursery, with an existing height. To support this, there are two constructors, one that takes no arguments (we call constructors that take no arguments default constructors and one that takes the existing height).

You might also want to call the **info( )** method in more than one way. For example, with a **String** argument if you have an extra message you want printed, and without if you have nothing more to say. It would seem strange to give two separate names to what is obviously the same concept. Fortunately, method overloading allows you to use the same name for both.

7.1.1 Distinguishing Overloaded Methods

If the methods have the same name, how can Java know which method you mean? There's a simple rule: Each overloaded method must take a unique list of argument types.

If you think about this for a second, it makes sense: how else could a programmer tell the difference between two methods that have the same name, other than by the types of their arguments?

Even differences in the ordering of arguments are sufficient to distinguish two methods: (Although you don't normally want to take this approach, as it produces difficult-to-maintain code.)

> 每个被重载的方法必须带有不同的参数列表。两个方法名称相同但所带参数的排列次序不同，也可称其为两个不同的方法，它们实现了方法的重载。由于这种方式的代码不易维护，通常情况下很少使用。

```java
//: OverloadingOrder.Java
// Overloading based on the order of
// the arguments.
public class OverloadingOrder {
    static void print(String s, int i) {         //成员方法的定义
        System.out.println("String: " + s +", int: " + i);
    }
    static void print(int i, String s) {
    //实现方法的重载，其参数的类型排列次序与上不同
        System.out.println("int: " + i +", String: " + s);
    }
    public static void main(String[] args) {
        print("String first", 11);               //调用的是第一个成员方法
        print(99, "Int first");                  //调用的是第二个成员方法
    }
}
```

The two **print( )** methods have identical arguments, but the order is different, and that's what makes them distinct.

7.1.2 Overloading with Primitives

Primitives can be automatically promoted from a smaller type to a larger one and this can be slightly confusing in combination with overloading. The following example demonstrates what happens when a primitive is handed to an overloaded method:

> 基本数据类型能自动从数据范围小的类型转换成数据范围大的类型，但这对方法的重载会引起一定的歧义。

```java
//: PrimitiveOverloading.Java
// Promotion of primitives and overloading
public class PrimitiveOverloading {
    // boolean can't be automatically converted
    static void prt(String s) {
        System.out.println(s);
    }
    void f1(char x) { prt("f1(char)"); }      //成员方法的定义，参数类型为 char
    void f1(byte x) { prt("f1(byte)"); }      //方法的重载，参数类型为 byte
    void f1(short x) { prt("f1(short)");}     //方法的重载，参数类型为 short
    void f1(int x) { prt("f1(int)");}         //方法的重载，参数类型为 int
    void f1(long x) { prt("f1(long)");}       //方法的重载，参数类型为 long
    void f1(float x) { prt("f1(float)");}     //方法的重载，参数类型为 float
    void f1(double x) { prt("f1(double)");}   //方法的重载，参数类型为 double
    void f2(byte x) { prt("f2(byte)"); }
    void f2(short x) { prt("f2(short)"); }
    void f2(int x) { prt("f2(int)"); }
    void f2(long x) { prt("f2(long)"); }
    void f2(float x) { prt("f2(float)"); }
    void f2(double x) { prt("f2(double)"); }
    void f3(short x) { prt("f3(short)"); }
    void f3(int x) { prt("f3(int)"); }
    void f3(long x) { prt("f3(long)"); }
    void f3(float x) { prt("f3(float)"); }
    void f3(double x) { prt("f3(double)"); }
    void f4(int x) { prt("f4(int)"); }
    void f4(long x) { prt("f4(long)"); }
    void f4(float x) { prt("f4(float)"); }
    void f4(double x) { prt("f4(double)"); }
    void f5(long x) { prt("f5(long)"); }
    void f5(float x) { prt("f5(float)"); }
    void f5(double x) { prt("f5(double)"); }
    void f6(float x) { prt("f6(float)"); }
    void f6(double x) { prt("f6(double)"); }
    void f7(double x) { prt("f7(double)"); }
    void testConstVal() {              //测一个常量的方法
        prt("Testing with 5");
        f1(5);f2(5);f3(5);f4(5);f5(5);f6(5);f7(5);//针对常量，调用相应的方法
    }
    void testChar() {                  //测 char 类型的方法
        char x = 'x';
        prt("char argument:");   //针对 char 类型调用直接相应的方法
        f1(x);f2(x);f3(x);f4(x);f5(x);f6(x);f7(x);
                        //自动转换成 int 类型，然后去匹配相应的方法
    }
    void testByte() {                  //测 byte 类型的方法
        byte x = 0;
        prt("byte argument:");   //针对 byte 类型调用直接相应的方法
        f1(x);f2(x);f3(x);f4(x);f5(x);f6(x);f7(x);
        //采用逐步拓宽自动转换功能，调用相应方法
    }
    void testShort() {                 //测 short 类型的方法
        short x = 0;
```

```java
        prt("short argument:");
        f1(x);f2(x);f3(x);f4(x);f5(x);f6(x);f7(x);
    }
    void testInt() {                //测 int 类型的方法
        int x = 0;
        prt("int argument:");
        f1(x);f2(x);f3(x);f4(x);f5(x);f6(x);f7(x);
    }
    void testLong() {               //测 long 类型的方法
        long x = 0;
        prt("long argument:");
        f1(x);f2(x);f3(x);f4(x);f5(x);f6(x);f7(x);
    }
    void testFloat() {              //测 float 类型的方法
        float x = 0;
        prt("float argument:");
        f1(x);f2(x);f3(x);f4(x);f5(x);f6(x);f7(x);
    }
    void testDouble() {             //测 double 类型的方法
        double x = 0;
        prt("double argument:");
        f1(x);f2(x);f3(x);f4(x);f5(x);f6(x);f7(x);
    }
    public static void main(String[] args) {
        PrimitiveOverloading p = new PrimitiveOverloading();//实例化一个对象
        p.testConstVal();           //调用对象的方法
        p.testChar();
        p.testByte();
        p.testShort();
        p.testInt();
        p.testLong();
        p.testFloat();
        p.testDouble();
    }
}
```

If you view the output of this program, you'll see that the constant value 5 is treated as an **int**, so if an overloaded method is available that takes an **int** it is used. In all other cases, if you have a data type that is smaller than the argument in the method, that data type is promoted. **char** produces a slightly different effect, since if it doesn't find an exact **char** match, it is promoted to **int**.

> 在方法的调用中，Java 优先采用参数类型完全一致的方法，但如果实参的数据类型比重载方法所带的参数类型小，为匹配该方法，此实参将进行自动类型转换。但 char 类型的实参有点不同，若找不到 char 类型的方法进行匹配，将自动转换为 int 类型。

What happens if your argument is bigger than the argument expected by the overloaded method? A modification of the above program gives the answer:

```java
//: Demotion.Java
// Demotion of primitives and overloading
public class Demotion {
```

```java
        static void prt(String s) {
            System.out.println(s);
        }
        void f1(char x) { prt("f1(char)"); }
        void f1(byte x) { prt("f1(byte)"); }
        void f1(short x) { prt("f1(short)"); }
        void f1(int x) { prt("f1(int)"); }
        void f1(long x) { prt("f1(long)"); }
        void f1(float x) { prt("f1(float)"); }
        void f1(double x) { prt("f1(double)"); }
        void f2(char x) { prt("f2(char)"); }
        void f2(byte x) { prt("f2(byte)"); }
        void f2(short x) { prt("f2(short)"); }
        void f2(int x) { prt("f2(int)"); }
        void f2(long x) { prt("f2(long)"); }
        void f2(float x) { prt("f2(float)"); }
        void f3(char x) { prt("f3(char)"); }
        void f3(byte x) { prt("f3(byte)"); }
        void f3(short x) { prt("f3(short)"); }
        void f3(int x) { prt("f3(int)"); }
        void f3(long x) { prt("f3(long)"); }
        void f4(char x) { prt("f4(char)"); }
        void f4(byte x) { prt("f4(byte)"); }
        void f4(short x) { prt("f4(short)"); }
        void f4(int x) { prt("f4(int)"); }
        void f5(char x) { prt("f5(char)"); }
        void f5(byte x) { prt("f5(byte)"); }
        void f5(short x) { prt("f5(short)"); }
        void f6(char x) { prt("f6(char)"); }
        void f6(byte x) { prt("f6(byte)"); }
        void f7(char x) { prt("f7(char)"); }
        void testDouble() {
            double x = 0;
            prt("double argument:");
            f1(x);f2((float)x);f3((long)x);f4((int)x);
            //先将double类型的数据进行强制类型转换，再调用匹配的方法
            f5((short)x);f6((byte)x);f7((char)x);
        }
        public static void main(String[] args) {
            Demotion p = new Demotion();
            p.testDouble();
        }
    }
```

Here, the methods take narrower primitive values. If your argument is wider then you must cast to the necessary type using the type name in parentheses. If you don't do this, the compiler will issue an error message.

> 在方法调用中，若实参的数据类型比方法本身所带的参数类型大，则必须使用带圆括号的类名将其强制转换成需要的类型，否则编译时会出错。

You should be aware that this is a narrowing conversion, which means you might lose information during the cast. This is why the compiler forces you to do it – to flag the narrowing conversion.

7.1.3 Overloading on Return Values

It is common to wonder "Why only class names and method argument lists? Why not distinguish between methods based on their return values?" For example, these two methods, which have the same name and arguments, are easily distinguished from each other:

```
void f() {}
int f() {}
```

This works fine when the compiler can unequivocally determine the meaning from the context, as in **int x = f( )** . However, you can call a method and ignore the return value; this is often referred to as calling a method for its side effect since you don't care about the return value but instead want the other effects of the method call. So if you call the method this way: f( );

How can Java determine which **f( )** should be called? And how could someone reading the code see it? Because of this sort of problem, you cannot use return value types to distinguish overloaded methods.

> 在 Java 中，不能依据方法返回值的不同实现方法的重载，其编译时会报错。

- **Fields** cannot be overridden but they can be **hidden**, ie if you declare a field in a subclass with the same name as one in the superclass, the superclass field can only be accessed using **super** or the superclasses type.

> 成员变量不能被覆盖但可以被隐藏。在子类对父类的继承中，如果子类的成员变量和父类的成员变量同名，此时子类会隐藏父类的成员变量。如果要访问父类的同名变量，则必须借助关键字 super 或父类名。

- A subclass can override methods in it's superclass and change it's implementation.
- It must have the same **return type**, **name**, and **parameter list** and can only throw **exceptions** of the same class/subclass as those declared in the original method.

> 可以在子类中覆盖父类的同名方法并重新实现此方法。方法的覆盖必须具有相同的返回类型、方法名、参数列表，并且只能抛出与被覆盖的原方法中声明抛出的异常相同的异常。

```
class Super {
    void test() {
        System.out.println("In Super.test()");
    }
}
class Sub extends Super {
    void test() {  //overrides test() in Super
        System.out.println("In Sub.test()");
    }
}
```

Overriding method cannot have weaker access rights than the original method.

> 覆盖方法不能比被覆盖方法的访问控制权限低。

In Sub class:
```
// compile-error, original has package access
private void test() {}
protected void test() {}      // compiles ok
public void test() {}         // compiles ok
```
- You can have multiple **overloaded** methods in a class but **only one** overriding method.

In Sub class:
```
void test() {}              //overrides test() in Super
 public void test(){}
//compile-error:test() already declared, 编译出错,覆盖不可重复定义
//different access modifiers not part of method signature for naming purposes
void test(String str) {}// compiles ok, overloads test(),可以实现多个重载
```

- Only accessible **non-static** methods can be overridden.
- **Static** methods can be **hidden**, ie you can declare a static method in the subclass with the same signature as a static method in the superclass. The superclass method will **not be** accessible from a subclass reference.
- Any class can override methods from its superclass to declare them **abstract**, turning a concrete method into an abstract one at that point in the type tree. Useful when a class's default implementation is invalid for part of the class hierarchy.

> 只有非静态的方法才可被覆盖。静态方法可以被隐藏,若子类中声明了一个与父类同标识的 static 方法,通过子类的引用无法访问父类中同名的 static 方法。任何类都可以覆盖父类中的抽象方法。在类的层次结构中,某些类的方法无法有效实现时,用抽象方法表示更适合。

7.1.4 Overriding with Constructors

- You cannot override a constructor in a superclass as they are not inherited.
- You cannot override a constructor in the same class as they would both have the same signatures; get an 'already declared' compile-error.
- If you're instantiating a Subclass object and if the Superclass constructor calls a method that is overridden in the Subclass, the Subclass method will called from the superclass constructor —NOT the one in the superclass.

> 父类的构造方法不能被它的子类继承,子类也不能覆盖父类的构造方法。同一个类中构造方法不能被覆盖,即两个构造方法不能有相同的签名,否则会引发已知的编译器错误。如果在父类的构造方法中所要调用的成员方法已在子类被覆盖了,在实例化一个子类对象时,父类的构造方法实际调用的将是子类的成员方法,而不是在父类中定义的那个成员方法。

```
class Super {                   //父类
    Super(){
        System.out.println("In Super constructor");
        test();                 //调用成员方法
    }
    void test() {               //Super 类的成员方法
        System.out.println("In Super.test()");
```

```
    }
}
class Sub extends Super { //子类
    Sub() {
        System.out.println("In Sub constructor");
    }
    void test() { // overrides test() in Super test()成员方法的覆盖
        System.out.println("In Sub.test()");
    }
}
```

Output if Sub sb = new Sub( ) is invoked:

In Super Constructor

In Sub.test( )

In Sub Constructor

7.2 Parameter Passing in Java - By Reference or By Value

This is another common question on Java newsgroups, made worse by the fact that people who should know better still perpetuate the following myth:

"Objects are passed by reference, primitives are passed by value"

Everything in Java is passed by value. Objects, however, are never passed at all.

That needs some explanation - after all, if we can't pass objects, how can we do any work? The answer is that we pass references to objects. That sounds like it's getting dangerously close to the myth, until you look at truth.

> 在 Java 中，传递参数时都是以传值的方式进行。对象传的是该对象的引用，即引用的副本；基本数据类型传的是值，即数据的副本。

The values of variables are always primitives or references, never objects.

This is probably the single most important point in learning Java properly. It's amazing how far you can actually get without knowing it, in fact, but vast numbers of things suddenly make sense when you grasp it.

When people hear the words "pass by reference", they may understand different things by the words. There are some pretty specific definitions of what it should mean. This would mean that if you wrote the following code:

```
Object x = null;
giveMeAString (x);
System.out.println (x);
...
void giveMeAString (Object y){
    y = "This is a string";
}
```

The result (if Java used pass-by-reference semantics) would be:

This is a string

Instead of the actual result:

null

Explaining the two truths above eliminates all of this confusion.

So what does passing a reference by value actually mean?

It means you can think of references how you think of primitives, to a large extent. For instance, the equivalent to the above bit of code using primitives would be:

```
int x = 0;
giveMeATen (x);
System.out.println (x);
...
void giveMeATen (int y)
{
    y = 10;
}
```

Now, the above doesn't print out "10". Why not? Because the value "0" was passed into the method giveMeTen, not the variable itself. Exactly the same is true of reference variables - the value of the reference is passed in, not the variable itself. It's the same kind of copying that happens on variable assignment. The first code snippet, if inlined, is equivalent to:

```
// Before the method call
Object x = null;
// Start of method call - parameter copying
Object y = x;   //对象 x 的引用即首地址赋值给引用类型变量 y
// Body of method call
y = "This is a piece of string.";
// End of method call
System.out.println (x);
```

If you want to think pictorially, you might find my "objects are balloons, references are pieces of string" analogy helpful.

> 在调用带形参的方法时，所传递的参数可分为基本数据类型和引用类型两种。传递的参数若是基本数据类型，传递的是数据的副本；传递的参数若是引用类型，传递的是引用的副本。

7.2.1 Passing Named Arguments to Java Programs

Here's an example of a better way to pass arguments to main() methods of Java classes than using String[] args. First, pass two named parameters, user and level, that come into a GamePlayer class:

```
Java  -Duser=champ  -Dlevel=expert  GamePlayer
```

Classes will use them as follows:

```
public class GamePlayer {
    public static void main( String[] args ) {
    /*得到参数名为 user 的参数值，否则采用所给的默认值'unknown'*/
        String user = System.getProperty( "user", "unknown" );
        // get user name or use 'unknown'
        System.out.println( "Welcome to Inter-Galactic Rooster
                    Race: " + user );
        GameStarter.start();         直接调用类的静态方法
        // start game which uses the start level
```

```
            }
    }
    class GameStarter { // use arguments in other classes as well
        static void start() {
            /*得到参数名为level的参数值,否则采用所给的默认值' beginner '*/
            System.out.println("Your level"+System.getProperty("
                level","beginner"));
        }
    }
```

The parameters are preceded by a -D, to indicate that they are named properties. The arguments will not be available in the String argument array of the main( ) method. With -D parameters, you do not have to worry about the positions, and can refer to the arguments by name. You can use the arguments in other methods and even other classes, without having to pass them around. A second parameter to System.getProperty( ) is used as a default value in case the named parameter has not been passed. You can use this code to list all passed properties, including system properties, to standard output:

> 在要传递的命名参数前需添加一个前缀-D,表示其为命名参数。这些命名参数会被传递给Java应用程序。参数是按名称访问的,与存放的位置无关。这些命名的参数不会被保存在main(String args[ ])方法的 args数组中。在其他的方法或类中,也可访问这些命名的参数。注意System.getProperty( )的用法,可通过它来获取命名参数的参数值。

```
System.getProperties().list( System.out );
// get properties from system and list them to stdout
```

7.2.2　Passing Information into a Method

When you write your method, you declare the number and type of the arguments required by that method. You declare the type and name for each argument in the method signature. For example, the following is a method that computes the monthly payments for a home loan based on the amount of the loan, the interest rate, the length of the loan (the number of periods), and the future value of the loan (presumably the future value of the loan is zero because at the end of the loan, you've paid it off):

```
double computePayment(double loanAmt, double rate, double futureValue,
int numPeriods) {
    double I, partial1, denominator, answer;
    I = rate / 100.0;
    partial1 = Math.pow((1 + I), (0.0 - numPeriods));
    denominator = (1 - partial1) / I;
    answer=((-1*loanAmt)/denominator)-
        ((futureValue*partial1)/denominator);
    return answer;
}
```

This method takes four arguments: the loan amount, the interest rate, the future value and the number of periods. The first three are double-precision floating point numbers, and the fourth is an integer.

As with this method, the set of arguments to any method is a comma-delimited list of variable declarations where each variable declaration is a type/name pair:

```
type name
```

As you can see from the body of the computePayment method, you simply use the argument name to refer to the argument's value.

> 在定义一个带形参的方法时，在形参列表中必须声明每个参数的类型和名称。在方法的实现中，参数名指向参数的值。

Argument Types　　// 参数的类型

In Java, you can pass an argument of any valid Java data type into a method. This includes primitive data types such as doubles, floats and integers as you saw in the computePayment method, and reference data types such as objects and arrays. Here's an example of a constructor that accepts an array as an argument. In this example, the constructor initializes a new Polygon object from a list of Points (Point is a class that represents an x, y coordinate):

```
Polygon polygonFrom(Point[] listOfPoints) {
    ...
}
```

Unlike some other languages, you cannot pass methods into Java methods. But you can pass an object into a method and then invoke the object's methods.

> 在调用带形参的方法时，Java 不可以将某一方法视为参数进行传递，但可以将某对象作为参数传递，在方法中对该对象的成员方法进行调用。

Argument Names　　// 参数的名称

When you declare an argument to a Java method, you provide a name for that argument. This name is used within the method body to refer to the item.

A method argument can have the same name as one of the class's member variables. If this is the case, then the argument is said to hide the member variable. Arguments that hide member variables are often used in constructors to initialize a class. For example, take the following Circle class and its constructor:

```
class Circle {
    int x, y, radius;
    public Circle(int x, int y, int radius) {
        ...
    }
}
```

The Circle class has three member variables: x, y and radius. In addition, the constructor for the Circle class accepts three arguments each of which shares its name with the member variable for which the argument provides an initial value.

The argument names hide the member variables. So using x, y or radius within the body of the constructor refers to the argument, not to the member variable. To access the member variable, you must reference it through this—the current object:

```
class Circle {
    int x, y, radius;
    public Circle(int x, int y, int radius) {
        this.x = x;
```

```
        this.y = y;
        this.radius = radius;
    }
}
```

Names of method arguments cannot be the same as another argument name for the same method, the name of any variable local to the method, or the name of any parameter to a catch clause within the same method.

> 在类的定义中，成员方法中形参的参数名称可以与类的成员变量同名，只是在此方法的内部，类的成员变量会被隐藏。使用 this（代表当前对象）访问被隐藏的成员变量。在同一方法中参数不能同名，参数和同一个方法中的局部变量不能同名，参数和同一个方法中的 catch 语句中的参数不能同名。

7.2.3 Pass by Value

In Java methods, arguments are passed by value. When invoked, the method receives the value of the variable passed in. When the argument is of primitive type, pass-by-value means that the method cannot change its value. When the argument is of reference type, pass-by-value means that the method cannot change the object reference, but can invoke the object's methods and modify the accessible variables within the object.

> 在 Java 中，传递参数时都是以传值的方式进行。若传递的参数类型是基本数据类型，其传递的是数值，在方法内部无法改变实参的值。若传递的参数类型是引用类型，值传递意味着无法改变对象的引用，但可以通过调用该对象的方法来修改对象中可以访问的变量。

This is often the source of confusion—a programmer writes a method that attempts to modify the value of one its arguments and the method doesn't work as expected. Let's look at such method and then investigate how to change it so that it does what the programmer originally intended.

Consider this series of Java statements which attempts to retrieve the current color of a Pen object in a graphics application:

```
...
int r = -1, g = -1, b = -1;
pen.getRGBColor(r, g, b);
System.out.println("red = " + r + ", green = " + g + ", blue = " + b);
...
```

At the time when the getRGBColor method is called, the variables r, g, and b all have the value −1. The caller is expecting the getRGBColor method to pass back the red, green and blue values of the current color in the r, g, and b variables.

However, the Java runtime passes the variables' values (-1) into the getRGBColor method; not a reference to the r, g, and b variables. So you could visualize the call to getRGBColor like this: getRGBColor(-1, -1, -1).

When control passes into the getRGBColor method, the arguments come into scope (get allocated) and are initialized to the value passed into the method:

```
class Pen {
    int redValue, greenValue, blueValue;
    void getRGBColor(int red, int green, int blue) {
        // red, green, and blue have been created
        // and their values are -1
        ...
    }
}
```

So getRGBColor gets access to the values of r, g, and b in the caller through its arguments red, green, and blue, respectively. The method gets its own copy of the values to use within the scope of the method. Any changes made to those local copies are not reflected in the original variables from the caller.

> 调用 getRGBColor 方法时通过参数 red、green 和 blue 访问到实参 r、g 和 b 值，在该方法调用中获取的是实参值的副本。在此方法中任何对形参值的改变都不会影响调用时传递的实参的值。

Now, let's look at the implementation of getRGBColor within the Pen class that the method signature above implies:

```
class Pen {
    int redValue, greenValue, blueValue;
    ...
    // this method does not work as intended
    void getRGBColor(int red, int green, int blue) {
        red = redValue;
        green = greenValue;
        blue = blueValue;
    }
}
```

This method will not work as intended. When control gets to the println statement in the following code, which was shown previously, getRGBColor's arguments, red, green, and blue, no longer exist. Therefore the assignments made to them within the method had no effect; r, g, and b are all still equal to -1.

```
...
int r = -1, g = -1, b = -1;
pen.getRGBColor(r, g, b);
System.out.println("red = " + r + ", green = " + g + ", blue = " + b);
...
```

Passing variables by value affords the programmer some safety: Methods cannot unintentionally modify a variable that is outside of its scope. However, you often want a method to be able to modify one or more of its arguments. The getRGBColor method is a case in point. The caller wants the method to return three values through its arguments. However, the method cannot modify its arguments, and, furthermore, a method can only return one value through its return value. So, how can a method return more than one value, or have an effect (modify some value) outside of its scope?

For a method to modify an argument, it must be of a reference type such as an object or array. Objects and arrays are also passed by value, but the value of an object is a reference. So the effect is that arguments of reference types are passed in by reference. Hence a reference to an object is the address of the object in memory. Now, the argument in the method is referring to the same memory location as the caller.

> 在Java中，可以通过引用类型来实现参数的修改，比如对象或数组类型。对象和数组在方法的参数传递中都是按值的方式进行传递，只不过其传递的值是对象的引用值。因此传递的是以引用方式传递的引用值。在方法调用时，方法的形参和实参都指向内存中的同样的地址。

Let's rewrite the getRGBColor method so that it actually does what you want. First, you must introduce a new type of object, RGBColor, that can hold the red, green and blue values of a color in RGB space:

```
class RGBColor {
    public int red, green, blue;
}
```

Now, we can rewrite getRGBColor so that it accepts an RGBColor object as an argument. The getRGBColor method returns the current color of the pen by setting the red, green and blue member variables of its RGBColor argument:

```
class Pen {
    int redValue, greenValue, blueValue;
    void getRGBColor(RGBColor aColor){//引用类型为变量
        aColor.red = redValue;
        aColor.green = greenValue;
        aColor.blue = blueValue;
    }
}
```

And finally, let's rewrite the calling sequence:

```
...
RGBColor penColor = new RGBColor();   //先实例化一个对象
pen.getRGBColor(penColor);            //传递一个对象的引用值
System.out.println("red = " + penColor.red + ", green
= " +penColor.green + ", blue = " + penColor.blue);
//将结果通过对象来输出
...
```

The modifications made to the RGBColor object within the getRGBColor method affect the object created in the calling sequence because the names penColor (in the calling sequence) and aColor (in the getRGBColor method) refer to the same object.

All parameters to methods in Java are pass-by-value.

What does this mean? How do I change a parameter? This short tutorial will help you figure out how parameter passing works in Java, and will help you avoid some common mistakes.

First let's get some terminology straight. An example will help set the stage.

```
// Method definition
public int mult(int x, int y){
    return x * y;
}
```

```
// Where the method mult is used
int length = 10;                    //给定义的实参赋初值
int width = 5;
int area = mult(length, width);
//调用方法，并将实参的值传给方法中的形参
```

We use the term formal parameters to refer to the parameters in the definition of the method. In the example, x and y are the formal parameters. You can remember to call them "formal" because they are part of the method's defintion, and you can think of a definition as being formal.

We use the term actual parameters to refer to variables in the method call, in this case length and width. They are called "actual" because they determine the actual values that are sent to the method.

You may have heard the term "argument" used or just "parameter" (without specifying actual or formal). You can usually tell by the context which sort of parameter is being referred to.

What Pass-by-Value Means

Pass-by-value means that when you call a method, a copy of the value of each actual parameter is passed to the method. You can change that copy inside the method, but this will have no effect on the actual parameter.

> 一般用形参表示方法中定义的参数，称为是形式上的；可以用实参引用方法调用时传递的变量，由于它们给方法传递了实际的值故看做是实际的。在方法的调用中，所传递的参数都是以传值的方式进行，即实参数值的副本。

Unlike many other languages, Java has no mechanism for changing the value of an actual parameter. Isn't this very restrictive? Not really. In Java, we can pass a reference to an object (also called a "handle")as a parameter. We can then change something inside the object; we just can't change what object the handle refers to. For primitives, the story is a little different.

> 跟其他多数语言不一样，Java没有改变实参值的机制。但Java中可以传递一个对象的引用，通过引用可以改变所指对象内部的值，不能改变的是对象的引用。

Let's look closely at some examples of passing different kinds of things to methods: primitives, objects, strings, and array.

7.2.4 Passing Primitive Types

Java has eight primitive data types: six number types, character and boolean. Consider the following example:

```java
public static void tryPrimitives(int i, double f, char c, boolean test) {
    i += 10;     //This is legal, but the new values
    c = 'z';     //won't be seen outside tryPrimitives.
    if(test)
        test = false;
    else
        test = true;
    f = 1.5;
}
```

If tryPrimitives is called within the following code, what will the final print statement produce?

```
int ii = 1;
double ff = 1.0;
char cc = 'a';
boolean bb = false;
//Try to change the values of these primitives:
tryPrimitives(ii, ff, cc, bb);
System.out.println("ii="+ii+",ff="+ff+",cc="+cc+",bb="+bb);
//输出结果: ii=1,ff=1.0,cc=a, bb=false
```

> 传递基本数据类型的实参到方法的形参中，不论形参在成员方法中发生何种变化，实参值不会受到影响及改变。

7.2.5 Return Values

Okay, so how do we change the values of variables inside methods? One way to do it is simply to return the value that we have changed. In the simple example below, we take two integers, a and b, as parameters and return a to the power of b.

```
public static int power(int a, int b){
    int i;
    int total = 1;
    for(i = 0; i < b; i++)
        total = total * a;
    return total;
}
```

Now when we call the method power, number gets assigned the value that power returns.

```
int number = 2;
int exponent = 4;
number = power(number, exponent);
System.out.println("New value of number is " + number);
```

Since methods should be designed to be simple and to do one thing, you will often find that returning a value is enough.

7.2.6 Passing Object References

What if a parameter to a method is an object reference? We can manipulate the object in any way, but we cannot make the reference refer to a different object.

Suppose we have defined the following class:

```
class Record{
    int num;
    String name;
}
```

Now we can pass a Record as a parameter to a method:

```
public static void tryObject(Record r){     //引用类型作为形参
    r.num = 100;
    r.name = "Fred";
}
```

In some other code we can create an object of our new class Record, set its fields, and call the method tryObject.

```
Record id = new Record();                    //创建一个实例化对象
id.num = 2;
id.name = "Barney";
tryObject(id);                               //传递实例化对象的引用
System.out.println(id.name + " " + id.num);  //输出 Fred 100
```

The print statement prints out "Fred 100"; the object has been changed in this case. Why? The reference to id is the parameter to the method, so the method cannot be used to change that reference; i.e., it can't make id reference a different Record. But the method can use the reference to perform any allowed operation on the Record that it already references.

Side Note: It is often not good programming style to change the values of instance variables outside an object. Normally, the object would have a method to set the values of its instance variables.

> 注意：一般在对象的外部修改其成员变量的值是一种不好的编程风格。通常情况下，对象都有设置内部成员变量值的方法。

We cannot make the object parameter refer to a different object by reassigning the reference or calling new on the reference. For example the following method would not work as expected:

```
public void createRecord(Record r, int n, String name){
    r = new Record();
    r.num = n;
    r.name = name;
}
```

> 不可以通过重新赋值或 new 操作修改引用类型的参数。

We can still encapsulate the initialization of the Record in a method, but we need to return the reference.

```
public Record createRecord(int n, String name){
    Record r = new Record();
    r.num = n;
    r.name = name;
    return r;
}
```

7.2.7 Passing Strings

When we write code with objects of class String, it can look as if strings are primitive data types. For example, when we assign a string literal to a variable, it looks no different than, say, assigning an number to an int variable. In particular, we don't have to use new.

```
String str = "This is a string literal.";
```

But in fact, a string is an object, not a primitive. A new is indeed required; it's just that Java does it behind the scenes. Java creates a new object of class String and initializes it to contain the string literal we have given it.

> 在用字符串对象编写代码时，可以将字符串当做一种基本数据类型。可以像给一个整型数据赋值一样将字符串直接赋值给声明为 String 类型的变量。但事实上，字符串是一个对象，不是一个基本的数据类型。本质上创建 String 类的对象是需要 new 操作的，Java 创建一个 String 类对象并用指定的文本初始化此对象。

Because str is an object, we might think that the string it contains can be changed when we pass str as a parameter to a method. Suppose we have the method tryString:

```
public static void tryString(String s){
    s = "a different string";
}
```

When the following code is executed, what does the print statement produce?

```
String str = "This is a string literal.";
tryString(str);
System.out.println("str = " + str) ;
```

7.2.8 Passing Arrays

Arrays are references. This means that when we pass an array as a parameter, we are passing its handle or reference. So, we can change the contents of the array inside the method.

```
public static void tryArray(char[] b){
    b[0] = 'x';
    b[1] = 'y';
    b[2] = 'z';
}
```

When the following code is executed, the array a does indeed have the new values in the array.

```
char[] a = {'a', 'b', 'c'};
tryArray(a);
System.out.println("a[0]="+a[0]+",a[1]="+a[1]+",a[2]=" + a[2]);
```

The print statements produces "a[0] = x, a[1] = y, a[2] = z".

> 数组是引用类型，可以在方法调用中将数组视为一个参数，传递数组的引用，实现改变数组中元素的值。

7.3 Recursion

In Java, a method may call itself. This is called **recursion**. You do not need to make any special declarations to make this possible. The trick is used mainly in exercises given in computer science class. In the real world, recursion has a bad name because you can overflow the stack if you call yourself too many times without returning. If you recurse too deeply you will get a StackOverflowError. Further, most cute recursive algorithms can be specified more efficiently with loops. The classic practical recursive algorithm is QuickSort. The classic problem is to calculate n! with or without recursion.

递归是指一种自己调用本身的方法。由于多次调用本身而不返回会使栈溢出，因此递归一直被认为是不好的机制。递归过深会出现 StackOverFlowError 的异常。大部分递归算法可以转换成更高效的循环算法。QuickSort 是比较典型而实用的递归算法。利用递归和非递归算法进行求解的典型例题如求 n!。

Simply put, recursion is when a function calls itself. That is, in the course of the function definition there is a call to that very same function. At first this may seem like a never ending loop, or like a dog chasing its tail. It can never catch it. So too it seems our method will never finish. This might be true in some cases, but in practice we can check to see if a certain condition is true and in that case exit (return from) our method. The case in which we end our recursion is called a base case. Additionally, just as in a loop, we must change some value and incrementally advance closer to our base case.

递归看起来像一个永不终止的循环。但在实践中会去检查是否已满足了某个限定条件，若达到则会退出此方法。导致递归结束的条件称为边界条件。就像循环一样需要修改某些值让其不断地接近边界值。

Consider this function.
```java
void myMethod( int counter){   //实现数据逐一递减输出，递减至 0 结束递归
    if(counter == 0)
        return;
    else{
        System.out.println(""+counter);
        myMethod(--counter);
        return;
    }
}
```

This recursion is not infinite, assuming the method is passed a positive integer value. What will the output be?

Consider this method:
```java
void myMethod( int counter){
    if(counter == 0)
        return;
    else    {
        System.out.println("hello" + counter);
        myMethod(--counter);
        System.out.println(""+counter);
        return;
    }
}
```

If the method is called with the value 4, what will the output be? Explain.

hello4
hello3
hello2
hello1
0

1
2
3

The above recursion is essentially a loop like a for loop or a while loop. When do we prefer recursion to an iterative loop? We use recursion when we can see that our problem can be reduced to a simpler problem that can be solved after further reduction.

Every recursion should have the following characteristics.

(1) A simple base case which we have a solution for and a return value.

(2) A way of getting our problem closer to the base case. I.e. a way to chop out part of the problem to get a somewhat simpler problem.

(3) A recursive call which passes the simpler problem back into the method.

> 递归的特性:用以返回的边界条件;每次调用使问题越来越简单,越来越接近边界条件;每次递归都应该是一次更简单的调用。

7.4　Controlling Access to Members of a Class

An access level determines whether other classes can use a particular member variable or call a particular method. The Java programming language supports four access specifiers for member variables and methods: private, protected, public, and, if left unspecified, package private. Table 7.1 shows the access permitted by each specifier.

> 访问控制能限制其他类访问该类中特定的数据成员和方法。Java 语言提供了四种关于类数据成员与成员方法的访问控制符: private, protected, public, 以及没有访问控制修饰符的缺省访问方式。表 7.1 所列为针对不同的情况下的访问控制权限, Y 表示可以被访问, N 表示不可以被访问。

Table 7.1　Access Permitted of Each Specifier

Specifier	Access Levels			
	Class	Package	Subclass	World
Private	Y	N	N	N
No Specifier	Y	Y	N	N
Protected	Y	Y	Y	N
Public	Y	Y	Y	Y

The first column indicates whether the class itself has access to the member defined by the access level. As you can see, a class always has access to its own members. The second column indicates whether classes in the same package as the class (regardless of their parentage) have access to the member. A package groups related classes and interfaces and provides access protection and namespace management. You'll learn more about packages in the Creating and Using Packages section. The third column indicates whether subclasses of the class — declared outside this package—have access to the member. The fourth column indicates whether all classes have access to the member.

Access levels affect you in two ways. First, when you use classes that come from another source, such as the classes in the Java platform, access levels determine which members of those classes your classes can use. Second, when you write a class, you need to decide what access level every member variable and every method in your class should have. One way of thinking about access levels is in terms of the API: access levels directly affect the public API of a class and determine which members of the class can be used by other classes. You need to put as much effort into deciding the access level for a member as you put into making other decisions about your class's API, such as naming methods.

> 类的内部访问权限决定了类的成员可否让类进行访问，可以从两个方面理解访问控制。一方面，类可以访问其他类中的哪些成员，例如访问 Java 中本身的类；另一方面，控制类中的数据成员和方法如何被外部访问。一种比较好的访问控制方式是通过在类中的公有 API 接口控制外部类访问类中的成员。

Let's look at a collection of classes and see access levels in action. The following figure shows the four classes that comprise this example and how they are related.

Classes and Packages of the Example Used to Illustrate Access Levels

7.4.1 Class Access Level

Here's a listing of a class, Alpha, whose members other classes will be trying to access. Alpha contains one member variable and one method per access level. Alpha is in a package called one:

```java
package one;
public class Alpha {
   //member variables    4 种成员变量的定义
   private   int privateVariable = 1;
   int packageVariable = 2;  //default access
   protected int protectedVariable = 3;
   public    int publicVariable = 4;
   //methods    4 种类型的成员方法的定义
   private void privateMethod() {
      System.out.format("privateMethod called%n");
   }
   void packageMethod() { //default access
      System.out.format("packageMethod called%n");
   }
   protected void protectedMethod() {
      System.out.format("protectedMethod called%n");
   }
   public void publicMethod() {
      System.out.format("publicMethod called%n");
   }
   public static void main(String[] args) {
      Alpha a = new Alpha();
      a.privateMethod();       //legal private 成员变量的调用
      a.packageMethod();       //legal default 成员变量的调用
      a.protectedMethod();     //legal protected 成员变量的调用
      a.publicMethod();        //legal public 成员变量的调用
      /*4 种类型的成员方法的调用 */
      System.out.format("privateVariable: %2d%n",
```

```
                a.privateVariable);      //legal
            System.out.format("packageVariable: %2d%n",
                a.packageVariable);      //legal
            System.out.format("protectedVariable: %2d%n",
                a.protectedVariable);    //legal
            System.out.format("publicVariable: %2d%n",
                a.publicVariable);       //legal
        }
    }
```

As you can see, Alpha can refer to all its member variables and all its methods, as shown by the Class column in the preceding table. The output from this program is:

  privateMethod called
  packageMethod called
  protectedMethod called
  publicMethod called
  privateVariable: 1
  packageVariable: 2
  protectVariable: 3
  publicVariable: 4

A member's access level determines which classes have access to that member, not which instances have access. So, for example, instances of the same class have access to one another's private members. Thus, we can add to the Alpha class an instance method that compares the current Alpha object (this) to another object, based on their privateVariables:

```
    package one;
    public class Alpha {
        ...
        public boolean isEqualTo(Alpha anotherAlpha) {
            if (this.privateVariable == anotherAlpha.privateVariable) {
            //legal
                return true;
            }
            else {
                return false;
            }
        }
    }
```

7.4.2 Package Access Level

Now consider the following class, DeltaOne, which is in the same package as Alpha. The methods and the variables this class can use are predicted by the Package column in the preceding table.

```
    package one;
    public class DeltaOne {
        public static void main(String[] args) {
            Alpha a = new Alpha();          //实例化一个对象
            /*4 种类型的成员方法的调用 */
            //a.privateMethod();             //illegal
            a.packageMethod();               //legal
```

```
            a.protectedMethod();        //legal
            a.publicMethod();           //legal
            /*输出4种类型的成员变量的值 */
            //System.out.format("privateVariable: %2d%n",
            //    a.privateVariable);  //illegal
            System.out.format("packageVariable: %2d%n",
                a.packageVariable);     //legal
            System.out.format("protectedVariable: %2d%n",
                a.protectedVariable);   //legal
            System.out.format("publicVariable: %2d%n",
                a.publicVariable);      //legal
        }
    }
```

DeltaOne cannot refer to privateVariable or invoke privateMethod but can access the other members of Alpha. If you remove the comment from the lines of code that are commented out and try to compile the class, the compiler will generate errors. Here's the output from the program when you run it as shown:

packageMethod called
protectedMethod called
publicMethod called
packageVariable: 2
protectedVariable: 3
publicVariable: 4

> 由上例测试可得，在同一个包当中，不同类之间，private的成员变量与成员方法都不能被访问。

7.5 Static Import

Many classes, including many in the Java core libraries, contain static constants that are used within the class and are also useful outside the class. For example, the Java.lang.Math class contains the constants PI and E for pi and e, respectively.

As another example, we see in Chapter 16 that the BorderLayout class contains constants such as NORTH, CENTER, etc. that are useful for laying out graphics components.

Prior to Java version 5.0, the only way to access those constants was by fully spelling out the names Math.PI, Math.E, BorderLayout.NORTH, etc. in your code. With static imports, you can use just PI, E, and NORTH without all the extra typing. To use static imports, add the static keyword to the import statement as follows:

```
        import static Java.lang.Math.PI;    //static 引入 Java.lang.Math.PI
```

Then in the code body you can use

```
        double area = PI * radius * radius;  //用 PI 取代 Math.PI
```

instead of

```
        double area = Math.PI * radius * radius;
```

Wildcards also work in the import static statement:

```
import static Java.lang.Math.*;     //static 引入 Java.lang.Math.* 包
```

For occasional use, simply typing the Math.PI is probably easier, but if constants from another class are used extensively, then the static import feature will reduce the code and also make it easier to read.

> Java 5.0 以前的版本，在代码中要对类中的静态常量进行访问，必须输入完整的名称，如 Math.PI、Math.E、BorderLayout.NORTH 等。在 5.0 版本后可以通过在程序的开头采用静态导入方式访问这些常量而无需加入类前缀。静态导入增强了程序的可读性，减少了代码的冗余。

In addition to the constants, all the many static methods in the Math class are also available when you use the wildcard static import line above. For example,

```
double logarithm = log (number);
```

instead of

```
double logarithm = Math.log (number);
```

For extensive use of the mathematical functions, the static import feature is a great addition to the Java language.

7.6　Arrays

7.6.1　Array Overview

Like a record, an array is a sequence of items. However, where items in a record are referred to by name, the items in an array are numbered, and individual items are referred to by their position number. Furthermore, all the items in an array must be of the same type. The definition of an array is: a numbered sequence of items, which are all of the same type. The number of items in an array is called the length of the array. The position number of an item in an array is called the index of that item. The type of the individual items in an array is called the base type of the array.

The base type of an array can be any Java type, that is, one of the primitive types, or a class name, or an interface name. If the base type of an array is int, it is referred to as an "array of ints." An array with base type String is referred to as an "array of Strings." However, an array is not, properly speaking, a list of integers or strings or other values. It is better thought of as a list of variables of type int, or of type String, or of some other type. As always, there is some potential for confusion between the two uses of a variable: as a name for a memory location and as a name for the value stored in that memory location. Each position in an array acts as a variable. Each position can hold a value of a specified type (the base type of the array). The value can be changed at any time. Values are stored in an array. The array is the container, not the values.

> 数组是由一组类型相同的元素的集合。数组中数据项的个数称为数组的长度。一个数据项在数组中的位置称为这个数据项的索引。数组中各个数据项的类型称为数组的基本类型。

7.6.1.1 Array Item

The items in an array—really, the individual variables that make up the array—are more often referred to as the elements of the array. In Java, the elements in an array are always numbered starting from zero. That is, the index of the first element in the array is zero. If the length of the array is N, then the index of the last element in the array is N−1. Once an array has been created, its length cannot be changed.

7.6.1.2 Array Object

Java arrays are objects. This has several consequences. Arrays are created using a form of the new operator. No variable can ever hold an array; a variable can only refer to an array. Any variable that can refer to an array can also hold the value null, meaning that it doesn't at the moment refer to anything. Like any object, an array belongs to a class, which like all classes is a subclass of the class Object. The elements of the array are, essentially, instance variables in the array object, except that they are referred to by number rather than by name.

Nevertheless, even though arrays are objects, there are differences between arrays and other kinds of objects, and there are a number of special language features in Java for creating and using arrays.

Suppose that A is a variable that refers to an array. Then the item at index k in A is referred to as A[k]. The first item is A[0], the second is A[1], and so forth. "A[k]" is really a variable, and it can be used just like any other variable. You can assign values to it, you can use it in expressions, and you can pass it as a parameter to subroutines. All of this will be discussed in more detail below. For now, just keep in mind the syntax：

```
array-variable [ integer-expression ]
```

for referring to an element of an array.

> 假如变量A指向一个数组。A中元素下标为k的就是A[k]。第一个是A[0]，第二个为A[1]，……。A[k]是一个变量，它可以像其他变量一样使用。可以给它赋值，也可以引用它，还可以把它作为一个子程序的参数。

Although every array, as an object, is a member of some class, array classes never have to be defined. Once a type exists, the corresponding array class exists automatically. If the name of the type is BaseType, then the name of the associated array class is BaseType[ ]. That is to say, an object belonging to the class BaseType[ ] is an array of items, where each item is a variable of type BaseType. The brackets, "[ ]", are meant to recall the syntax for referring to the individual items in the array. "BaseType[ ]" is read as "array of BaseType" or "BaseType array." It might be worth mentioning here that if ClassA is a subclass(子类)of ClassB, then ClassA[ ] is automatically a subclass of ClassB[ ].

> 每个数组都是某个类的实例化对象，然而不需要定义数组类。一旦某种类型产生，其相对应的数组类也自动产生。一个属于 BaseType[]类的对象是数组的元素，而每个数组元素是 BaseType 类型变量。数组中访问元素是通过中括号[]来表示。值得提的是，假如 ClassA 是 ClassB 的一个子类，那么 ClassA[ ]就自动是 ClassB[ ]的子类。

7.6.1.3 Array Types

Java arrays are reference objects that are dynamically created, much like a class. Like a class, they can refer only to instances of arrays or to the *null* reference, as the following snippet demonstrates:

```
int[] myIntArray = new int[5];
int[] anotherIntArray = null;
```

An array is a descendant of the Object class and, thus, all methods of class Object may be invoked on an array. An array object contains elements. The number of elements may be zero, in which case the array is said to be empty. All arrays are zero-based, meaning the first element is indexed by the number 0. All accesses on the elements of an array are checked at run time, and an attempt to use an index that is less than zero or greater than or equal to the length of the array causes an ArrayIndexOutOfBoundsException to be thrown.

> 数组是由 Object 类派生的，因此 Object 类中的所有方法可以被数组调用。

The elements of an array are referenced by integer index values, as follows:

```
int[] myIntArray = { 9, 5, 6 };
int int1 = myIntArray[0];
int int2 = myIntArray[1];
int int3 = myIntArray[2];
```

An array object's length can never change. To change the length of an array variable, another array must be created and assigned to the variable, as we see here:

```
int[] myIntArray = { 9, 5, 6 };
System.out.println("myIntArray length = " +myIntArray.length);//output is 3
myIntArray = new int[] { 3, 6, 4, 2, 8 };
System.out.println("myIntArray length = " +myIntArray.length);//output is 5
```

7.6.1.4 Create Array

Let's try to get a little more concrete(具体的) about all this, using arrays of integers as our first example. Since int[ ] is a class, it can be used to declare variables. For example,

```
int[] list;
```

creates a variable named list of type int[ ]. This variable is capable of referring to an array of ints, but initially its value is null (if it is a member variable in a class) or undefined (if it is a local variable in a method). The new operator is used to create a new array object, which can then be assigned to list. The syntax for using new with arrays is different from the syntax you learned previously. As an example,

```
list = new int[5];
```

creates an array of five integers. More generally, the constructor "new BaseType[N]" is used to create an array belonging to the class BaseType[ ]. The value N in brackets specifies the length of the array, that is, the number of elements that it contains. Note that the array "knows" how long it is. The length of the array is an instance variable in the array object. In fact, the length of an array, list, can be referred to as list.length. (However, you are not allowed to change the value of list.length, so it's really a "final" instance variable, that is, one whose value cannot be changed after it has been initialized.)

The situation produced by the statement "list = new int[5];" Figure7.1 shows the procedure of array initialization.

Note that the newly created array of integers is automatically filled with zeros. In Java, a newly created array is always filled with a known, default value: zero for numbers, false for boolean, the character with Unicode number zero for char, and null for objects.

> 注意：这个新创建的整型数组自动填入的 0 值。在 Java 中，一个新创建的数组总是用一些已知的默认值来填入：数值型数组用 0，布尔型数组用 false，unicode 字符用 0 表示，对象数组用空。

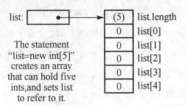

Figure 7.1 Array Initialization

The elements in the array list, are referred to as list[0], list[1], list[2], list[3], and list[4]. (Note again that the index for the last item is one less than list.length.) However, array references can be much more general than this. The brackets in an array reference can contain any expression whose value is an integer. For example if indx is a variable of type int, then list[indx] and list[2*indx+7] are syntactically correct references to elements of the array list. Thus, the following loop would print all the integers in the array, list, to standard output:

```
for (int i = 0; i < list.length; i++) {
    System.out.println( list[i] );
}
```

The first time through the loop, i is 0, and list[i] refers to list[0]. So, it is the value stored in the variable list[0] that is printed. The second time through the loop, i is 1, and the value stored in list[1] is printed. The loop ends after printing the value of list[4], when i becomes equal to 5 and the continuation condition "i<list.length" is no longer true. This is a typical example of using a loop to process an array. I'll discuss more examples of array processing throughout this chapter.

Every use of a variable in a program specifies a memory location. Think for a moment about what the computer does when it encounters a reference to an array element, list[k], while it is executing a program. The computer must determine which memory location is being referred to. To the computer, list[k] means something like this: "Get the pointer that is stored in the variable, list. Follow this pointer to find an array object. Get the value of k. Go to the k-th position in the array, and that's the memory location you want." There are two things that can go wrong here. Suppose that the value of list is null. If that is the case, then list doesn't even refer to an array. The attempt to refer to an element of an array that doesn't exist is an error. This is an example of a "null pointer" error. The second possible error occurs if list does refer to an array, but the value of k is outside the legal range of indices for that array. This will happen if k < 0 or if k >= list.length. This is called an "array index out of bounds" error. When you use arrays in a program, you should be mindful that both types of errors are possible. However, array index out of bounds errors are by far the most common error when working with arrays.

For an array variable, just as for any variable, you can declare the variable and initialize it in a single step. For example,

```
int[] list = new int[5];
```

If list is a local variable in a subroutine, then this is exactly equivalent to the two statements:
```
int[] list;
list = new int[5];
```

If list is an instance variable, then of course you can't simply replace "int[ ] list = new int[5];" with "int[ ] list; list = new int[5];" since the assignment statement "list = new int[5];" is only legal inside a subroutine.

7.6.1.5 Initial Array

The new array is filled with the default value appropriate for the base type of the array—zero for int and null for class types, for example. However, Java also provides a way to initialize an array variable with a new array filled with a specified list of values. In a declaration statement that creates a new array, this is done with an array initializer. For example,

```
int[] list = { 1, 4, 9, 16, 25, 36, 49 };
```

creates a new array containing the seven values 1, 4, 9, 16, 25, 36, and 49, and sets list to refer to that new array. The value of list[0] will be 1, the value of list[1] will be 4, and so forth. The length of list is seven, since seven values are provided in the initializer.

An array initializer takes the form of a list of values, separated by commas and enclosed between braces. The length of the array does not have to be specified, because it is implicit in the list of values. The items in an array initializer don't have to be constants. They can be variables or arbitrary expressions, provided that their values are of the appropriate type. For example, the following declaration creates an array of eight Colors. Some of the colors are given by expressions of the form "new Color(r,g,b)":

```
Color[] palette =
    {
        Color.black,
        Color.red,
        Color.pink,
        new Color(0,160,0),        //dark green
        Color.green,
        Color.blue,
        new Color(160,160,255),    //light blue
        Color.white
    };
```

> 数组的初始化以一组值的形式表示，由逗号和括号分开。数据的长度可以不必指出，因为在这组值中长度是隐含的。数组初始化的值不一定为常数，可以为变量和任意的表达式，只要它们的值与数组的类型一致。例如，下面语句创建了一个 8 色数组。一些颜色是以 new Color(r,g,b) 来表示。

A list initializer of this form can be used only in a declaration statement, to give an initial value to a newly declared array variable. It cannot be used in an assignment statement to assign a value to a variable that has been previously declared. However, there is another, similar notation for creating a new array that can be used in an assignment statement or passed as a parameter to a subroutine. The notation uses another form of the new operator to create and initialize a new array object. (This rather odd syntax is reminiscent of the syntax for anonymous classes, For example to assign a new value to an array variable, list, that was declared previously, you could use:

```
list = new int[] { 1, 8, 27, 64, 125, 216, 343 };
```

The general syntax for this form of the new operator is

```
new base-type [ ] { list-of-values }
```

This is an expression whose value is an array object. It can be used in any context where an object of type base-type[ ] is expected. For example, if makeButtons is a method that takes an array of Strings as a parameter, you could say:

```
makeButtons( new String[] { "Stop", "Go", "Next", "Previous" } );
```

One final note: For historical reasons, the declaration

```
int[] list;
```

can also be written as

```
int list[];
```

which is a syntax used in the languages C and C++. However, this alternative syntax（选择性语法）does not really make much sense in the context of Java, and it is probably best avoided. After all, the intent is to declare a variable of a certain type, and the name of that type is "int[ ]". It makes sense to follow the "type-name variable-name;" syntax for such declarations.

7.6.2 Java Arrays

As we saw in Java Variables, you can store program data in variables. Each variable has an identifier, a type, and a scope. When you have closely related data of the same type and scope, it is often convenient to store it together in a data structure instead of in individual variables. The most common data structure is the array. Arrays are fixed-length structures for storing multiple values of the same type. An array implicitly extends Java.lang.Object so an array is an instance of Object. But arrays are directly supported language features. This means that their performance is on par with primitives and that they have a unique syntax that is different than objects.

> 数组具备如下特性：有着固定长度，可以存放多个相同类型的变量，可通过一个共同的名称进行引用。数组是从 Java.lang.Object 中派生出来的，数组是 Object 类的一个实例。数组有自己的语法规则。

7.6.2.1 Structure of Java Arrays

The Java array depicted in figure 7.2 has 7 elements. Each element in an array holds a distinct value. In the figure, the number above each element shows that element's index. Elements are always referenced by their indices. The first element is always index 0. This is an important point, so I will repeat it! The first element is always index 0. Given this zero-based numbering, the index of the last element in the array is always the array's length minus one. So in the array pictured above, the last element would have index 6 because 7 minus 1 equals 6.

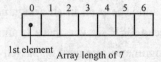

Figure 7.2   the Structer of Array

> 可以通过下标来访问数组中的特定元素，数组下标都是从 0 开始。

7.6.2.2 Java Array Declaration

An array variable is declared the same way that any Java variable is declared. It has a type and a valid Java identifier. The type is the type of the elements contained in the array. The [] notation is used to denote that the variable is an array. Some examples:

```
int[] counts;
String[] names;
int[][] matrix; //this is an array of arrays
```

> 数组变量的声明和 Java 其他类型的变量声明方式一样。数组定义也需要一个类型和合法的标识符。数组变量的类型就是数组中所有元素的数据类型。[]符号用来标识该变量是一个数组变量。

7.6.2.3 Java Array Initialization

Once an array variable has been declared, memory can be allocated to it. This is done with the new operator, which allocates memory for objects. (Remember arrays are implicit objects.) The new operator is followed by the type, and finally, the number of elements to allocate. The number of elements to allocate is placed within the [] operator. Some examples:

```
counts = new int[5];
names = new String[100];
matrix = new int[5][];
```

An alternate shortcut syntax is available for declaring and initializing an array. The length of the array is implicitly defined by the number of elements included within the {}. An example:

```
String[] teacher = {"Michael", "james", "hunter"};
```

> 数组变量一旦被声明，会立即给其分配内存。使用 new 操作符可以为数组的每个对象分配内存空间。注意 new 操作符后要紧跟数据类型，将数组中元素的个数放至[]内。可以用一种简单的方式声明并初始化数组，直接在花括号中指定值给数组初始化。数组的长度取决于花括号内元素的个数。

7.6.2.4 Java Array Usage

To reference an element within an array, use the [] operator. This operator takes an int operand and returns the element at that index. Remember that array indices start with zero, so the first element is referenced with index 0.

```
int month = months[3]; //get the 4th month (April)
```

> 通过使用[]运算符引用数组元素，该运算符以一个整数作为其操作数，返回对应下标的元素。Java 中数组元素下标从 0 开始计数，故数组中的第一个元素下标为 0。

In most cases, a program will not know which elements in an array are of interest. Finding the elements that a program wants to manipulate requires that the program loop through the array with the for construct and examine each element in the array.

```
String months[] =
    {"Jan", "Feb", "Mar", "Apr", "May", "Jun",
```

```
            "July", "Aug", "Sep", "Oct", "Nov", "Dec"};
    //use the length attribute to get the number of elements in an array
    for(int i = 0; i < months.length; i++ ) {
        System.out.println("month: " + month[i]);    //打印输出数组中的元素
    }
```

Using Java 5.0, the enhanced **for** loop makes this even easier:

```
String months[] =
    {"Jan", "Feb", "Mar", "Apr", "May", "Jun",
     "July", "Aug", "Sep", "Oct", "Nov", "Dec"};
// Shortcut syntax loops through array months
// and assigns the next element to variable month
// for each pass through the loop
for(String month: months) {
    System.out.println("month: " + month);    //打印输出数组中的元素
}
```

Arrays are a great way to store several to many values that are of the same type and that are logically related to one another: lists of invoices, lists of names, lists of Web page hits, etc. But, being fixed-length, if the number of values you are storing is unknown or changes, there are better data structures available in the Java

7.7 String

7.7.1 Creating a String

Strings are created like any other object in Java using the new operator. For example one of the String constructors allows an array of characters to be taken as input

```
char letters[] = {'J','a','v','a'};
String str=new String (letters);//通过定义字符串数组,去构造 String 对象
System.out.println(str);        //will print Java
```

> 在 Java 中与构造其他对象一样采用 new 操作,运用 String 类提供的构造函数实现 String 对象的构造。

However, there is a special syntax that can be used when creating objects of the String class. This doesn't use the new operator but simply an assignment, for example:

```
String str = "Java";              //note the ""s around the letters
System.out.println (str);         //will print Java
```

There is another special piece of string syntax using the + operator. This allows strings to be added together, at the moment it is convenient to know how to use this. The way it works is to use the StringBuffer append method and convert back to a string, this will become clearer after the next lesson. An example of the plus operator in string syntax is:

```
String s = "Hello";
String t = s + "there" + ", how are you" + "today";
// t = "Hello there, how are you today"
```

7.7.2 Strings Operation

Once you have created your strings it is nice to be able to do something useful with them.

The equals method

Comparing two Strings references to see if they hold identical strings is easily achieved with the equals method. This method is called on one string object and takes another as an input parameter, if the they are equal it returns the boolean value true, otherwise it returns false. For example:

```
String str = "hello";
String str2 = "hello";
String str3 = "goodbye";
System.out.println ("str and str2" + str.equals(str2)); // true
System.out.println ("str and str3" + str.equals(str2)); // false
```

7.7.3 Alter Strings

```
class altStr
{
    public static void main (String args[])
    {
    String str = "Hello";
    String str2 = "Java";
    str = str.toUpperCase ();      //将字符串中的所有字符从小写转换为大写
    str2 = str2.toLowerCase ();    //将字符串中的所有字符从大写转换为小写
    System.out.println (str);      //HELLO
    System.out.println (str2);     //Java
    str=str.concat(str2);          //str now equals "HELLO Java" 连接两个字符串
    System.out.println (str);
    str=str.trim ();//str now equals "HELLOJava"    //删除字符串中的空格
    System.out.println (str);
    str=str.substring (5,str.length());              // str = "Java"
      //提取[5,str.length())区间的字符串
    System.out.println (str);
    str = str.replace ('a', 'i');
      //str = "jivi", 用'i'字符替代 str 字符串中出现的'a'字符
    System.out.println (str);
    str = String.valueOf (3.141);
      //str = "3.141", 将数据类型转换为字符串
    System.out.println (str);
    }
}
```

To run this example cut and past it, save as altStr.Java and compile and run as normal.

7.8 Command Line Arguments

Your Java application can accept any number of arguments from the command line. Command line arguments allow the user to affect the operation of an application. For example, an application might allow the user to specify verbose mode—that is, specify that the application display a lot of trace information—with the command line argument -verbose.

When invoking an application, the user types the command line arguments after the application name. For example, suppose you had a Java application, called Sort, that sorted lines in a file, and that the data you want sorted is in a file named friends.txt. If you were using Windows 95/NT, you would invoke the Sort application on your data file like this:

C:\> Java Sort friends.txt

In the Java language, when you invoke an application, the runtime system passes the command line arguments to the application's main method via an array of Strings. Each String in the array contains one of the command line arguments. In the previous example, the command line arguments passed to the Sort application is an array that contains a single string: "friends.txt".

> 命令行参数是在程序执行时紧跟在命令行上程序名称后面的信息，Java 应用程序能接收从命令行传递过来的任何参数。在 Java 中，在执行程序时，命令行的参数将通过一个字符串数组传递到 main 方法中。如上例中传递给 Sort 应用程序的命令行参数就是一个简单字符串 "friends.txt"。

Echo Command Line Arguments

This simple application displays each of its command line arguments on a line by itself:

```
class ECIT {
    public static void main (String[] args) {
        for (int i = 0; i < args.length; i++)
            System.out.println(args[i]);   //输出命令行参数
    }
}
```

Try this: Invoke the Echo application. Here's an example of how to invoke the application using Windows 95/NT:

C:\> **Java ECIT always mean quality**

Always
mean
quality

7.9　Sample Examples

Examples 7.1

```
//此例用来练习类中构造方法的定义
//Constructor in a class
class Box
{
    double width;
    double height;
    double depth;
    Box()              //构造方法
    {
        System.out.println("Constructing a box");
        width = 10;
        height = 15;
        depth = 5;
```

```java
    }
    double volume()        //返回盒子的体积
    {
        return width*height*depth;
    }
}
class Constructor
{
    public static void main(String args[])
    {
        Box mybox1 = new Box();
        double vol;
        vol = mybox1.volume();
        System.out.println(vol);
    }
}
```

Examples 7.2

```java
//此例用来练习构造方法的重载
//Overloading constructor
class Box
{
    double width;
    double height;
    double depth;
    Box(double w,double h,double d) //构造方法的重载,可用来构建长方体
    {
        width = w;
        height = h;
        depth = d;
    }
    Box()                           //默认的构造方法
    {
        width = -1;
        height = -1;
        depth = -1;
    }
    Box(double len)                 //构造方法的重载,可用来构建正方体
    {
        width = height = depth = len;
    }
    double volume()                 //求体积的成员方法
    {
        return width*height*depth;
    }
}
class Constructor_overloading
{
    public static void main(String args[])
    {
        Box mybox1 = new Box(10,20,25);
        Box mybox2 = new Box();
        Box mycube = new Box(7);
```

```
        double vol;
        vol = mybox1.volume();     //得到长方体的体积
        System.out.println(vol);
        vol = mybox2.volume();
        System.out.println(vol);
        vol = mycube.volume();     //得到正方体的体积
        System.out.println(vol);
    }
}
```

Examples 7.3

```
//此例用来练习方法的重载
//Method overloading
class Overload
{
    void test()               //不带任何参数的成员方法
    {
        System.out.println("No parameters");
    }
    void test(int a)          //成员方法的重载,带一个整型参数
    {
        System.out.println("a:" + a);
    }
    void test(int a,int b)    //成员方法的重载,带两个整型参数
    {
        System.out.println("a and b:" + a + ' ' + b);
    }
    double test(double a)     //成员方法的重载,带一个双精度类型的参数
    {
        return a*a;
    }
}
class Method_overloading
{
    public static void main(String args[])
    {
        Overload O = new Overload();
        //针对所带的参数的个数及参数的类型去调用相应的成员方法
        O.test();
        O.test(1);
        O.test(1,2);
        double result = O.test(123.4);
        System.out.println("the result is:" + result);
    }
}
```

Examples 7.4

```
//此例用来练习堆栈的实现,栈的原理是先进后出
//Practical implementation of stack
class Stack
{        /* 创建一个数组对象,用来存放栈的元素*/
    private int Stack[] = new int [10];
    private int tos;                    //tos 用来表明栈的下标
```

```java
        Stack()                         //构造方法
        {
            tos = -1;                   //给 tos 赋初值
        }
        void Push(int i)                //入栈方法
        {
            if(tos == 9)                //若下标为 9，则输出栈已满
            {
                System.out.println("The stack is full");
                return ;
            }
            else                        //否则将元素存放入数组中
            {
                Stack[++tos] = i;
            }
        }
        int Pop()                       //出栈方法
        {
            if(tos == 0)                //若下标为 0，则输出栈溢出
            {
                System.out.println("The stack is overflow");
                return 0;
            }
            else                        //否则根据栈下标从数组中取出元素
            {
                return Stack[tos--];
            }
        }
    }
    class newStack
    {
       public static void main(String args[])
       {
            Stack s = new Stack();   //创建一个空栈
            int i;
            for(i=0; i<10; ++i)         //将 0 至 9 逐个入栈
            {
                s.Push(i);
            }
            /*将栈中的元素按后进先出的原理逐个出栈*/
            for(i=0; i<10; ++i)
            {
                System.out.println("Pop out of the stack:" + s.Pop());
            }
       }
    }
```

Examples 7.5
```java
//此例用来再次练习堆栈的实现
//Stack class
class Stack
{
     public static void main(String args[])
```

```
        {
            int i;
            /*创建了两个空栈*/
            Stack mystack1 = new Stack();
            Stack mystack2 = new Stack();
            for(i=0; i<10; ++i)        //将 0 至 9 逐个入栈
            {
                mystack1.push(i);
            }
            for(i=10; i<20; ++i)       //将 10 至 19 逐个入栈
            {
                mystack2.push(i);
            }
            /*将 mystack1 栈中的元素按后进先出的原理逐个出栈*/
            System.out.println("Pop out mystack1:");
            for(i=0; i<10; ++i)
            {
                System.out.println(mystack1.pop() + ' ');
                /* 输出的结果将是出栈的元素与空格对应的 ASCII 值之和*/
            }
            /*将 mystack2 栈中的元素按后进先出的原理逐个出栈*/
            System.out.println("Pop out mystack2:");
            for(i=10; i<20; ++i)
            {
                System.out.println(mystack2.pop() + ' ');
                /* 输出的结果将是出栈的元素与空格对应的 ASCII 值之和*/
            }
        }
    }
```

7.10 Exercise for you

1. Define one object and use the object as parameter.
2. What is a constructor? how you can overload a constructor? write one program.
3. What is recursive synonym? Write one example to show one recursive synonym.

Chapter 8   Inheritance

Object-oriented programming (OOP) is a popular and powerful programming philosophy. One of the main techniques of OOP is known as inheritance. Inheritance means that a very general form of a class can be defined and compiled. Later, more specialized versions of that class may be defined by starting with the already defined class and adding more specialized instance variables and methods. The specialized classes are said to inherit the methods and instance variables of the previously defined general class.

> 面向对象程序设计(OOP)是一种流行且强大的编程思想。OOP 最主要的技术之一就是继承。继承首先定义和编译类的最一般的形式，之后在已经定义的类的基础上定义更具体的类，并为之增加更多特殊的实例成员变量和成员方法。这个特定的类就称为从先前已定义的普通类继承了方法和实例变量。

8.1   Derived Classes

Inheritance is the process by which a new class—known as a derived class—is created from another class, called the base class. A derived class automatically has all the instance variables and all the methods that the base class has, and can have additional methods and/or additional instance variables.

> 继承是一个新的类(称为派生类)从另一个称为基类的类创建产生的过程。派生类自动拥有基类的所有实例变量和成员方法，同时该新类也能定义额外的方法和实例成员变量。

Suppose we are designing a record-keeping program that has records for salaried employees and hourly employees. There is a natural hierarchy for grouping these classes.

Employees who are paid an hourly wage are one subset of employees. Another subset consists of salaried employees who are paid a fixed wage each month. Although the program may not need any type corresponding to the set of all employees, thinking in terms of the more general concept of employees can be useful. For example, all employees have a name and a hire date (when they started working for the company), and the methods for setting and changing names and hire dates will be the same for salaried and hourly employees. The classes for hourly employees and salaried employees may be further subdivided as diagrammed in figure 8.1.

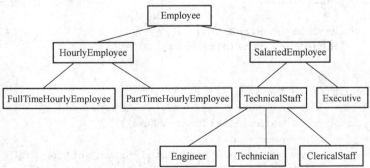

Figure 8.1 Inheritance Hierarchy of Employee Classes

Within Java you can define a class called Employee that includes all employees, whether salaried or hourly, and then use this class to define classes for hourly employees and salaried employees. You can then, in turn, use classes like HourlyEmployee to define classes like PartTimeHourlyEmployee, and so forth.

> 可以用Java语言定义一个Employee类，该类涵盖了所有员工的共同特性，并利用这个类创建其他的如按小时付费的员工类和受薪员工类；然后还可以使用已定义的类如HourlyEmployee类定义兼职按小时付费员工类PartTimeHourlyEmployee等。

The following example shows our definition for the class Employee. The class Employee is a pretty ordinary class much like earlier classes we have seen. What will be interesting about the class Employee is how we use it to create a class for hourly employees and a class for salaried employees. It is legal to create an object of the class Employee, but our reason for defining the class Employee is so that we can define derived classes for different kinds of employees.

```java
public class Employee{
    private String name;
    private Date hireDate;
    public Employee(){
        name = "No name";
        hireDate = new Date("January", 1, 1000); //Just a placeholder.
    }
    /**
    Precondition: Neither theName nor theDate is null.
    */
    public Employee(String theName, Date theDate){
        if (theName == null || theDate == null){
            System.out.println("Fatal Error creating employee.");
            System.exit(0);
        }
        name = theName;
        hireDate = new Date(theDate);
    }
    public Employee(Employee originalObject){
        name = originalObject.name;
        hireDate = new Date(originalObject.hireDate);
    }
    public String getName(){
        return name;
    }
    public Date getHireDate(){
        return new Date(hireDate);
    }
    /**
    Precondition newName is not null.
    */
    public void setName(String newName){
        if (newName == null){
            System.out.println("Fatal Error setting employee name.");
            System.exit(0);
        }
        else
```

```
            name = newName;
    }
    /**
    Precondition newDate is not null.
    */
    public void setHireDate(Date newDate){
        if (newDate == null){
            System.out.println("Fatal Error setting employee hire date.");
            System.exit(0);
        }
        else
            hireDate = new Date(newDate);
    }
    public String toString(){
        return (name + " " + hireDate.toString());
    }
    public boolean equals(Employee otherEmployee){
        return (name.equals(otherEmployee.name)  && hireDate.equals
                (otherEmployee.hireDate));
    }
}
```

The following example contains the definition of a class for hourly employees. An hourly employee is an employee, so we define the class HourlyEmployee to be a derived class of the class Employee. A Derived class is a class defined by adding instance variables and methods to an existing class. The existing class that the derived class is built upon is called the base class. In our example, Employee is the base class and HourlyEmployee is the derived class. As you can see in the example, the way we indicate that Hourly - Employee is a derived class of Employee is by including the phrase extends Employee on the first line of the class definition, like so:

> 派生类是一种在已知类的基础上增加了实例变量和成员方法的类。这个已知类被称为基类。在例子中，Employee 类是基类，而 HourlyEmployee 是派生类。从例子中可以看到，HourlyEmployee 是 Employee 的派生类，是通过在首行包含了 extends Employee 短语体现的。

```
    public class HourlyEmployee extends Employee{
        private double wageRate;
        private double hours; //for the month
        public HourlyEmployee(){
            super();
            wageRate = 0;
            hours = 0;
        }
    /**
    Precondition: Neither theName nor theDate is null;
    theWageRate and theHours are nonnegative.
    */
        public HourlyEmployee(String theName, Date theDate,
                    double theWageRate, double theHours){
            super(theName, theDate);
            if ((theWageRate >= 0) && (theHours >= 0)){
                wageRate = theWageRate;
                hours = theHours;
```

```
            }
            else{
                System.out.println("Fatal Error: creating an illegal hourly
                                        employee.");
                System.exit(0);
            }
        }
        public HourlyEmployee(HourlyEmployee originalObject){
            super(originalObject);
            wageRate = originalObject.wageRate;
            hours = originalObject.hours;
        }
        public double getRate(){
            return wageRate;
        }
        public double getHours(){
            return hours;
        }
/**
Returns the pay for the month.
*/
        public double getPay(){
            return wageRate*hours;
        }
/**
Precondition: hoursWorked is nonnegative.
*/
        public void setHours(double hoursWorked){
            if (hoursWorked >= 0)
                hours = hoursWorked;
            else{
                System.out.println("Fatal Error: Negative hours worked.");
                System.exit(0);
            }
        }
/**
Precondition: newWageRate is nonnegative.
*/
        public void setRate(double newWageRate){
            if (newWageRate >= 0)
                wageRate = newWageRate;
            else{
                System.out.println("Fatal Error: Negative wage rate.");
                System.exit(0);
            }
        }
}
```

    A derived class is also called a subclass, in which case the base class is usually called a superclass. However, we prefer to use the terms derived class and base class.

    When you define a derived class, you give only the added instance variables and the added methods. For example, the class HourlyEmployee has all the instance variables and all the methods of the class Employee, but you do not mention them in the definition of HourlyEmployee. Every object of the class HourlyEmployee has instance variables called name and hireDate, but you do not specify the instance variable name or the instance variable hireDate in the definition of the class HourlyEmployee. The class HourlyEmployee (or any other derived class) is said to inherit the instance variables and methods of the base class that it extends.

> 派生类也称为子类，与之相对应的基类通常称为超类。本书更愿意称之为派生类和基类。当定义派生类的时候，只需要给出增加的实例成员变量和成员方法的定义即可。

Just as it inherits the instance variables of the class Employee, the class HourlyEmployee inherits all the methods from the class Employee. So, the class HourlyEmployee inherits the methods getName, getHireDate, setName, and setHireDate, from the class Employee.

> 与从 Employee 继承实例成员变量一样，HourlyEmployee 类从 Employee 继承了所有的成员方法。因此，HourlyEmployee 类从 Employee 继承了getName、getHireDate、setName 和 setHireDate 方法。

For example, suppose you create a new object of the class HourlyEmployee as follows:

```
HourlyEmployee joe = new HourlyEmployee();
```

Then, the name of the object joe can be changed using the method setName, which the class HourlyEmployee inherited from the class Employee. The inherited method setName is used just like any other method; for example:

```
joe.setName("Josephine");
```

Parent and Child Classes

A base class is often called the parent class. A derived class is then called a child class. A class that is a parent of a parent of a parent of another class (or some other number of "parent of" iterations) is often called an ancestor class. If class A is an ancestor of class B, then class B is often called a descendent of class A.

> 基类通常也称为父类。派生类通常也称为子类。另外一个类的父类的父类的父类(或更多级父类)通常称为祖先类。如果 A 类是 B 类的祖先类，则 B 类又称为 A 类的子孙类。

Extend

When you derive a class from a base class we say the derived subclass **extends** the base class. An interface can also **extend** another interface. In contrast, when a class provides the methods (possibly abstract) necessary to conform to some interface we say that class **implements** the interface. Extending a class is closely related to implementing an interface. A new class can extend at most one superclass, but it may implement several interfaces.

> 从一个类中可以派生出另一个类，一个接口也可以派生出另一个接口。接口声明中的方法全都是抽象的，这些方法要靠使用接口的类去具体实现。一个类只能有一个超类，但可以实现多个接口。

Implement

When a class provides the methods (possibly abstract) necessary to conform to some interface we say that class **implements** the interface. Implementing an interface is closely related to extending a class. A new class can extend at most one superclass, but it may implement several interfaces.

> 当一个类定义了和接口一致的方法实现或是抽象方法，则称此类实现了接口。接口的实现和类的继承很相似。一个类最多只能有一个超类，却可以实现多个接口。

Multiple Inheritances

Java does not support multiple inheritances, allowing a class to extend more than one base class. It does however support implementing multiple "interfaces", which is similar to multiple inheritance but simpler and safer. Inheritance describes an IS-A relationship. Dalmatian IS-A Dog. You can also have HAS-A relationships. A Body HAS-A Leg. HAS-A relationships are implemented with embedded references to other objects. HAS-A relationships are more flexible. An object can have multiple HAS-A relationships. It can even have a variable number of them, by having a reference to an array or ArrayList of references.

> Java中不支持多继承，即一个类不能同时从多个类扩展。但Java支持实现多接口，这和多继承非常相似，但更为安全。多重接口类似多重继承。继承描述的是一个"IS-A"的关系，例如"Dalmatian IS-A Dog"。同时还有一种"HAS-A"的关系，例如"A Body HAS-A Leg"。"HAS-A"的关系是指在一个对象中定义其他类对象的引用作为其内部成员。"HAS-A"的关系使用时更为灵活，一个对象可以和多个对象建立"HAS-A"的关系。可以通过在内部定义指向对象数组或ArrayList对象的引用建立该对象和其他任意多个对象之间的"HAS-A"关系。

Key Points
- Note the use of the **extends** keyword in the Whole_example and Natural_example classes, indicating inheritance.
- Note the use of the **super** keyword in the constructors for the Whole_example and Natural_example classes. To create an instance of Natural_example, one must create an instance of Whole_example, and to create an instance of Whole_example, one must create an instance of Integer_example. The super keyword within the constructor does this.
- Note the use of the **super** keyword in the toString methods of the Whole_example and Natural_example classes. In each case, we return the result of the toString method from the superclass with another message appended to that indicating the nature of the subclass.

8.2 Abstract Classes

Abstract classes are those which can be used for creation of objects. However their methods and constructors can be used by the child or extended class. The need for abstract classes is that you can generalize the super class from which child classes can share its methods. The subclass of an abstract class which can create an object is called as "concrete class". Following is an example.

```
Abstract class A {                //抽象类A
    abstract void method1();      //抽象类方法method1()
    void method2(){
        System.out.println("this is concrete method");
```

```
        }
    }
    class B extends A {
        void method1(){                    //覆盖父类的抽象方法 method1()
            System.out.println("B is implementation of method1");
        }
    }
    class demo{
        public static void main(String arg[]){
            B b=new B();                   //创建子类B的实例b
            b.method1();                   //输出B is implementation of method1
            b.method2();                   //输出this is concrete method
        }
    }
```

For another example, all card objects are one of the three types (Valentine, Holiday, or Birthday) and the parent class Card is used only to group them into a hierarchy. There will never be an object that is just a card. This is useful to do, just as a store has all its various greeting cards in one display, arranged into several categories.

An **abstract class** in Java is a class that is never instantiated. Its purpose is to be a parent to several related classes. The child classes inherit from the abstract parent class.

> 抽象类可以被用来创建对象。抽象类的方法和构造方法可以被它们的派生类使用。抽象类是用来被继承的，抽象类的子类可以创建一个对象，将它实例化，即"实体类"。Java 中的抽象类是不能被实例化的。它可以同时是几个相关类的父类，这些子类从这个抽象类继承。

In hierarchy drawings (such as on the previous page), abstract classes are drawn with dotted lines. An abstract class is defined like this:

```
Abstract class Classname
{
…// definitation to methods and variables.
}
```

Access modifiers such as public can be placed before abstract. Even though it can not be instantiated, an abstract class can define methods and variables that children classes inherit.

> 在抽象类的 abstract 前面可以加上 public，虽然不能被实例化，但抽象类可以为它的子类定义方法和变量。

An abstract class is missing definitions for one or more methods. You can't thus create an object of that class.

You must first create a subclass and provide definitions for the abstract methods. Unlike interfaces, abstract classes may implement some of the methods. Though you can't instantiate an abstract class, you can invoke its static methods.

> 抽象类不能直接实例化对象，必须先创建抽象类的子类并在子类中实现这些抽象方法，然后对子类进行实例化。与接口不同，抽象类可以实现其中的一些方法。虽然无法对一个抽象类创建实例化对象，但可以调用它的 static 方法。

8.3 Keyword "final"

Java's final keyword has slightly different meanings depending on the context, but in general it says "This cannot be changed". You might want to prevent changes for two reasons: design or efficiency. Because these two reasons are quite different, it's possible to misuse the final keyword.

> Java 中的 final 关键字在不同的上下文中意思稍有不同，但一般来说意思是"它是不能被改变的"。不希望改变的原因有两个：设计考虑角度或是效率因素。这两个原因区别很大，很可能会错用 final 关键字。

The following sections discuss the three places where final can be used: for data, methods, and classes.

8.3.1 Final Data

Many programming languages have a way to tell the compiler that a piece of data is "constant." A constant is useful for two reasons:
- It can be a compile-time constant that won't ever change.
- It can be a value initialized at run-time that you don't want changed.

> 很多程序设计语言的编译器都有将某个数据设置成常量的方式，使用常量的情况有两种：一种是编译时就不能改变的常量；一种是运行时值被初始化后就不能被改变。

In the case of a compile-time constant, the compiler is allowed to "fold" the constant value into any calculations in which it's used; that is, the calculation can be performed at compile-time, eliminating some run-time overhead. In Java, these sorts of constants must be primitives and are expressed using the final keyword. A value must be given at the time of definition of such a constant.

> 对于编译时的常量，编译器允许常量通过计算得到，也就是说，某些计算可以在编译时进行，从而减少一些运行时的开销。Java 中，编译时常量必须是基本类型数据，并且用 final 关键字定义，且在定义该常量时必须给其赋值。

A field that is both static and final has only one piece of storage that cannot be changed.

> 同时用 static 和 final 关键字定义的数据成员对类的所有对象都只有一份副本，且该数据成员的值是不能改变的。

When using final with object references rather than primitives the meaning gets a bit confusing. With a primitive, final makes the value a constant, but with an object reference, final makes the reference a constant. Once the reference is initialized to an object, it can never be changed to point to another object. However, the object itself can be modified; Java does not provide a way to make any arbitrary object a constant. (You can, however, write your class so that objects have the effect of being constant.) This restriction includes arrays, which are also objects.

> 基本类型数据定义为 final 表示该数据为常量，但是对于一个对象引用定义为 final，该对象引用则变为引用常量，即当一个引用被初始化指向一个对象，则该引用再也不能指向其他的对象，但是被指向的对象本身是能被修改的。

Here's an example that demonstrates final fields:

```java
class Value
{
    int i = 1;
}
public class FinalData
{
    // Can be compile-time constants
    final int i1 = 9;
    static final int VAL_TWO = 99;
    // Typical public constant:
    public static final int VAL_THREE = 39;
    // Cannot be compile-time constants:
    final int i4 = (int)(Math.random()*20);
    static final int i5 = (int)(Math.random()*20);
    Value v1 = new Value();
    final Value v2 = new Value();
    static final Value v3 = new Value();
    // Arrays
    final int[] a = { 1, 2, 3, 4, 5, 6 };
    public void print(String id){
        System.out.println(id + ": " + "i4 = " + i4 + ", i5 = " + i5);
    }
    public static void main(String[] args)
    {
        FinalData fd1 = new FinalData();
        //! fd1.i1++; // Error: can't change value
        fd1.v2.i++; // Object isn't constant!
        fd1.v1 = new Value(); // OK -- not final
        for(int i = 0; i < fd1.a.length; i++)
            fd1.a[i]++; // Object isn't constant!
        //! fd1.v2 = new Value(); // Error: Can't
        //! fd1.v3 = new Value(); // change reference
        //! fd1.a = new int[3];
        fd1.print("fd1");
        System.out.println("Creating new FinalData");
        FinalData fd2 = new FinalData();
        fd1.print("fd1");
        fd2.print("fd2");
    }
}
```

Since i1 and VAL_TWO are final primitives with compile-time values, they can both be used as compile-time constants and are not different in any important way. VAL_THREE is the more typical way you'll see such constants defined: public so they're usable outside the package, static to emphasize that there's only one, and final to say that it's a constant. Note that final static primitives with constant initial values (that is, compile-time constants) are named with all capitals by convention, with words separated by underscores. Also note that i5 cannot be known at compile-time, so it is not capitalized.

Just because something is final doesn't mean that its value is known at compile-time. This is demonstrated by initializing i4 and i5 at run-time using randomly generated numbers. This portion of the example also shows the difference between making a final value static or non-static. The difference is shown in the output from one run:

fd1: i4 = 15, i5 = 9
Creating new FinalData
fd1: i4 = 15, i5 = 9
fd2: i4 = 10, i5 = 9

Note that the values of i4 for fd1 and fd2 are unique, but the value for i5 is not changed by creating the second FinalData object. That's because it's static and is initialized once upon loading and not each time a new object is created.

The variables v1 through v3 demonstrate the meaning of a final reference. As you can see in main( ), just because v2 is final doesn't mean that you can't change its value. However, you cannot rebind v2 to a new object, precisely because it's final. That's what final means for a reference. You can also see the same meaning holds true for an array, which is just another kind of reference.

8.3.1.1 Blank Finals

Java allows the creation of blank finals, which are fields that are declared as final but are not given an initialization value. In all cases, the blank final must be initialized before it is used, and the compiler ensures this. However, blank finals provide much more flexibility in the use of the final keyword since, for example, a final field inside a class can now be different for each object and yet it retains its immutable quality. Here's an example:

> Java 允许定义未赋值的 final 变量，即成员变量被定义为 final 但没有给出初始值。在任何情况下使用未赋值的 final 变量，都必须先对其初始化，编译器会进行检查。然而，未赋值的 final 变量在 final 的使用上提供了更大的灵活性。例如，在一个类内部定义的 final 类型的成员变量，对每个对象来说可以不同，尽管它仍然有不可以被改变的特性。

```java
class Poppet{
}
public class BlankFinal {
    final int i = 0;     // Initialized final
    final int j;         // Blank final
    final Poppet p;      // Blank final reference
    // Blank finals MUST be initialized
    // in the constructor:
    BlankFinal() {
        j = 1;           // Initialize blank final
        p = new Poppet();
    }
    BlankFinal(int x){
        j = x;           // Initialize blank final
        p = new Poppet();
    }
    public static void main(String[] args){
        BlankFinal bf = new BlankFinal();
    }
}
```

You're forced to perform assignments to finals either with an expression at the point of definition of the field or in every constructor. This way it's guaranteed that the final field is always initialized before use.

8.3.1.2 Final Arguments

Java allows you to make arguments final by declaring them as such in the argument list. This means that inside the method you cannot change what the argument reference points to:

> Java 允许将参数列表中的参数声明为 final，表示在这个方法中不能修改该参数所指向数据的值。

```
class Gizmo {
    public void spin() {
    }
}
public class FinalArguments{
    void with(final Gizmo g){
        //! g = new Gizmo(); // Illegal -- g is final
    }
    void without(Gizmo g){
        g = new Gizmo(); // OK -- g not final
        g.spin();
    }
    // void f(final int i) { i++; } // Can't change
    // You can only read from a final primitive:
    int g(final int i){
        return i + 1;
    }
    public static void main(String[] args){
        FinalArguments bf = new FinalArguments();
        bf.without(null);
        bf.with(null);
    }
}
```

Note that you can still assign a null reference to an argument that's final without the compiler catching it, just like you can with a non-final argument.

The methods f( ) and g( ) show what happens when primitive arguments are final: you can read the argument, but you can't change it.

8.3.2 Final Methods

There are two reasons for final methods. The first is to put a "lock" on the method to prevent any inheriting class from changing its meaning. This is done for design reasons when you want to make sure that a method's behavior is retained during inheritance and cannot be overridden.

The second reason for final methods is efficiency. If you make a method final, you are allowing the compiler to turn any calls to that method into inline calls. When the compiler sees a final method call it can (at its discretion) skip the normal approach of inserting code to perform the method call mechanism (push arguments on the stack, hop over to the method code and execute it, hop back and clean off the stack arguments, and deal with the return value) and instead replace the method call with a copy of the actual code in the method body. This eliminates the overhead of the method call. Of course, if a method is big, then your code begins to bloat and you probably won't see any performance gains from inlining, since any improvements will be dwarfed by the amount of time spent inside the method. It is implied that the Java compiler is able to detect these situations and choose wisely whether to inline a final method. However, it's better to not trust that the compiler is able to do this and make a method final only if it's quite small or if you want to explicitly prevent overriding.

> 有两种原因需要定义 final 方法。一是给某个方法加上锁以防止继承类去修改该方法。这是从设计上考虑的，可以确保该方法的行为在继承的过程中保持不变且不被覆盖。二是考虑到 final 方法的效率。如果将一个方法定义为 final，意味着可以允许编译器将对该方法的任意调用变为内联访问的形式。当编译器发现一个 final 方法被调用，可以将该方法的方法体插入到调用方法中而不执行常规的方法调用机制。这样减少了方法调用的开销。当然，如果一个方法很复杂，使用内联方式将使代码变得膨大而使效率可能无法得到提升，在这个方法中花费的时间使得效率的提升相形见拙。这也说明 Java 编译器能根据具体情况进行最优的选择，即是否选用内联方式。

Any private methods in a class are implicitly final. Because you can't access a private method, you can't override it (even though the compiler doesn't give an error message if you try to override it, you haven't overridden the method, you've just created a new method). You can add the final specifier to a private method but it doesn't give that method any extra meaning.

This issue can cause confusion, because if you try to override a private method (which is implicitly final) it seems to work:

```java
class WithFinals {
    // Identical to "private" alone:
    private final void f() {
        System.out.println("WithFinals.f()");
    }
    // Also automatically "final":
    private void g() {
        System.out.println("WithFinals.g()");
    }
}
class OverridingPrivate extends WithFinals {
    private final void f() {
        System.out.println("OverridingPrivate.f()");
    }
    private void g() {
        System.out.println("OverridingPrivate.g()");
    }
}
class OverridingPrivate2 extends OverridingPrivate {
    public final void f() {
        System.out.println("OverridingPrivate2.f()");
    }
    public void g() {
        System.out.println("OverridingPrivate2.g()");
    }
```

```
}
public class FinalOverridingIllusion {
    public static void main(String[] args) {
        OverridingPrivate2 op2 =
            new OverridingPrivate2();
        op2.f();
        op2.g();
        // You can upcast:
        OverridingPrivate op = op2;
        // But you can't call the methods:
        //! op.f();
        //! op.g();
        // Same here:
        WithFinals wf = op2;
        //! wf.f();
        //! wf.g();
    }
}
```

"Overriding" can only occur if something is part of the base-class interface. That is, you must be able to upcast an object to its base type and call the same method (the point of this will become clear in the next chapter). If a method is private, it isn't part of the base-class interface. It is just some code that's hidden away inside the class, and it just happens to have that name, but if you create a public, protected or "friendly" method in the derived class, there's no connection to the method that might happen to have that name in the base class. Since a private method is unreachable and effectively invisible, it doesn't factor into anything except for the code organization of the class for which it was defined.

> "覆盖"一般只发生在基类的接口部分。也就是说，必须将一个对象转换成它的基类，并需要调用同样的方法。如果一个方法是私有的，它就不是这个基类的接口部分，而只是这个类内部的隐藏代码，它和基类中碰巧同名的方法没有任何关系。

8.3.3 Final Classes

When you say that an entire class is final (by preceding its definition with the final keyword), you state that you don't want to inherit from this class or allow anyone else to do so. In other words, for some reason the design of your class is such that there is never a need to make any changes, or for safety or security reasons you don't want subclassing.

```
class SmallBrain {}
final class Dinosaur {
  int i = 7;
  int j = 1;
  SmallBrain x = new SmallBrain();
  void f() {}
}
//! class Further extends Dinosaur {}
// error: Cannot extend final class 'Dinosaur'
public class Jurassic {
```

```java
    public static void main(String[] args) {
        Dinosaur n = new Dinosaur();
        n.f();
        n.i = 40;
        n.j++;
    }
}
```

Note that the data members can be final or not, as you choose. The same rules apply to final for data members regardless of whether the class is defined as final. Defining the class as final simply prevents inheritance—nothing more. However, because it prevents inheritance all methods in a final class are implicitly final, since there's no way to override them.

> 当将一个完整的类定义为 final(在这个类定义的前面加上 final 关键字)，意味着不能从这个类继承。换句话说，由于类设计需要，这个类不需要做任何变化，或处于安全考虑的原因不需要派生其子类。将一个类定义为 final 类型仅意味着它不能被继承，没有任何其他含义。由于 final 类不能被继承，它所有的方法也隐含被声明为 final，无法覆盖这些方法。

You can add the final specifier to a method in a final class, but it doesn't add any meaning.

8.4 Sample Example

Example 8.1

```java
//use of super: using super to overcome harm hiding 通过super调用被隐藏的成员
class A
{
    int i;
}
class B extends A          //类B是类A的子类
{
    int i;
    B(int a, int b)
    {
        super.i = a;       //将变量a赋值给父类的成员变量i
        i = b;             //将变量b赋值给成员变量i
    }
    void show()
    {
        System.out.println("In super class A:" + super.i);
        System.out.println("In sub class B:" + i);
    }
}
class useSuper
{
    public static void main(String args[])
    {
        B subob = new B(1,2);  //创建子类的实例
        subob.show();
    }
}
//输出：In super class A:1    In sub class B:2
```

Example 8.2
```
//Practised overriding 覆盖
class A
{
    int i,j;
    A(int a, int b)//父类的构造方法
    {
        i = a;
        j = b;
    }
    void show()
    {
        int  k;
        System.out.println(i+j);
    }
}
class B extends A
{
    int k;
    B(int a, int b, int c)//子类覆盖父类的构造方法
    {
        super(a,b);
        k = c;
    }
    void show()
    {
        System.out.println(i+j+k);
    }
}
class Overide
{
    public static void main(String args[])
    {
        B subob = new B(1,2,3);
        subob.show();
    }
}
//输出结果：6
```

Example 8.3
```
//Allocation of an object to inheritise another 根据方法调用的实际参数创建不同对象
    class Box//这个类重载了4个不同的构造方法
    {
        double height;
        double width;
        double depth;
        Box()
        {
            width = height = depth = -1;
        }
        Box(Box ob)
        {
```

```java
            width = ob.width;
            height = ob.height;
            depth = ob.depth;
        }
        Box(double w, double h, double d)
        {
            width = w;
            height = h;
            depth = d;
        }
        Box(double len)
        {
            width = height = depth = len;
        }
        double volume()
        {
            return width*height*depth;
        }
    }
    class Obj_inheritence
    {
        public static void main(String args[])
        {
            Box mybox1 = new Box(10,20,15);
            //创建Box(double w, double h, double d)的实例
            Box mybox2 = new Box();              //创建Box()的实例
            Box mycube = new Box(7);             //创建Box(double len)的实例
            Box myclone = new Box(mybox1);       //创建Box(Box ob)的实例
            double vol;
            vol = mybox1.volume();
            System.out.println(vol);             //输出3000.0
            vol = mybox2.volume();
            System.out.println(vol);             //输出-1.0
            vol = mycube.volume();
            System.out.println(vol);             //输出343.0
            vol = myclone.volume();
            System.out.println(vol);             //输出3000.0
        }
    }
```

Example 8.4

```java
    //Writing a object as a parameter 通过对象传递参数
    class Test
    {
       int a,b;
       Test(int i, int j)//constructor
       {
            a = i;
            b = j;
       }
       boolean equal(Test t)
       {
            if(t.a==a && t.b==b)
```

```
            {
            return true;
            //如果参数对象的成员变量与自己的成员变量相等则返回"true"
            }
            else
            {
            return false;    //否则返回"false"
            }
        }
}
class Obj_parameter
{
    public static void main(String args[])
    {
        Test ob1 = new Test(100,22);
        Test ob2 = new Test(100,22);
        Test ob3 = new Test(-1,-1);
        System.out.println("ob1 == ob2 is " + ob1.equal(ob2));
        //输出 ob1 == ob2 is true
        System.out.println("ob2 == ob3 is " + ob2.equal(ob3));
        //输出 ob2 == ob3 is false
    }
}
```

Example 8.5
```
//Inheritance
class Box
{
    double width;
    double height;
    double depth;
    //这个类重载了 4 个不同的构造方法
    Box()
    {
        width = height = depth = -1;
    }
    Box(double len)
    {
        width = height = depth = len;
    }
    Box(double w, double h, double d)
    {
        width = w;
        height = h;
        depth = d;
    }
    double volume()
    {
        return width*height*depth;
    }
}
class boxWeight extends Box        //类 boxWeight 是类 Box 的子类
{
```

```
            double weight;
            boxWeight(double w, double h, double d, double wh)//重载了父类的构造方法
            {
                width = w;
                height = h;
                depth = d;
                weight = wh;
            }
            double volume()
            {
                return width*height*depth*weight;
            }
        }
        class Sample
        {
            public static void main(String args[])
            {
                boxWeight mybox1 = new boxWeight(10,11,12,13); //创建一个子类的实例
                double vol;
                vol = mybox1.volume();
                System.out.println(vol);                       //输出 17160.0
            }
        }
```

8.5 Exercise for you

1. How a super class variable can reference a subclass object. Give an example.

2. How you can use the super class constructor-using super? Give an example.

3. Your grand father is tall, your father is handsome and your mother is white. Make one program to show that you are tall, handsome and white.

Chapter 9 Packages and Interfaces

The most innovative features of Java are the Package and the Interfaces. Packages are used to contain classes in a logical, hierarchical and streamlined manner so that no class name collides with other one.

> Java 中最具有创新性的特性就是包和接口。包被用来按一定的逻辑结构、分层次、高效地将类组织在一起，以确保类名不会冲突。

In Java the multiple inheritance is not possible but that you can do by using an interface. Through the use of the interface keyword, Java allows you to fully abstract the interface from its implementation. Using interface you can specify a set of methods which can be implemented by one or more classes.

> Java 不支持多重继承，通过接口可以实现多重继承。通过使用 interface 关键字，可以从类的实现中很好地抽象出接口。接口中可以定义一组方法，这组方法可以被多个类实现。

9.1 Package

9.1.1 Packages Overview

A package is a collection of classes and interfaces. Each package has its own name and organizes its top-level (that is, nonnested) classes and interfaces into a separate namespace, or name collection. Although same-named classes and interfaces cannot appear in the same package, they can appear in different packages because a separate namespace assigns to each package.

> 包是类和接口的集合。每个包都有包名，它将非内部类接口组织在独立的名称空间或名称集合。名称相同的类和接口不能出现在同一包中，由于每个包具有独立的名称空间，这些同名的类和接口可以定义在不同的包中。

From an implementation perspective, equating a package with a directory proves helpful, as does equating a package's classes and interfaces with a directory's classfiles. Keep in mind other approaches—such as the use of databases—to implementing packages, so do not get into the habit of always equating packages with directories. But because many JVMs use directories to implement packages, this article equates packages with directories. The Java 2 SDK organizes its vast collection of classes and interfaces into a tree-like hierarchy of packages within packages, which is equivalent to directories within directories. That hierarchy allows Sun Microsystems to easily distribute (and you to easily work with) those classes and interfaces.

> 从实现的角度出发，包和目录是等同的，也可以将一个包中的类和接口等同于一个目录下的 class 文件。但是也不要总是养成把包等同于目录的习惯。由于许多 JVMs（Java 虚拟机）使用目录来实现包。Java 2 SDK 将它数量庞大的类和接口的集合组织成一个包中含包的树形结构，等同于目录中包含目录的层次结构。这种树形结构使得 Sun 公司很容易发布它的类和包。

Examples of Java's packages include:
- **Java.lang:** A collection of language-related classes, such as Object and String, organized in the Java package's lang subpackage
- **Java.lang.ref:** A collection of reference-related language classes, such as SoftReference and ReferenceQueue, organized in the ref sub-subpackage of the Java package's lang subpackage
- **Javax.swing:** A collection of Swing-related component classes, such as JButton, and interfaces, such as ButtonModel, organized in the Javax package's swing subpackage

Period characters separate package names. For example, in Javax.swing, a period character separates package name Javax from subpackage name swing. A period character is the platform-independent equivalent of forward slash characters (/), backslash characters (\), or other characters to separate directory names in a directory-based package implementation, database branches in a hierarchical database-based package implementation, and so on.

> 包名用点字符隔开。例如：在 Javax.swing 中，点字符将包名 Javax 与子包名 swing 隔开。点字符是平台无关的字符，类似于斜杠字符/，反斜杠字符\，或在基于目录的包生成时用于分隔目录名称的其他字符等。

9.1.2 Packages in Java

Many implementations of Java use a hierarchical file system to manage source and class files. Figure 9.1 shows the relationship between package name and file system

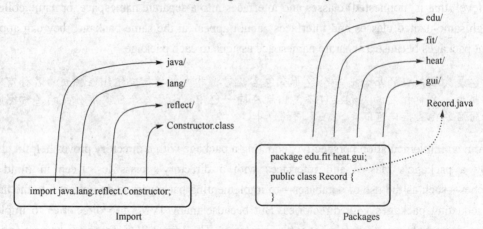

Figure 9.1 Relationship Between Package Name and file System

9.1.3 Access Specifiers

- Private: accessible only in the class
- No modifier: so-called "package" access—accessible only in the same package
- Protected: accessible (inherited) by subclasses, and accessible by code in same package
- Public: accessible anywhere the class is accessible, and inherited by subclasses

Notice that private protected is not syntactically legal. Table 9.1 shows different access to fields:

Chapter 9 Packages and Interfaces

Table 9.1 Access to Fileds

Summary of access to fields in Java				
access by	Private	"package"	protected	public
the class itself	yes	yes	yes	yes
a subclass in same package	no	yes	yes	yes
non-subclass in same package	no	yes	yes	yes
a subclass in other package	no	no	yes	yes
non-subclass in other package	no	no	no	yes

Classes in the same package can be in different directories.

dir1/dir2/A/B/C.class

dir3/A/B/D.class

Java -classpath dir1/dir2:dir3 ...

Packages are simple-minded in Java and dangerous. You can accidentally use the wrong classes. So if you misspell a class name and there happens to be a class by that name in the same directory, Java will not detect it.

```
import java.util.*;  // imports java.util.Date
import java.sql.*;   // also import java.util.Date
public class Collision {
    HashMap m;
    Date d;
}
```

Basically, files in one directory (or package). would have different functionality from those of another directory. For example, files in java.io package do something related to I/O, but files in Java.net package give us the way to deal with the network . In GUI applications, it's quite common for us to see a directory with a name "ui" (user interface), meaning that this directory keeps files related to the presentation part of the application. On the other hand, we would see a directory called "engine", which stores all files related to the core functionality of the application instead. Figure 9.2 shows the basic structer of JDK Package in file system:

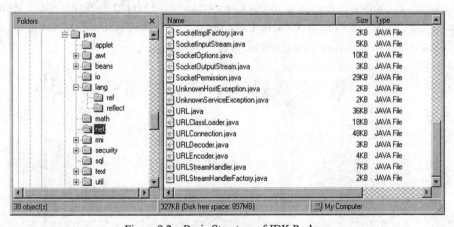

Figure 9.2 Basic Structure of JDK Package

基本上在一个目录(或包)中的文件与在另一个目录中的文件具有不同的功能。例如，在java.io包中的文件一般处理与I/O相关的操作，而在Java.net包中的文件则处理网络相关的操作。在GUI(图形用户界面)应用中，经常会看到一个命名为"ui"(用户界面)的目录，意味着这个目录中的文件与应用程序的界面表示相关。另一方面，还能看到"engine"目录，该目录主要存储与应用程序内核功能相关的文件。

Packaging also helps us to avoid class name collision when we use the same class name as that of others. For example, if we have a class name called "Vector", its name would crash with the Vector class from JDK. However, this never happens because JDK use Java.util as a package name for the Vector class (Java.util.Vector). So our Vector class can be named as "Vector" or we can put it into another package like com.mycompany.Vector without conflicting with anyone. The benefits of using package reflect the ease of maintenance, organization, and increase collaboration among developers. Understanding the concept of package will also help us manage and use files stored in jar files in more efficient ways.

在使用相同类名时，包有助于避免相同类名冲突。使用包有利于类的维护和组织，同时也使开发人员更容易协作。对包的理解有利于更有效地管理和使用存储在jar包中的文件。

9.1.4 How to Create a Package

Suppose we have a file called HelloWorld.Java, and we want to put this file in a package **world**. First thing we have to do is to specify the keyword **package** with the name of the package we want to use (**world** in our case) on top of our source file, before the code that defines the real classes in the package, as shown in our HelloWorld class below:

```
// only comment can be here
package world;
public class HelloWorld {
    public static void main(String[] args) {
        System.out.println("Hello World");
    }
}
```

One thing you must do after creating a package for the class is to create nested subdirectories to represent package hierarchy of the class. In our case, we have the **world** package, which requires only one directory. So, we create a directory **world** and put our HelloWorld.Java into it. Figure 9.3 shows class HelloWorld in package world is located in directory c:\world.

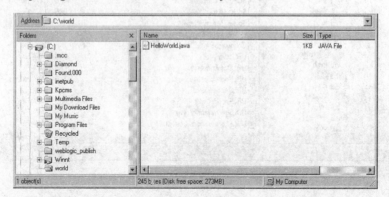

Figure 9.3　HelloWorld in world Package (C:\world\HelloWorld.Java)

That's it!!! Right now we have HelloWorld class inside world package. Next, we have to introduce the world package into our **CLASSPATH**

> 创建包的两个步骤：在文件源代码前加上 Package 关键字和包名称；创建与包名相同的子目录结构，并把文件放到子目录中去。

9.1.5 Setting Up the CLASSPATH

From figure 9.2 we put the package world under C:. So we just set our CLASSPATH as:
>   **set CLASSPATH=.;C:\;**

We set the CLASSPATH to point to 2 places, . (dot) and C:\ directory.

Note: If you used to play around with DOS or UNIX, you may be familiar with . (dot) and .. (dot dot). We use . as an alias for the current directory and .. for the parent directory. In our CLASSPATH we include this . for convenient reason. Java will find our class file not only from C: directory but from the current directory as well. Also, we use ; (semicolon) to separate the directory location in case we keep class files in many places.

When compiling HelloWorld class, we just go to the world directory and type the command:

C:\world\Javac HelloWorld.Java

If you try to run this HelloWorld using **Java HelloWorld**, you will get the following error:

C:\world>Java HelloWorld
Exception in thread "main" Java.lang.NoClassDefFoundError: HelloWorld (wrong name: world/HelloWorld)
>   **at Java.lang.ClassLoader.defineClass0(Native Method)**
>   **at Java.lang.ClassLoader.defineClass(ClassLoader.Java:442)**
>   **at Java.security.SecureClassLoader.defineClass(SecureClassLoader.Java:101)**
>   **at Java.net.URLClassLoader.defineClass(URLClassLoader.Java:248)**
>   **at Java.net.URLClassLoader.access$1(URLClassLoader.Java:216)**
>   **at Java.net.URLClassLoader$1.run(URLClassLoader.Java:197)**
>   **at Java.security.AccessController.doPrivileged(Native Method)**
>   **at Java.net.URLClassLoader.findClass(URLClassLoader.Java:191)**
>   **at Java.lang.ClassLoader.loadClass(ClassLoader.Java:290)**
>   **at sun.misc.Launcher$AppClassLoader.loadClass(Launcher.Java:286)**
>   **at Java.lang.ClassLoader.loadClass(ClassLoader.Java:247)**

The reason is right now the HelloWorld class belongs to the package world. If we want to run it, we have to tell JVM about its **fully-qualified class name** (world.HelloWorld) instead of its plain class name (HelloWorld).

>   **C:\world>Java world.HelloWorld**
>   **C:\world>Hello World**

Note: **fully-qualified class name** is the name of the Java class that includes its package name

To make this example more understandable, let's put the HelloWorld class along with its package (world) be under **C:\myclasses** directory instead. The new location of our HelloWorld should be as shown in table

We just changed the location of the package from C:\world\HelloWorld.Java to C:\myclasses\world\HelloWorld.Java as shown in figure 9.4. Our CLASSPATH then needs to be changed to point to the new location of the package world accordingly.

```
set CLASSPATH=.;C:\myclasses;
```

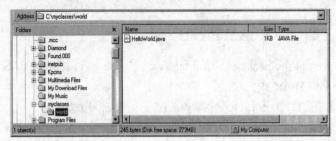

Figure 9.4   HelloWorld class (in world package) under **myclasses** directory

Thus, Java will look for Java classes from the current directory and **C:\myclasses** directory instead.

Someone may ask "Do we have to run the HelloWorld at the directory that we store its class file everytime?". The answer is NO. We can run the HelloWorld from anywhere as long as we still include the package world in the CLASSPATH. For example,

C:\>set CLASSPATH=.;C:\;
C:\>set CLASSPATH // see what we have in CLSSPATH
CLASSPATH=.;C:\;
C:\>cd world
C:\world>Java world.HelloWorld
Hello World
C:\world>cd ..
C:\>Java world.HelloWorld
Hello World

9.1.6   Subpackage (Package inside Another Package)

Assume we have another file called **HelloMoon.Java**. We want to store it in a subpackage **"moon"**, which stays inside package **world**. The HelloMoon class should look something like this:

```
package world.moon;
public class HelloMoon {
    private String holeName = "rabbit hole";
    public getHoleName() {
        return hole;
    }
    public setHole(String holeName) {
        this.holeName = holeName;
    }
}
```

If we store the package world under C: as before, the **HelloMoon.Java** would be **c:\world\moon\HelloMoon.Java** as shown in figure 9.5:

Although we add a subpackage under package world, we still don't have to change anything in our CLASSPATH. However, when we want to reference to the HelloMoon class, we have to use world.moon.HelloMoon as its fully-qualified class name.

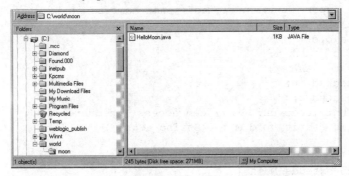

Figure 9.5   HelloMoon in world.moon package

9.1.7   How to Use Package

There are 2 ways in order to use the public classes stored in package.
(1) Declare the fully-qualified class name.
For example,

```
...
world.HelloWorld helloWorld = new world.HelloWorld();
world.moon.HelloMoon helloMoon = new world.moon.HelloMoon();
String holeName = helloMoon.getHoleName();
...
```

(2) Use an "import" keyword.

```
import world.*;   //we can call any public classes inside the world package
import world.moon.*;
//we can call any public classes inside the world.moon package
import java.util.*;
//import all public classes from Java.util package
import java.util.Hashtable;
// import only Hashtable class(not all classes in Java.util package)
```

Thus, the code that we use to call the HelloWorld and HelloMoon class should be

```
...
   HelloWorld helloWorld = new HelloWorld();
// don't have to explicitly specify world.HelloWorld anymore
   HelloMoon helloMoon = new HelloMoon();
// don't have to explicitly specify world.moon.HelloMoon anymore
...
```

Note that we can call public classes stored in the package level we do the import only. We can't use any classes that belong to the subpackage of the package we import. For example, if we import package world, we can use only the HelloWorld class, but not the HelloMoon class.

> 只能调用 import 引入的包中的公有类，不能使用其子包中的任何类。例如，如果通过 import 引入 world 包，只能使用 HelloWorld 类，而不能使用 HelloMoon 类。

Example-one/Main.java
```
package one;
import another.Unrelated;
// Can't import "another.AlsoSub"; it is not public
// Can't import "one.Unrelated"; same name as in "another"
public class Main  {
    int a,b,c;
    public    int x;
    private   int y;
    protected int z;
    Unrelated u;
    one.Unrelated uu;      // can access "one.Unrelated"
    public static void main (String args[])  {
    }
}
class Sub extends Main {
    private int d = a;     // can access "package" variables
    private int e = z;     // can access protected variables of "Main"
}
class AlsoUnrelated  {
    private Main m = new Main();
    private int f = m.a;   // can access "package" variables
    private int g = m.z;   // can access protected variables of "Main"
}
```

Important Classes:

Java.awt.Graphics

You will use this class for drawing on your applet. It represents a graphics context, or "surface" for drawing. The class has several methods for drawing on the "surface" it represents such as "drawString," "drawRect," and "fillRect." All of these methods can be found in the API reference. It is important to note that the Graphics class is abstract so you may not use the "new" operator with it. You may only fill variables of type "Graphics" with methods that have the return type "Graphics."

Java.awt.Image

This class is obvious enough; it holds a gif or jpeg picture. It contains several methods for manipulating the image it stores. As with the graphics class, it is abstract and an "Image" variable may only be obtained from a method with return type "Image."

Java.lang.Math

This class contains important mathematical functions such as square roots and logarithms. You do not need to declare a variable of type "Math" to use the methods of the class. Here is an example:

```
Double N=60;
Double SINN=Math.sin(N).
```

Because it is in "Java.lang," the math class is part of the Java language and needs no import statement.

Java.net.URL

This class contains a Universal Resouce Locator(URL), or the location of a document on the Internet. An example URL would be "HTTP://Java.Sun.Com" the home of the makers of Java. The URL of this page is probably written in a text box at the top of your browser window.

9.2 Interface

9.2.1 Interface Overview

An interface is a named collection of method definitions, without implementations.

In general, an interface is a device or a system that unrelated entities use to interact. According to this definition, a remote control is an interface between you and a television set, the English language is an interface between two people, and the protocol of behavior enforced in the military is the interface between individuals of different ranks.

Within the Java programming language, an **interface** is a type, just as a class is a type. Like a class, an interface defines methods. Unlike a class, an interface never implements methods; instead, classes that implement the interface implement the methods defined by the interface. A class can implement multiple interfaces.

> 接口是方法定义的命名集合，没有提供任何具体实现。在 Java 编程语言中，接口是一种类型，和类类型一样，接口也定义方法。与类不同的是，接口从不实现方法。然而，实现接口的类需要实现接口中定义的方法。一个类可以实现多个接口。

The bicycle class and its class hierarchy define what a bicycle can and cannot do in terms of its "bicycleness." But bicycles interact with the world on other terms. For example, a bicycle in a store could be managed by an inventory program. An inventory program doesn't care what class of items it manages as long as each item provides certain information, such as price and tracking number. Instead of forcing class relationships on otherwise unrelated items, the inventory program sets up a communication protocol. This protocol comes in the form of a set of method definitions contained within an interface. The inventory interface would define, but not implement, methods that set and get the retail price, assign a tracking number, and so on.

To work in the inventory program, the bicycle class must agree to this protocol by implementing the interface. When a class implements an interface, the class agrees to implement all the methods defined in the interface. Thus, the bicycle class would provide the implementations for the methods that set and get retail price, assign a tracking number, and so on.

You use an interface to define a protocol of behavior that can be implemented by any class anywhere in the class hierarchy. Interfaces are useful for the following:

- Capturing similarities among unrelated classes without artificially forcing a class relationship
- Declaring methods that one or more classes are expected to implement
- Revealing an object's programming interface without revealing its class
- Modeling multiple inheritance, a feature of some object-oriented languages that allows a class to have more than one superclass

> 接口被用来定义行为规范，类的层次结构中的任意类都可以实现接口中定义的行为规范。接口一般用于：在无关的类之间找到相似点，而无需人工强加类之间的关系；声明一个或多个需要被实现的方法；对外提供一个程序的接口而不是类的实现；构建多重继承模型，某些面向对象语言允许的一个类可以有多个父类(超类)的特性。

9.2.2　Creating and Using Interfaces

An interface defines a protocol of behavior that can be implemented by any class anywhere in the class hierarchy; it also defines a set of methods but does not implement them. A class that implements the interface agrees to implement all the methods defined in the interface, thereby agreeing to certain behavior.

> 接口定义了可被类层次结构中任何类实现的行为规范，它定义了一系列方法，但不提供方法的实现。一个实现接口的类需要实现接口中定义的所有方法，因此该类符合接口定义的行为规范。

Because an interface is simply a list of unimplemented, and therefore abstract, methods, you might wonder how an interface differs from an abstract class. The differences are significant.
- An interface cannot implement any methods, whereas an abstract class can.
- A class can implement many interfaces but can have only one superclass.
- An interface is not part of the class hierarchy.
- Unrelated classes can implement the same interface.

> 接口只是简单的一系列未实现的、抽象的方法列表。接口和抽象类的区别是明显的，包括：接口不能实现任何方法，而抽象类可以有实现的方法；类能实现多个接口而只能有一个父类；接口不是类层次结构中的一部分；无关的类能实现同一接口。

Let's set up the example we'll be using in this section. Suppose you have written a class that can watch stock prices coming over a data feed. This class allows other classes to register to be notified when the value of a particular stock changes. First, your class, StockApplet, would implement a method that lets other objects register for notification, as follows.

```
public class StockMonitor {
    public void watchStock(StockWatcher watcher,
                           TickerSymbol tickerSymbol,
                           BigDecimal delta) {
        ...
    }
}
```

The first argument(参数) to this method is a StockWatcher object. StockWatcher is the name of an interface whose code you will see in the next section. That interface declares one method(方法): valueChanged. An object that wants to be notified of stock changes must be an instance of a class that implements this interface and thus implements the valueChanged method.

The other two arguments provide the symbol of the stock to watch and the amount of change that the watcher considers interesting enough to be notified of. When the StockMonitor class detects an interesting change, it calls the valueChanged method of the watcher.

The watchStock method ensures, through the data type of its first argument, that all registered objects implement the valueChanged method. It makes sense to use an interface data type here because it only matters whether registrants implement a particular method. If StockMonitor uses a class name as the data type, which would artificially force a class relationship on its users. Because a class can have only one superclass, it would also limit what type of objects can use this service. By using an interface, the registered objects class could be anything (for example, Applet or Thread), thus allowing any class anywhere in the class hierarchy to use this service.

9.2.3 Defining an Interface

Following program shows that an interface definition has two components: the interface declaration and the interface body. The interface declaration defines various attributes of the interface, such as its name and whether it extends other interfaces. The interface body contains the constant and the method declarations for that interface.

> 以下代码显示了接口的定义由两部分组成：接口声明和接口体。接口声明定义了接口的不同属性，如接口的名称和是否继承其他接口；接口体由常量和接口的方法定义组成。

```
public interface Stock {
    void valueChanged( TickerSymbol tickersymbol, BigDecimal newValue);
}
         ↑                    ↑                              ↑
interface declaration   Method declaration            interface body

public interface StockWatcher {
    void valueChanged(TickerSymbol tickerSymbol,
                BigDecimal newValue);
}
```

The StockWatcher interface declares, but does not implement, the valueChanged method. Classes that implement this interface provide the implementation for that method.

9.2.4 The Interface Body

The interface body contains method declarations for all the methods included in the interface. A method declaration within an interface is followed by a semicolon (;) because an interface does not provide implementations for the methods declared within it. All methods declared in an interface are implicitly public and abstract.

An interface can contain constant declarations in addition to method declarations. All constant values defined in an interface are implicitly public, static, and final.

Member declarations in an interface prohibit the use of some declarations; you cannot use transient, volatile, or synchronized in a member declaration in an interface. Also, you cannot use the private and protected specifiers when declaring members of an interface.

> 接口体包含了接口中所有方法的定义。由于接口不提供方法的实现，在接口中一个方法的声明后接一个分号(;)结束。接口中声明的所有方法都隐式定义为公有的和抽象的。接口除了方法声明，还能包含常量声明。所有定义在接口中的常量也都隐式地定义为公有的、静态的和final的。接口中的成员声明不允许某些声明；不能使用transient、volatile或synchronized关键字声明接口中的成员。同时接口中声明成员时不能使用private和protected修饰符。

9.2.5　Implementing an Interface

An interface defines a protocol of behavior. A class that implements an interface adheres to the protocol defined by that interface. To declare a class that implements an interface, include an implements clause in the class declaration. Your class can implement more than one interface (the Java platform supports multiple inheritance for interfaces), so the implements keyword is followed by a comma-separated list of the interfaces implemented by the class.

> 接口定义了行为规范。实现接口的类依附于接口所定义的规范。声明一个类实现接口，需要使用 implements 关键字。类可以实现多个接口（Java 平台支持接口的多继承），implements 关键字后面接逗号分隔的被类实现的接口列表。

By Convention: The implements clause follows the extends clause, if it exists.

> 约定：如果出现 implements 子句和 extends 子句，则 implements 子句在 extends 子句之后。

The following is a partial example of an applet that implements the StockWatcher interface.

```
public class StockApplet extends Applet  implements StockWatcher {
    public void valueChanged(TickerSymbol tickerSymbol,BigDecimal newValue) {
        switch (tickerSymbol) {
            case SUNW:
                ...
                break;
            case ORCL:
                ...
                break;
            case CSCO:
                ...
                break;
            default:
                //handle unknown stocks
                ...
                break;
        }
    }
}
```

When a class implements an interface, it is essentially signing a contract. Either the class must implement all the methods declared in the interface and its superinterfaces, or the class must be declared abstract. The method signature — the name and the number and type of arguments — in the class must match the method signature as it appears in the interface. The StockApplet implements the StockWatcher interface, so the applet provides an implementation for the valueChanged method. The method ostensibly updates the applet's display or otherwise uses this information.

> 当一个类实现一个接口，它必然遵循一个约定。这个类需要实现接口及其父接口定义的所有方法，或这个类必须声明为抽象类。类中方法的定义必须与接口中要实现方法的名称、参数个数和参数类型一致。

9.2.6 Using an Interface as a Type

When you define a new interface, you are defining a new reference data type. You can use interface names anywhere you can use any other data type name. Recall that the data type for the first argument to the watchStock method in the StockMonitor class is StockWatcher.

```java
public class StockMonitor {
    public void watchStock(StockWatcher watcher,
                           TickerSymbol tickerSymbol,
                           BigDecimal delta) {
    ...
    }
}
```

Only an instance of a class that implements the interface can be assigned to a reference variable whose type is an interface name. So only instances of a class that implements the StockWatcher interface can register to be notified of stock value changes. StockWatcher objects are guaranteed to have a valueChanged method.

9.3 Sample Example

Example 9.1
```java
/*Break to exit a loop in side the package Mypack. You must create a package
*"Mypack" and keep the program in side that to run as described before.
*/
package Mypack;
class Break_loop
{
    public static void main(String args[])
    {
        for(int i=0; i<100; ++i)
        {
            if(i == 10)
                break;
        }
        System.out.println("Loop completed");
    }
}
```

Example 9.2
```java
import java.util.Random;
interface Showme
{
    int No = 0;
    int Yes = 1;
    int Maybe = 2;
    int Later = 3;
    int Soon = 4;
    int Never = 5;
}
class Question implements Showme
```

```java
    {
        Random Rand = new Random();
        int ask()
        {
            int prob = (int)(100 * Rand.nextDouble());
            if(prob < 30)
                return No;        //30%
            else if(prob < 60)
                return Yes;       //30%
            else if(prob < 75)
                return Later;     //15%
            else if (prob < 98)
                return Soon;      //13%
            else
                return Never;     //2%
        }
    }
    class ask_me implements Showme
    {
        static void answer(int result)
        {
            switch(result)
            {
                case No:
                    System.out.println("No");
                    break;
                case Yes:
                    System.out.println("Yes");
                    break;
                case Later:
                    System.out.println("Later");
                    break;
                case Soon:
                    System.out.println("Soon");
                    break;
                case Never:
                    System.out.println("Never");
                    break;
            }
        }
        public static void main(String args[])
        {
            Question q = new Question();
            answer(q.ask());
        }
    }
```

Example 9.3

```java
    //Interface
    interface callback
    {
        void callback(int param);
    }
```

```java
class client implements callback {
    public void callback(int p) {
        System.out.println("Callback called with " + p) ;
    }
}
class clientDemo {
    public static void main(String args[]) {
        callback c = new client();
        c.callback(42);
    }
}
```

Example 9.4
```java
//one interface can extend another.
interface a {
    void meth1();
    void meth2();
}
//b now includes meth1() and meth2()-----it adds meth3().
interface b extends a {
    void meth3();
}
//now this class must implement all of class a and b.
class myclass implements b {
    public void meth1 () {
        System.out.println("implement meth1().");
    }
    public void meth2 () {
        System.out.println("implement meth2().");
    }
    public void meth3 () {
        System.out.println("implement meth3().");
    }
}
class extend {
    public static void main (String args[]) {
        myclass ob = new myclass();
        ob.meth1();
        ob.meth2();
        ob.meth3();
    }
}
```

Example 9.5
```java
//Interface
interface callback
{
    void callback(int param);
}
class client2 implements callback {
    public void callback (int p) {
        System.out.println("Another version of callback");
        System.out.println("p Squared is " + (p*p));
    }
}
```

```
class testface2 {
    public static void main(String args[]) {
        callback c = new client();
        client2 c2 = new client2();
        c.callback(42);
        c = c2; // Now c refers to client2 object.
        c.callback(42);
    }
}
```

9.4 Exercise for you

1. Create a package "A1" in side, which create a class "AA1". Create another package B2 and inside that a class BB2. Get the class properties of AA1 to BB1.

2. Write a program to show all the utility of public, private, protected and default access modifies.

3. Define and apply an interface to test how multiple inheritances are possible.

Chapter 10 Exception Handling

Errors and exceptions are very common in a program practice. This chapter will cover the common exception handling mechanism in Java. Error or "exception" is an abnormal condition that arises in a run time. The errors must be checked, detected and handled. This is a troublesome approach however Java makes this practice easy by exception handling and avoids problems during the run time. It brings a run time error management concept to the OOP concept.

> 错误与异常是程序运行中非常普遍的问题。本章将涵盖Java中各种常见的异常处理机制。错误或异常通常是程序运行时出现的非正常情况。各种错误必须能被检测、发现并及时处理。这本是一个麻烦的过程,然而Java通过异常处理机制使这个过程变得简单容易,以避免程序运行过程中出现问题。Java将运行时错误管理概念引入了面向对象概念之中。

10.1 Definition of Exception

10.1.1 What is an Exception

An **exception** is an error thrown by a class or method reporting an error in operation. For example, **dividing by zero** is undefined in mathematics, and a calculation can fail if this winds up being the case because of **an error in user input**. In this particular case an *Arithmetic Exception* is thrown, and unless the programmer looks for this exception and manually puts in code to handle it, the program will crash stating the exception thrown and a stack trace, which would be unhelpful to a casual user of a Java program. If the programmer handles the exception, he could deliver a useful error to the user and return the user to the beginning of the program so that they could continue to use it.

> 异常是指程序运行过程中由类或方法抛出的错误,需要程序员编写代码检测异常情况并进行处理,否则程序将崩溃,并显示异常抛出的轨迹。处理异常时,需要向用户传递有用的错误信息,并返回到程序的开始位置,以便用户继续运行该程序。

10.1.2 Common Exceptions

There are many different exceptions that can be thrown by a program, and the Java API contains **quite a few**. A lot are contained in the default package, *Java.lang*; however, when you start using more functionality such as AWT, Swing, or *Java.io*, the packages may also contain additional exceptions thrown by those libraries. As you start **expanding the functionality**, it might be a good idea to look at potential exceptions in the package and when they might be thrown in the course of your application.

Here is a primer of some:

ArithmeticException: thrown if a program attempts to perform division by zero.

ArrayIndexOutOfBoundsException: thrown if a program attempts to access an index of an array that does not exist.

StringIndexOutOfBoundsException: thrown if a program attempts to access a character at a non-existent index in a String.

NullPointerException: thrown if the JVM attempts to perform an operation on an Object that points to no data, or null.

NumberFormatException: thrown if a program is attempting to convert a string to a numerical datatype, and the string contains inappropriate characters (i.e. 'z' or 'Q').

ClassNotFoundException: thrown if a program can not find a class it depends at runtime (i.e., the class's ".class" file cannot be found or was removed from the CLASSPATH).

IOException: actually contained in Java.io, but it is thrown if the JVM failed to open an I/O stream.

As I said before, many different exceptions exist, and you should probably use your API documentation to learn about additional exceptions that may apply to your **particular application**.

"Catching" Exceptions

The Java language contains keywords used specifically for testing for and handling exceptions. The ones we will be using here are *try* and *catch*, and they must be used in conjunction with one another. They sort of work like if-else:

```
try{
  /*   包含被检测的代码    */
}catch(Exception e){
  /*   包含当异常实例被捕获的时候执行的代码 */
}
```

The *catch* statement can look for all exceptions, using the *Exception* **superclass**(超类), or it can catch a specific exception that you think could be thrown in the code you are testing with the *try* block. You can even have multiple *catch* blocks to catch and execute custom code for a number of different errors. A good thing to note would be that any particular exception that is caught is compared with each *catch* statement sequentially; so it is a good idea to put more generic exceptions, like *Exception*, towards the bottom of the list.

> Java 语言中有专门用于检测和处理异常的机制，采用了 try…catch 语句块。通过使用 Exception 超类，catch 块能捕获所有的异常，或者捕获有可能在检测块中抛出的特殊异常。该语句块中可以包含多个 catch 块来对多种不同类型异常和错误进行处理。需要注意的是，任何特定的异常捕获都会顺序和每个 catch 块进行逐一匹配，因而好的建议是将更一般的异常类尽可能得放置在异常捕获列表的后面。

10.1.3 The Throwable Superclass

The *catch* statement also stores an instance of the exception that was caught in the variable that the programmer uses, in the previous example *Exception e*. While all exceptions are subclasses of *Exception*, *Exception* itself is a subclass of *Throwable*, which contains a nice suite of methods that you can use to get all kinds of information to report about your exceptions:

- getMessage( ): returns the error message reported by the exception in a String

- printStackTrace( ): prints the stack trace of the exception to standard output, useful for debugging purposes in locating where the exception occurred
- printStackTrace(PrintStream s): prints the stack trace to an alternative output stream
- printStackTrace(PrintWriter s): prints the stack trace to a file, this way you can log stack traces transparent to the user or log for later reference
- toString( ): if you just decide to print out the exception it will print out this: NAME_OF_EXCEPTION: getMessage( ).

Using the catch you now have control over what the error message is and where it goes.

10.1.4 Effectively Using try-catch

In this section, I'm going to give you a few code samples on using try-catch blocks to give you an idea of the flexibility you as a programmer have over exceptions in your programs. The scenario is a user has input a file name to a program of a file that does not exist. In this scenario we are going to be using a text-based program; however, in a graphical environment you can easily use the catch block to draw dialog boxes.

In the first example, we want to print a useful error message and exit the program gracefully, saving the user from the confusing stack trace with something useful:

```
try{
    BufferedReader in = new BufferedReader(new FileReader(userInput));
    System.out.println(in.readLine())
}catch(IOException e){
    System.err.println(e.getMessage());
    System.err.println("Error: " + userInput + " is not a valid file. " +
    "Please verify that the file exists and that you have access to it.");
    System.exit(1);
}
```

Here, *System.err.println*( ) prints the error message to **standard error output** which is a high priority(高优先级) buffer that is used to report error messages to the console. If you're being a good programmer, you have separate methods that handle the different functionality of your programs; this way you can easily start the program from a previous place in the program before an exception occurs. In this next example, the user inputs the filename through the function *getUserInput*( ) elsewhere in the program; we want to report the "helpful error message" to the user and return the user to a place where he can enter a new filename.

> System.err.println( )将错误消息输出到标准错误输出中去，标准错误输出具有高优先级将出错信息报告给控制台的缓冲区。作为一个好的程序员，应该用独立的方法处理程序中的不同功能，以便异常发生时容易从程序前一位置开始执行。

```
public String getFileInput(String userInput){
    try{
        BufferedReader in = new BufferedReader(new FileReader(userInput));
        return in.readLine();
    }catch(IOException e){
        System.err.println(e.getMessage());
        System.err.println("Error: " + userInput + " is not a valid file.
```

```
        Please verify that the file exists and that you have access to it.");
        return getFileInput(getUserInput());
    }
}
```

Now you have an idea of how you can control your programs once an exception is thrown, and this should give you the basic idea of how to work with exceptions in the Java language.

Final Thoughts on Exceptions

The best way to prevent your application from crashing from exceptions is to avoid exceptions in the first place. It is much more costly for a system to catch and handle exceptions than to account for potential errors directly in your code. A lot of times, an exception can point to a problem in the programmer's logic rather than in the user's use of the program. You should look for the cause of exceptions, and try to understand why it occurred and make sure that you really had considered everything. You will be making far more resilient and robust applications.

> 防止应用程序由于异常发生而崩溃的最好方式就是一开始就避免异常发生。对一个系统来说，捕获并处理异常比直接在代码中处理各种可能的异常情况开销更大。很多时候，异常的产生大多是由于程序逻辑错误而并非用户使用程序造成的。

Java projects rarely feature a consistent and thorough exception-handling strategy. Often, developers add the mechanism as an afterthought or an as-you-go addition. Significant reengineering during the coding stage can make this oversight an expensive proposition indeed. A clear and detailed error-and-exception-handling strategy pays off in the form of robust code, which in turn, enhances user value.

Java's exception-handling mechanism offers the following benefits:
- It separates the working/functional code from the error-handling code by way of try-catch clauses.
- It allows a clean path for error propagation. If the called method encounters a situation it can't manage, it can throw an exception and let the calling method deal with it.
- By enlisting the compiler to ensure that "exceptional" situations are anticipated and accounted for, it enforces powerful coding.

In order to develop a clear and consistent strategy for exception handling, examine these questions that continually plague Java developers:
- Which exceptions should I use?
- When should I use exceptions?
- How do I best use exceptions?
- What are the performance implications?

When trying to design APIs and applications that can cross system boundaries or be implemented by third parties, these issues only exacerbate.

Let's delve deeper into the various aspects of exceptions.

Use of Exception

Which exceptions should you use?

Exceptions are of two types:

(1) Compiler-enforced exceptions, or checked exceptions

(2) Runtime exceptions, or unchecked exceptions

Compiler-enforced (checked) exceptions are instances of the Exception class or one of its subclasses—excluding the RuntimeException branch. The compiler expects all checked exceptions to be appropriately handled. Checked exceptions *must* be declared in the throws clause of the method throwing them—assuming, of course, they're not being caught within that same method. The calling method *must* take care of these exceptions by either catching or declaring them in its throws clause. Thus, making an exception checked forces the programmer to pay heed to the possibility of it being thrown.

> 异常有两种类型：编译强制性异常或检测性异常；运行时异常或非检测异常。编译强制性异常是 Exception 类或 Exception 类子类（除 RuntimeException 子类外）的实例。编译器希望所有被检测的异常都能被适当地处理。强制性异常必须在抛出它们的 throws 子句声明。假定它们不在同一方法中被捕获，调用方法必须认真处理这些异常或在 throws 子句中声明异常抛出。因此，编译强制性异常促使程序员尽可能减少异常抛出的可能性。

An example of a checked exception is Java.io.IOException. As the name suggests, it throws whenever an input/output operation is abnormally terminated.

Examine the following code:

```java
try
{
    BufferedReader br = new BufferedReader(new FileReader("MyFile.txt"));
    String line = br.readLine();
}
catch(FileNotFoundException fnfe)
{
    System.out.println("File MyFile.txt not found.");
}
catch(IOException ioe)
{
    System.out.println("Unable to read from MyFile.txt");
}
```

The constructor of FileReader throws a FileNotFoundException—a subclass of IOException—if the said file is not found. Otherwise, if the file exists but for some reason the readLine( ) method can't read from it, FileReader throws an IOException.

Runtime(unchecked) exceptions are instances of the RuntimeException class or one of its subclasses. You need not declare unchecked exceptions in the throws clause of the throwing method. Also, the calling method doesn't have to handle them—although it may. Unchecked exceptions usually throw only for problems arising in the Java Virtual Machine (VM) environment. As such, programmers should refrain from throwing these, as it is more convenient for the Java VM to manage this part.

> 运行时(非检测)异常是 RuntimeException 类或其子类的实例。非检测性异常不需要在 throwing 子句中声明。调用方法也没有必要一定要处理这些异常，尽管这些异常会发生。非检测性异常通常只抛出 Java 虚拟机出现的问题。

Java.lang.ArithmeticException is an example of an unchecked exception thrown when an exceptional arithmetic condition has occurred. For example, an integer "divide by zero" throws an instance of this class. The following code illustrates how to use an unchecked exception:

```java
    public static float fussyDivide(float dividend, float divisor) throws
        FussyDivideException
    {
        float q;
        try
        {
            q = dividend/divisor;
        }
        catch(ArithmeticException ae)
        {
            throw new FussyDivideException("Can't divide by zero.");
        }
    }
    public class FussyDivideException extends Exception
    {
        public FussyDivideException(String s)
        {
            super(s);
        }
    }
```

fussyDivide( ) forces the calling method to ensure that it does not attempt to divide by zero. It does this by catching ArithmeticException — an unchecked exception — and then throwing FussyDivide Exception—a checked exception.

To help you decide whether to make an exception checked or unchecked, follow this general guideline: If the exception signifies a situation that the calling method must deal with, then the exception should be checked, otherwise it may be unchecked.

10.1.5 When should You Use Exceptions

The Java Language Specification states that "an exception will be thrown when semantic constraints are violated," which basically implies that an exception throws in situations that are ordinarily not possible or in the event of a gross violation of acceptable behavior. In order to get a clearer understanding of the kinds of behavior that can be classified as "normal" or exceptional, take a look at some code examples.

Case 1

```java
    Passenger getPassenger()
    {
        try
        {
            Passenger flier = object.searchPassengerFlightRecord("Ashok Dash");
        }catch(NoPassengerFoundException npfe)
        {
            //执行一些处理任务
        }
    }
```

Case 2

```java
    Passenger getPassenger()
    {
        Passenger flier = object.searchPassengerFlightRecord("Ashok Dash");
```

```
        if(flier == null)
            //执行一些处理任务
}
```

In Case 1, if the search for the passenger is not fruitful, then the NoPassengerFoundException throws; whereas in Case 2, a simple null check does the trick. Developers encounter situations similar to the preceding in their day-to-day work; the trick is to engineer a sound and efficient strategy.

So, following the general philosophy behind exceptions, should you dismiss the possibility that searches will return nothing? When a search comes up empty, is it not more a case of normal processing? Therefore, in order to use exceptions judiciously, choose the approach in Case 2 over Case 1. (Yes—We recognize the performance angle. If this code were in a tight loop, then multiple if evals would adversely affect performance. However, whether or not the statement lies in the critical path would be known only after profiling and extensive performance analysis. Empirical results show that trying to code for performance up front—ignoring sound design principles—tends to produce more harm than good. So, go ahead and design the system right in the first cut, and then change later if you must.)

A good example of an exceptional situation: If somehow the object instance—which the search method invokes—was null, this becomes a fundamental violation of the getPassenger methods' semantics. In order to understand the performance implications of exceptions, read the paragraph on performance.

10.1.6 How do You Best Use Exceptions

All Java developers must address the challenging task of catching different kinds of exceptions and knowing what to do with them. This grows even more complicated when the code must transform the error messages from cryptic system-level exceptions to more user-friendly application-level ones. This holds true particularly for API-type coding, where you plug your code into another application, and you don't own the GUI.

Typically, there are three approaches to handling exceptions:
(1) Catch and handle all the exceptions.
(2) Declare exceptions in the throws clause of the method and let them pass through.
(3) Catch exceptions and map them into a custom exception class and re-throw.

> 有三种处理异常的方式：捕获并处理所有异常；在方法的 throws 子句中声明异常，并将其向上抛出；捕获异常并将其映射成为自定义的异常处理类并再次抛出。

Let's look at some issues with each of those options and try to develop a practicable solution.

Case 1
```
Passenger getPassenger()
{
    try
    {
        Passenger flier = object.searchPassengerFlightRecord("Ashok Dash");
    }catch(MalformedURLException ume)
    {
        //执行一些处理任务
```

```
}catch(SQLException sqle)
{
    //执行一些处理任务
}
```

At one extreme, you could catch all exceptions and then find a way to signal the calling method that something is wrong. This approach, as illustrated in Case 1, needs to return null values or other special values to the calling method to signal the error.

As a design strategy, this approach presents significant disadvantages. You lose all compile-time support, and the calling method must take care in testing all possible return values. Also, the normal code and error-handling code blend together, which leads to cluttering.

> 作为一种设计模式，这种方式存在很大的缺陷。将失去所有编译时支持，调用方法必须处理测试中所有出现的可能值。而且常规代码和错误处理代码绑定在一起，将导致程序逻辑混乱。

Case 2
```
Passenger getPassenger() throws MalformedURLException,SQLException
{
    Passenger flier = object.searchPassengerFlightRecord("Ashok Dash");
}
```

Case 2 presents the other extreme. The getPassenger() method declares all exceptions thrown by the method it calls in its throws clause. Thus getPassenger(), though aware of the exceptional situations, chooses not to deal with them and passes them on to *its* calling method. In short, it acts as a pass-through for the exceptions thrown by the methods it calls. However this does not offer a viable solution, as all responsibility for error processing is "bubbled up"—or moves up the hierarchy—which can present significant problems particularly in cases where multiple system boundaries exist. Pretend, for example, that you are Sabre (the airline reservation system), and the searchPassengerFlightRecord() method is part of your API to the user of your system, Travelocity.com, for example. The Travelocity application, which includes getPassenger() as part of its system, would have to deal with every exception that your system throws. Also, the application may not be interested in whether the exception is a MalformedURLException or SQLException, as it only cares for something like "Search failed."

> 情况2描述的是另一种极端。getPassenger()方法在throws子句中声明了所有它调用的方法抛出的异常。getPassenger()尽管知道各种异常情况，但选择不进行处理，而是将其抛给调用它的方法。简而言之，它仅仅充当了传递它所调用的方法抛出异常的通道。这并不是一个可行的解决方案，所有错误处理都像冒泡一样向方法调用层次结构的上层传递，这将导致一些很严重的问题出现，尤其是多系统边界问题。

Let us investigate further by examining Case 3.

Case 3
```
Passenger getPassenger() throws TravelException
{
    try
    {
```

```
        Passenger flier = object.searchPassengerFlightRecord("Tang Bin");
    }catch(MalformedURLException ume)
    {
        //执行一些处理任务
        throw new TravelException("Search Failed", ume);
    }catch(SQLException sqle)
    {
        //执行一些处理任务
        throw new TravelException("Search Failed", sqle);
    }
}
```

Case 3 meets midway between the two extremes of Case 1 and 2 by using a custom exception class called TravelException. This class features a special characteristic that understands the actual exception thrown as an argument and transforms a system-level message into a more relevant application-level one. Yet, you retain the flexibility of knowing what exactly caused the exception by having the original exception as part of the new exception object instance, which is handy for debugging purposes. This approach provides an elegant solution to designing APIs and applications that cross system boundaries.

> 情况 3 处于情况 1 和情况 2 两个极端之间，使用自定义异常类 TravelException。这个类的特性是将抛出的真正异常作为一个参数，并把系统级信息转换成更接近应用程序级的信息。将源异常看成新异常对象实例的一部分，新异常对象以方便实现调试为目的。这种方式为设计 API 和跨平台应用提供了一个优秀的解决方案。

The following code shows the TravelException class, which we used as our custom exception in Case 3. It takes two arguments. One is a message, which can be displayed on the error stream; the other is the real exception, which caused the custom exception to be thrown. This code shows how to package other information within a custom exception. The advantage of this packaging is that, if the calling method really wants to know the underlying cause of the TravelException, all it has to do is call getHiddenException( ). This allows the calling method to decide whether it wants to deal with specific exceptions or stick with TravelException.

```
public class TravelException extends Exception
{
    private Exception hiddenException_;
    public TravelException(String error, Exception excp)
    {
        super(error);
        hiddenException_ = excp;
    }
    public Exception getHiddenException()
    {
        return(hiddenException_);
    }
}
```

10.2 The Throw Statement

All methods use the throw statement to throw an exception. The throw statement requires a single argument: a throwable object. Throwable objects are instances of any subclass of the Throwable class.

> 所有方法都使用 throw 声明来抛出异常。throw 表达式需要一个单独的参数：throwable 对象。throwable 对象可以是 throwable 类任意子类的实例。

Here's an example of a throw statement.

```
throw someThrowableObject;
```

Let's look at the throw statement in context. The following pop method is taken from a class that implements a common stack object. The method removes the top element from the stack and returns the object.

```java
public Object pop() throws EmptyStackException {
    Object obj;
    if (size == 0) {
        throw new EmptyStackException();
    }
    obj = objectAt(SIZE - 1);
    setObjectAt(SIZE - 1, null);
    size--;
    return obj;
}
```

The pop method checks to see whether any elements are on the stack. If the stack is empty (its size is equal to 0), pop instantiates a new EmptyStackException object (a member of Java.util) and throws it. For now, all you need to remember is that you can throw only objects that inherit from the Java.lang.Throwable class.

Note that the declaration of the pop method contains a throws clause. EmptyStackException is a checked exception, and the pop method makes no effort to catch it. Hence, the method must use the throws clause to declare that it can throw that type of exception.

10.3 The Finally Statement

The following example shows the use of a finally statement that is useful for guaranteeing that some code gets executed whether or not an exception occurs.

For example, the following code example:

```
try {
    //执行一些任务
} finally {
    //之后执行一些清理任务
}
```

is similar to

```
try {
    //执行一些任务
```

```
    } catch(Object e){
        //执行一些处理任务
        throw e;
    }
```

The finally statement is executed even if the try block contains a return, break, continue, or throw statement.

For example, the following code example always results in "finally" being printed, but "after try" is printed only if a != 10.

```
try {
    if (a == 10) {
        return;
    }
} finally {
    print("finally\n");
}
print("after try\n");
```

10.4 Runtime Exceptions

This section contains a list of the exceptions that the Java runtime throws when it encounters various errors.

10.4.1 ArithmeticException

Attempting to divide an integer by zero or take a modulus by zero throw the ArithmeticException —no other arithmetic operation in Java throws an exception. For information on how Java handles other arithmetic errors see Operators on Integers and Operators on Floating Point Values.

For example, the following code causes an ArithmeticException to be thrown:

```
class Arith {
    public static void main(String args[]) {
        int j = 0;
        j = j / j;
    }
}
```

10.4.2 NullPointerException

An attempt to access a variable or method in a null object or a element in a null array throws a NullPointerException. For example, the accesses o.length and a[0] in the following class declaration throws a NullPointerException at runtime.

```
class Null {
    public static void main(String args[]) {
        String o = null;
        int a[] = null;
        o.length();
        a[0] = 0;
    }
}
```

It is interesting to note that if you throw a null object you actually throw a NullPointerException.

10.4.3 IncompatibleClassChangeException

In general the IncompatibleClassChangeException is thrown whenever one class's definition changes but other classes that reference the first class aren't recompiled. Four specific changes that throw a IncompatibleClassChangeException at runtime are:
- A variable's declaration is changed from static to non-static in one class but other classes that access the changed variable aren't recompiled.
- A variable's declaration is changed from non-static to static in one class but other classes that access the changed variable aren't recompiled.
- A field that is declared in one class is deleted but other classes that access the field aren't recompiled.
- A method that is declared in one class is deleted but other classes that access the method aren't recompiled.

> 一般来说,当一个类的定义改变而其他相关类没有重新编译时,将抛出 IncompatibleClassChangeException 异常。运行时抛出 IncompatibleClassException 的4种特殊变化情况: 一个类中某一个变量的声明从 static 改为 non-static,而其他需要访问改变后变量的类没有重新编译; 一个类中某一变量的声明从 non-static 改为 static,而其他需要访问改变后变量的类没有重新编译; 一个类中声明的字段被删除,而其他需要访问此字段的类没有重新编译; 一个类中声明的方法被删除,而其他需要访问此方法的类没有重新编译。

10.4.4 ClassCastException

A ClassCastException is thrown if an attempt is made to cast an object O into a class C and O is neither C nor a subclass of C.

The following class declaration results in a ClassCastException at runtime:

```
class ClassCast {
    public static void main(String args[]) {
        Object o = new Object();
        String s = (String)o;           //类转换
        s.length();
    }
}
```

10.4.5 NegativeArraySizeException

A NegativeArraySizeException is thrown if an array is created with a negative size.

For example, the following class definition throws a NegativeArraySizeException at runtime:

```
class NegArray {
    public static void main(String args[]) {
        int a[] = new int[-1];
        a[0] = 0;
    }
}
```

10.4.6 OutOfMemoryException

An OutOfMemoryException is thrown when the system can no longer supply the application with memory. The OutOfMemoryException can only occur during the creation of an object, i.e., when new is called.

> 当系统内存不能满足应用程序运行需要时，将抛出 OutOfMemoryException 异常。OutOfMemoryException 异常只在创建对象过程中 new 方法被调用时发生。

For example, the following code results in an OutOfMemoryException at runtime:

```
//本例将引发 OutOfMemoryException 异常
class Link {
    int a[] = new int[1000000];
    Link l;
}
class OutOfMem {
    public static void main(String args[]) {
        Link root = new Link();
        Link cur = root;
        while(true) {
            cur.l = new Link();
            cur = cur.l;
        }
    }
}
```

10.4.7 NoClassDefFoundException

A NoClassDefFoundException is thrown if a class is referenced but the runtime system cannot find the referenced class.

For example, class NoClass is declared:

```
class NoClass {
    public static void main(String args[]) {
        C c = new C();   //将引发 NoClassDefFoundException 异常
    }
}
```

When NoClass is run, if the runtime system can't find C.class it throws the NoClassDefFoundException. C.class must have existed at the time NoClass is compiled.

10.4.8 IncompatibleTypeException

An IncompatibleTypeException is thrown if an attempt is made to instantiate an interface.
For example, the following code causes an IncompatibleTypeException to be thrown.

```
interface I {
}
class IncompType {
    public static void main(String args[]) {
        I r = (I)new("I");
    }
}
```

10.4.9 ArrayIndexOutOfBoundsException

An attempt to access an invalid element in an array throws an ArrayIndexOutOfBoundsException. For example:

```java
class ArrayOut {
    public static void main(String args[]) {
        int a[] = new int[0];
        a[0] = 0;
    }
}
```

10.4.10 UnsatisfiedLinkException

An UnsatisfiedLinkException is thrown if a method is declared native and the method cannot be linked to a routine in the runtime.

```java
class NoLink {
    static native void foo();
    public static void main(String args[]) {
        foo();
    }
}
```

10.5 Sample Examples

Example 10.1

```java
//简单的异常例子
class exceptionDemo1
{
    public static void main(String args[])
    {
        try
        {
            int a = 0,
            b = 42/a;
            System.out.println("a = " + a);
        }
        catch(ArithmeticException e)
        {
            System.out.println(e);
        }
    }
}
```

Example 10.2

```java
//简单的异常处理例子
class ExceptionClass
{
    String name = "Bob";
    public void show()
    {
```

```
            System.out.println("name is: " + name);
        }
    }
    class exceptionDemo3
    {
        public static void main(String args[])
        {
            try
            {
                int a=4,b=0,c;
                c = a/b;
                System.out.println("c = " + c);
                ExceptionClass obj = null;
                obj.show();
            }catch(ArithmeticException e)
            {
                System.out.println("Denominator can't be zero");
            }catch(NullPointerException e)
            {
                System.out.println("Null reference has been use to access a
                    member");
            }catch(Exception e)
            {
                System.out.println("Unexpected error");
            }
        }
    }
```

Example 10.3

```
    //使用了finally的简单异常例子
    class exceptionDemo5
    {
        public static void main(String args[])
        {
            try
            {
                int val = 0;
                val = Integer.parseInt(args[0]);  //将第一个参数转变为整数
                System.out.println("No exception is raised");
            }catch(NumberFormatException e)
            {
                System.out.println("Exception is caught");
            }finally
            {
                System.out.println("Executing finally block");
            }
        }
    }
```

Example 10.4

```
    //本例展示一些内置异常
    class exceptionDemo6
    {
```

```java
public static void func()
{
    for(int i=0; i<3; ++i)
    {
        int k;
        try
        {
            System.out.println("Exception handling");
            try
            {
                switch(i)
                {
                    case 0:
                        int zero = 0;
                        k = 6/0;
                        break;
                    case 1:
                        int[] array = new int[2];
                        k = array[9];
                        break;
                    case 2:
                        char ch = "Hello".charAt(66);
                        break;
                }
            }
            catch(ArithmeticException e)
            {
                System.out.println("Arithmetic    Exception:    " +
                    e.getMessage());
            }
            catch(IndexOutOfBoundsException e)
            {
                System.out.println("Index Out Of Bounds Exception: " +
                    e.getMessage());
                throw e;
            }
        }catch(ArrayIndexOutOfBoundsException e)
        {
            System.out.println("Array Index Out Of Bounds Exception: "
                + e.getMessage());
        }
    }
}
public static void main(String args[])
{
    try
    {
        func();
    }catch(RuntimeException e)
    {
```

```
            System.out.println("Main() Runtime Exception: "+e.getMessage());
        }
    }
}
```

Example 10.5

```
/*一个简单的用户自定义异常处理程序，请注意，return 语句可以用在 catch 语句块以使控制权
*返回给异常块。如果在 catch 块中使用 return，则 catch 块之后的语句将不会被显示出来。
*A simple user-defined exception handling program.pay more attention!
*"The return statement may be used in the catch block to return the control
* back to the exception block. If we use return in the catch block, the
* statement after the catch"
* block will not be displayed!"
*/
class InrangeException extends Exception
{
    InrangeException(int wrongNum)
    {
        super(wrongNum + " is not a valid number");
    }
}
class Validation
{
    public void Inrange(int num) throws Exception
    {
        if(num>0 && num<1111)
        {
            System.out.println("n is in the range");
        }else
        {
            throw new InrangeException(num);
        }
    }
}
class exceptionDemo8
{
    public static void main(String args[])
    {
        int num = 0;
        num = Integer.parseInt(args[0]);    //Input from command prompt!
        Validation obj = new Validation();
        try
        {
            obj.Inrange(num);
        }catch(Exception e)
        {
            System.out.println(e.toString());
            return; //num 将永远无法打印在屏幕上！
        }
         System.out.print(num);
    }
}
```

10.6 Exercise for you

1. How can you find error and catch them in multiple lines?
2. What's the use of "throw" and "throws". Give an example to use those.
3. What are Java's built in exceptions and how they can be used? Give some examples.

Chapter 11 Multithread

Java provides a built in support for multithread programming. Multithread means it supports two or more parts in a program that can run concurrently. Each part is called a thread and each one has a special path of performance. It could be called as the special feature of a multitasking. Actually in the program one thread runs at one time but the design of all the threads and their respective slipping time is managed so well that they seem to run at the same time and thus its seems everything is running at one time, hence called multithread programming concept.

> Java 提供了内建的多线程编程支持功能。多线程意味着它支持程序中两个到多个部分并行运行。每一部分为一个线程，并且每个都具有独立执行路径。它被称为多任务驱动的特殊功能调用。实际上程序中每个线程运行在某一时间片上，但是所有线程及其对应时间片管理很好，以至于看上去各个线程似乎同时运行，从而看上去所有事情都在同一时间运行，这就是多线程程序设计的概念。

11.1 Multithread Overview

Definition of Multithread

Multithread is one of the most important concepts of the Java language. You simply cannot do without multithread in real-world programming. Multithread basically enables your program to do more than one task at once and also to synchronize the various tasks. But before we launch into multithread we will briefly summarize the points about threads.

> 多线程是 Java 程序语言中最重要的概念之一。在实时程序设计中没有多线程，就什么都做不成。多线程从根本上保证程序能同时执行多个任务，并实现在不同任务间的同步。在深入多线程学习前，先简要概述一下线程的基本概念。

- Brief Recapitulation of threads(线程简要描述)
- Synchronization(同步)
- Inter-thread Communication(线程间通信)

Single thread and multithread life cycle in a program is shown in figure 11.1 and figure 11.2.

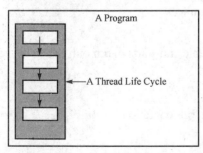

Figure 11.1　Single Thread Life Cycle

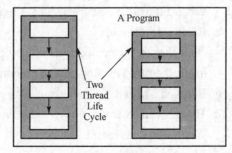

Figure 11.2　Multithread Life Cycle

There are two ways to create threads.
- Subclass Thread and override run( )
- Implement Runnable and override run( )

Either of these two approaches may be used. Since multiple inheritance doesn't allow us to extend more than one class at a time, implementing the Runnable interface may help us in this situation.

> 创建线程有两种方式:继承 Thread 类并重载 run( )函数;实现 Runnable 接口并重载 run( )函数。这两种方式都可使用。由于 Java 不允许同时继承 2 个以上的类,但可以通过接口来实现多重继承。

You call a thread by the start( ) method. And start calls the run( ) method. You never call run( ) directly. The stop( ) method is now deprecated and should be avoided. Threads have priorities between 1-10, the default being 5 i.e normal priority.

> 通过 start( )方法调用一个线程,start( )会调用 run( )方法。用户不必直接调用 run( )方法。Stop( )方法尽量避免使用。线程有 1~10 的优先级,缺省的优先级是 5,也是正常的优先级。

A daemon thread is a thread that has no other role other than to serve other threads. When only daemon threads remain, the program exits. When a new thread object is created, the new thread has priority equal to the creating thread, and is a daemon thread if and only if the creating thread is a daemon.

When the JVM starts, there is usually a single non-daemon thread which typically calls the main( ) method of the class.

> 和其他服务器线程相比,守护线程只是一个为其他线程提供服务的线程。只要守护线程还保留在内存中,程序就存在。当一个新线程对象创建时,相对于正在创建的线程来说具有优先权,一个守护线程之所以为守护线程是当且仅当创建的线程是一个后台服务。但 Java 虚拟机启动后,通常就有一个 main( )方法的非守护服务线程运行了。
>
> 注:Java 给每个线程安排优先级以决定与其他线程比较时该如何对待该线程,线程优先级是详细说明线程间优先关系的整数。作为绝对值,优先级是毫无意义的,当只有一个线程时,优先级高的线程并不比优先级低的线程运行得快。相反,线程的优先级用来决定何时从一个运行的线程切换到另一个。

Threads can be in one of four states.
- **New Threads**

  When a thread is first created, the thread is not yet running.
- **Runnable Threads**

  Once the start( ) method is invoked the thread is runnable and starts to run only when the code inside the run( ) method begins executing.
- **Blocked Threads**

  Threads can enter the blocked state when any of these four conditions occur.

  When *sleep*( ) is called.

  When *suspend*( ) is called.

  When *wait*( ) is called.

The thread calls an operation e.g. during input/output, which will not return until reading/writing is complete.

- **Dead Threads**

A thread dies because of two reasons.

It dies a natural death when the run( ) method exits.

It is killed because its stop( ) method was invoked.

Now it is time for some examples. Take a look at two examples below for creating more than one thread.

> 线程的 4 种状态：
> 　　新建状态，当线程第一次被创建，还没有运行时的状态。
> 　　运行状态，一旦 start( )方法被调用就标志着线程进行运行状态，同时调用 run( )方法，run( )方法内的代码开始执行。
> 　　中断状态，当以下 4 种情况之一发生时线程进入中断状态：当 sleep( )方法被调用；当 suspend( )方法被调用；当 wait( )方法被调用；线程调用了一个输入/输出操作，但读/写操作还没有完成。
> 　　死亡状态，线程由于 2 个原因进入死亡状态：run( )方法退出时线程自然死亡；调用 stop( )方法杀死线程。

Four state transition of thread is shown in figure 11.3.

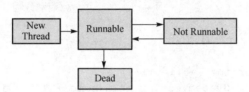

Figure 11.3　Thread State Transitions

```
class NewThread extends Thread {
    String name;
    Thread t;
    NewThread(String threadname) {
        name=threadname;//设置线程的 name 成员变量
        t=new Thread(this, name);//创建新线程
        System.out.println("New Thread: " + t );
        t.start(); //启动新线程
    }
    public void run() {
        try {
            for(int i=5; i>0;i--) {//循环 5 次，输出 i 的值
                System.out.println(name + ":" + i) ;
                Thread.sleep(1000);//线程休眠 1000 毫秒
            }
        }catch (InterruptedException e) {
            System.out.println(name + " Interrupted. ");
        }
        System.out.println(name + " Exiting.");
    }
}
```

```java
class MultiThreadDemo {
    public static void main (String args[]) {
        new NewThread("One");
        new NewThread("Two");
        new NewThread("Three");
        try {
            Thread.sleep(10000);
        }catch (InterruptedException e) {
            System.out.println("Main Thread Interrupted.");
        }
        System.out.println("main Thread Exiting.");
    }
}
```

And the second one.

Note: Suspend and resume are deprecated methods.

```java
class NewThread implements Runnable {
    String name;
    Thread t;
    NewThread(String threadname) {
        name=threadname;                        //设置变量 name
        t=new Thread(this, name);               //创建线程
        System.out.println("New Thread: " + t);
        t.start();                              //启动线程
    }
    public void run() {
        try {
            for(int i=5;i>0;i--) {
                System.out.println(name + ":" + i);
                Thread.sleep(200);
            }
        }catch (InterruptedException e) {
            System.out.println(name + "Interrupted. ");
        }
        System.out.println(name + " Exiting.");
    }
}
//下面创建两个线程,然后通过主线程main 分别对它们调用 suspend 和 resume 方法
class SuspendResume {
    public static void main(String args[]) {
        NewThread ob1 = new NewThread("One");
        NewThread ob2 = new NewThread("Two");
        try {
            Thread.sleep(1000);
            System.out.println("Suspending thread One");
            Thread.sleep(1000);
            ob1.t.suspend();
            System.out.println("Resuming thread One");
            ob1.t.resume();
            System.out.println("Suspending thread Two");
            Thread.sleep(1000);
            ob2.t.suspend();
```

```
            System.out.println("Resuming thread Two");
            ob2.t.resume();
        }catch (InterruptedException e) {
            System.out.println("main thread interrupted." );
        }
        try {
            ob1.t.join();
            ob2.t.join();
        }catch (InterruptedException e) {
            System.out.println("main thread interrupted.");
        }
        System.out.println("Main thread Exiting.");
    }
  }
}
```

11.2　Synchronization

When two or more threads need access to a shared resource, they need some way to ensure that the resource will be used by only one thread at a time. The process by which this is achieved is *synchronization*.

> 同步是指当两个或两个以上的线程需要访问共享资源时，需要通过一些方式确保资源在某一时刻只能被一个线程使用，实现这个目的的过程就叫同步。

Key to synchronization is the concept of the monitor. A monitor is an object that is used as a mutually exclusive lock. Only one thread can own the monitor at a given time. When a thread acquires a lock, it is said to have entered the monitor. The other threads attempting to enter the locked monitor will be suspended until the first exits the monitor.

> 同步的关键是监视器的概念。监视器是一个独占对象锁的对象。在给定时间段时只有一个线程能拥有监视器。但一个线程获得一个对象锁时，也就是说可具有监视器了。其他线程要进入中断状态时，会被挂起直至第一个线程退出监视器。有两种代码同步方式：synchronized()方法；synchronized 状态。

There are two ways you can synchronize your code.
(1) synchronized methods
(2) synchronized statement

Both involve the use of the *synchronized* keyword. See below for an example.

```
    import java.io.*;
    class Deposit {
        static int balance = 1000;
        public static void main(String args[]) {
            PrintWriter out = new PrintWriter(System.out, true);
            Account account = new Account(out);
            DepositThread first, second;
            first = new DepositThread(account, 1000, "#1");
            second=new DepositThread(account, 1000, "\t\t\t\t#2");
```

```java
            first.start();
            second.start();
            try {
                first.join();
                second.join();
            }catch (InterruptedException e) {  }
            out.println("*** Final balance is  " + balance);
        }//main 函数定义结束
}//类 Deposit 定义结束
class Account {
    PrintWriter out;
    Account(PrintWriter out) {
        this.out=out;
    }
    synchronized void deposit(int amount, String name ) {
        int balance;
        out.println(name + " trying to deposit " + amount);
        out.println(name + " getting balance... " );
        balance=getBalance();
        out.println(name + " balance got is " + balance);
        balance += amount;
        out.println(name + " setting balance...");
        setBalance(balance);
        out.println(name + " balance set to " + Deposit.balance);
    }
    int getBalance() {
        try {
            Thread.sleep(1000);
        }catch (InterruptedException e) {  }
        return Deposit.balance;
    }
    void setBalance(int balance) {
        try {
            Thread.sleep(1000);
        }catch(InterruptedException e) {  }
        Deposit.balance = balance;
    }
}//类 Account 定义结束
class DepositThread extends Thread {
    Account account;
    int deposit_amount;
    String message;
    DepositThread(Account account, int amount, String message) {
        this.message=message;
        this.account=account;
        this.deposit_amount=amount;
    }
    public void run() {
        account.deposit(deposit_amount, message);
    }
} //类 DepositThread 定义结束
```

11.2.1 Inter-thread Communication

Java's inter-thread communication process involves the use of *wait( )*, *notify( )* and *notifyall( )* methods. These methods are implemented as final methods in *Object,* so all classes have them. These methods can only be called from within synchronized code.

> Java 线程间同步是通过调用 wait( )、notify( )和 notifyall( )方法实现的。这些方法作为 Object 类的最终方法被实现,所有类都继承这些方法,而且这些方法只能在进行代码同步时调用。

Rules for using these methods:
- *wait*( ) tells the calling thread to give up the monitor and go to sleep until some other thread enters the same monitor and calls *notify*( ).
- *notify*( ) wakes up the first thread that called *wait*( ) on the object.
- *notifyall*( ) wakes up all the threads waiting on the object. The highest priority thread will run first.

> 方法使用规则
> wait( )方法:使处于运行状态线程放弃独占资源转入睡眠状态,让其他线程获得资源并调用 notify( )。
> notify( )方法: 唤醒等待队列中第一个调用 wait( )方法的线程。
> notifyall( )方法: 唤醒等待队列中所有处于等待状态的线程,优先级高的线程先返回。

See below for an incorrect implementation of a producer/consumer example.

```
//一个没有正确实现生产者消费者功能的程序
class Q {
    int n;
    synchronized int get() {
        System.out.println("Got: " + n);
        return n;
    }
    synchronized void put(int n) {
        this.n=n;
        System.out.println("Put: " + n);
    }
}
class Producer implements Runnable {
    Q q;
    Producer(Q q) {
        this.q=q;
        new Thread(this, "Producer").start();
    }
    public void run() {
        int i=0;
        while(true) {
            q.put(i++);
        }
    }
}
```

```java
class Consumer implements Runnable {
    Q q;
    Consumer(Q q) {
        this.q=q;
        new Thread(this, "Consumer").start();
    }
    public void run() {
        while(true) {
            q.get();
        }
    }
}
class PC {
    public static void main(String args[]) {
        Q q = new Q();
        new Producer(q);
        new Consumer(q);
        System.out.println("Press Control-C to stop");
    }
}
```

The correct way would be using *wait*( ) and *notify*( ) as shown here.

```java
//正确实现了生产者消费者功能的程序
class Q {
    int n;
    boolean valueset = false;
    synchronized int get() {
        if (!valueset)
        try {
            wait();
        }catch (InterruptedException e) {
            System.out.println("InterruptedException caught");
        }
        System.out.println("Got: " + n);
        valueset=false;
        notify();
        return n;
    }
    synchronized void put(int n) {
        if (valueset)
        try {
            wait();
        }catch(InterruptedException e) {
            System.out.println("InterruptedException caught");
        }
        this.n=n;
        valueset=true;
        System.out.println("Put: " + n);
        notify();
    }
}
class Producer implements Runnable {
```

```
    Q q;
    Producer(Q q) {
        this.q=q;
        new Thread(this, "Producer").start();
    }
    public void run() {
        int i=0;
        while(true) {
            q.put(i++);
        }
    }
}
class Consumer implements Runnable {
    Q q;
    Consumer(Q q) {
        this.q=q;
        new Thread(this, "Consumer").start();
    }
    public void run() {
        while(true) {
            q.get();
        }
    }
}
class PCFixed {
    public static void main(String args[]) {
        Q q = new Q();
        new Producer(q);
        new Consumer(q);
        System.out.println("Press Control-C to stop");
    }
}
```

Now to summarize the points about multithreading: thread synchronization, inter-thread communication, thread priorities, thread scheduling, and daemon threads.

11.2.2 Java Thread Scheduling

Features:

(1) The JVM schedules using a preemptive, priority based scheduling algorithm.

(2) All Java threads have a priority and the thread with the highest priority is scheduled to run by the JVM.

(3) In case two threads have the same priority a FIFO (first in first out) ordering is followed.

功能：
(1) Java 虚拟机(JVM)使用基于优先级的算法调度线程。
(2) 所有 Java 线程都具有各自的优先级，JVM 将安排最高优先级的线程运行。
(3) 当两个线程优先级相同时，遵循先进先出原则(FIFO)。

A different thread is invoked to run in case one of the following events occur:

(1) The currently running thread exits the Runnable state ie either blocks or terminates.

(2) A thread with a higher priority than the thread currently running enters the Runnable state. The lower priority thread is preempted and the higher priority thread is scheduled to run.

> 当发生以下事件之一时,另一不同的线程将被调用并运行:
> (1) 当前运行的线程因终止或阻塞退出。
> (2) 另一个比当前运行线程优先级高的线程进入运行状态。低优先级线程被中断运行,同时高优先级线程被安排运行。

Time Slicing is dependent on the implementation.

A thread can voluntarily yield control through the yield( ) method. Whenever a thread yields control of the CPU another thread of the same priority is scheduled to run. A thread voluntarily yielding control of the CPU is called Cooperative Multitasking.

> 线程通过 yield( )方法进入自动控制状态。此时任何一个具有相同优先级的线程将被安排运行。这种自动控制 CPU 的线程称做协同多任务处理。

11.2.3 Thread Priorities

JVM selects to run a Runnable thread with the highest priority. All Java threads have a priority in the range 1-10.Top priority is 10, lowest priority is 1.Normal priority by default is 5.

Thread.MIN_PRIORITY - minimum thread priority
Thread.MAX_PRIORITY - maximum thread priority
Thread.NORM_PRIORITY - normal thread priority

Whenever a new Java thread is created it has the same priority as the thread which created it.
Thread priority can be changed by the setpriority( ) method.

> 任何时候创建一个新的 Java 线程时,它具有和创建它的线程相同的优先级,优先级可以通过 setpriority( )方法来改变。

Java Based Round-Robin Scheduler(An example)

```
public class Scheduler extends Thread
{
    public Scheduler(){
        timeSlice = DEFAULT_TIME_SLICE;
        queue = new Circularlist();
    }
    public Scheduler(int quantum){
        timeSlice = quantum;
        queue = new Circularlist();
    }
    public addThread(Thread t) {
        t.setPriority(2);
        queue.additem(t);
    }
    private void schedulerSleep() {
        try{
            Thread.sleep(timeSlice );
```

```
            } catch (InterruptedException e){};
        }
        public void run(){
            Thread current;
            This.setpriority(6);
            While (true) {
                //得到下一个线程
                current = (Thread)qeue.getnext();
                if ( (current != null) && (current.isAlive()) ){
                    current.setPriority(4);
                    schedulerSleep();
                    current.setPriority(2)
                }
            }
        }
        private CircularList queue;
        private int timeSlice;
        private static final int DEFAULT_TIME_SLICE = 1000;
}
public class TesScheduler
{
    public static void main()String args[]) {
        Thread.currentThread().setpriority(Thread.Max_Priority);
        Schedular CPUSchedular = new Scheduler ();
        CPUSchedular.start()
        TestThread t1 = new TestThread("Thread 1");
        t1.start()
        CpuSchedular.addThread(t1);
        TestThread t2 = new TestThread("Thread 2");
        t2.start()
        CpuSchedular.addThread(t2);
        TestThread t3 = new TestThread("Thread 1");
        t3.start()
        CpuSchedular.addThread(t3);
    }
}
```

11.2.4 Java Synchronization

Every object in Java has a single lock associated with it. This lock is not used ordinarily. When a method is declared as SYNCHRONIZED, calling the method requires acquiring the lock for the object. When the lock is owned by a different thread the thread blocks and is put into the entry set of the object's lock.

> Java 中的每个对象都有一个相关的唯一性锁，只是这个锁平常用得少。若方法被定义为 SYNCHRONIZED，调用此方法就需要获得对象的锁。

Example:
```
public synchronized void enter(Object item){
    while (count == BUFFER_SIZE)
        Thread.yeild();
```

```
        ++count;
        buffer[in] = item;
        in = (in+1) % BUFFER_SIZE;
    }
    public synchronized void remove (){
        Object item;
        while (count == 0)
            Thread.yeild();
        --count;
        item = buffer[out]
        out = (out+1) % BUFFER_SIZE;
        return item
    }
```

Wait( ) & Notify( )

When a thread calls the wait( ) method:

(1) The thread releases the lock for the object.

(2) The state of the thread is set to blocked.

(3) The Thread is placed in the wait set for the object

> 以下情况时，线程调用 wait( )方法：线程释放对象的锁；线程进入中断状态；线程进入等待队列中。

When a thread calls the notify( ) method:

(1) An arbitrary thread is picked from the list of threads in the wait set.

(2) Moves the selected thread from the wait set to the entry set.

(3) Sets the state of the selected thread from blocked to runnable.

> 以下情况时，线程调用 notify( )方法：从线程等待队列中取出一个任意的线程；将此线程从等待队列送入就绪队列中；将此线程状态从中断转为运行。

Examples of Wait( ) and Notify( )

```
    public synchronized void enter(Object item){
        while (count == BUFFER_SIZE){
            try{
                wait();
            }catch (InterruptedException e) {}
        }
        // 添加一个项到缓冲区
        ++count;
        buffer[in] = item;
        in = (in+1) % BUFFER_SIZE;
        notify();
    }
    public synchronized void remove(Object item){
        while (count == 0){
            try{
                wait();
            }catch (InterruptedException e) {}
        }
```

```
//从缓冲区删除一个项
--count;
item = buffer[out];
out = (out+1) % BUFFER_SIZE;
notify();
return item;
}
```

11.3 The Life Cycle of a Thread

Now that you've seen how to give a thread something to do, let's review some details that were glossed over in the previous section. In particular, we look at how to create and start a thread, some of the special things it can do while it's running, and how to stop it.

> 如何给线程安排任务，具体来说，就是如何创建和开始一个线程，线程运行时工作任务如何安排，以及如何停止一个线程。

The following figure 11.4 shows the states a thread can be in during its life and illustrates which method calls cause a transition to another state. This figure is not a complete finite state diagram but rather an overview of the more interesting and common facets of a thread's life. The remainder of this section uses the previously introduced Clock applet to discuss a thread's life cycle in terms of its state.

> 图 11.4 显示了线程生命周期中的各种状态，并展示了引起状态转变的各种方法。这个图不是一个完全有限状态图，而只显示线程生命周期中有趣的和常用的方面。后面将使用前面介绍的 Clock applet 实例来讨论线程生命周期的不同状态。

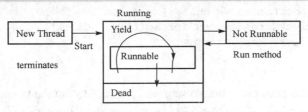

Figure 11.4　Thread States

11.3.1　Creating a Thread

The application in which an applet is running calls the applet's start method when the user visits the applet's page. The Clock applet creates a Thread, clockThread, in its start method with the code shown in boldface.

> 当用户访问 applet 所在页面时，applet 所在应用程序运行并调用 applet 的 start 方法。Clock applet 就创建一个 clockThread 线程。

```
public void start() {
    if (clockThread == null) {
        clockThread = new Thread(this, "Clock");
        clockThread.start();
    }
}
```

After the statement has been executed, clockThread is in the New Thread state. A thread in this state is merely an empty Thread object; no system resources have been allocated for it yet. When a thread is in this state, you can only start the thread. Calling any method besides start when a thread is in this state makes no sense and causes an IllegalThreadStateException. In fact, the runtime system throws an IllegalThreadStateException whenever a method is called on a thread and that thread's state does not allow for that method call.

> 在执行指定代码之后，clockThread 线程就处于新建状态。处于这种状态下的线程大多数是空线程对象，还没有为其分配系统资源。当线程处于这种状态时，只能开始线程。调用 start()外的任何方法都没有反应，并会引起 IllegalThreadStateException 异常。实际上，当调用处于新建状态线程方法时都会抛出这种异常。

Note that the Clock instance is the first argument to the thread constructor. The first argument to this thread constructor must implement the Runnable interface and becomes the thread's target. The clockThread gets its run method from its target Runnable object — in this case, the Clock instance. The second argument is just a name for the thread.

> 注意：Clock 实例是线程构造函数的首要参数，而且线程构造函数必须实现 Runnable 接口并成为线程对象。clockThread 通过可运行对象获得它运行的各种方法。第一要素就是线程的名称。

11.3.2 Starting a Thread

Now consider the next line of code in Clock's start method, shown here in boldface.

```
public void start() {
    if (clockThread == null) {
        clockThread = new Thread(this, "Clock");
        clockThread.start();
    }
}
```

The start method creates the system resources necessary to run the thread, schedules the thread to run, and calls the thread's run method. clockThread's run method is defined in the Clock class.

> start 方法为运行线程创建必要的系统资源，安排线程运行，并调用线程的 run 方法。clockThread 线程的 run 方法在 Clock 类定义。

After the start method has returned, the thread is "running"; however, it's somewhat more complex than that. As the previous figure 11.4 shows, a thread that has been started is in the Runnable state. Many computers have a single processor, thus making it impossible to run all running threads at the same time. The Java runtime system must implement a scheduling scheme that shares the processor among all running threads; see the Thread Scheduling section for more information about scheduling. So, at any given time, a running thread may be waiting for its turn in the CPU.

> 当 start 方法返回后，线程即开始"运行"。然而，有时比这更复杂。如图 11.4 所示，当一个线程已经开始后即处于运行状态。很多计算机都是单 CPU 的，因此要在同一时间运行所有运行状态的线程是不可能的。Java 运行时系统必须运行一个时间安排表，让所有处于运行状态的线程共享处理器。因此，在任何给定时间片，运行时线程都在等待属于它的 CPU 序列时间。

Here's another look at Clock's run method.

```
public void run() {
    Thread myThread = Thread.currentThread();
    while (clockThread == myThread) {
        repaint();
        try {
            Thread.sleep(1000);
        } catch (InterruptedException e) {
            //虚拟机要求不能再休息更多时间了，因此必须返回去工作
        }
    }
}
```

Clock's run method loops while the condition clockThread == myThread is true. This exit condition is explained in more detail in the Stopping a Thread section. For now, however, know that it allows the thread, and thus the applet, to exit gracefully.

Within the loop, the applet repaints itself and then tells the thread to sleep for one second (1,000 milliseconds). An applet's repaint method ultimately calls the applet's paint method, which does the actual update of the applet's display area. The Clock paint method gets the current time, formats it, and displays.

```
public void paint(Graphics g) {
    //得到时间，并将其转换为日期
    Calendar cal = Calendar.getInstance();
    Date date = cal.getTime();
    //格式化并显示
    DateFormat dateFormatter = DateFormat.getTimeInstance();
    g.drawString(dateFormatter.format(date), 5, 10);
}
```

11.3.3 Making a Thread Not Runnable

A thread enters Not Runnable state when one of the following events occurs:

- Its sleep method is invoked.
- The thread calls the wait method to wait for a specific condition to be satisfied.
- The thread is blocking on I/O.

> 当线程碰到下列情况之一时，就进入不可运行状态：sleep( )方法被调用；线程调用 wait( ) 方法后等待需满足的特殊条件；线程碰 I/O 阻塞。

The clockThread in the Clock applet becomes Not Runnable when the run method calls sleep on the current thread.

```
public void run() {
    Thread myThread = Thread.currentThread();
    while (clockThread == myThread) {
        repaint();
        try {
            Thread.sleep(1000);
        } catch (InterruptedException e) {
```

                }
            }
        }
//虚拟机要求不能再休息更多时间了，因此必须返回去工作

During the second that the clockThread is asleep, the thread does not run, even if the processor becomes available. After the second has elapsed, the thread becomes Runnable again; if the processor becomes available, the thread begins running again.

> 在 clockThread 处于睡眠期间，即使处理器空闲，线程也不运行。它休眠时间过去后，线程再进入可运行状态，如果 CPU 处理器可用，则线程重新开始运行。

For each entrance into the Not Runnable state, a specific and distinct exit returns the thread to the Runnable state. An exit works only for its corresponding entrance. For example, if a thread has been put to sleep, the specified number of milliseconds must elapse before the thread becomes Runnable again. The following list describes the exit for every entrance into the Not Runnable state:

> 对每个进入不可运行状态的入口，都有一个特定的返回到可运行状态的出口相对应。出口只和它相关联的入口一起工作。例如，如果一个线程已在休眠，则特定数量的毫秒必须用完线程才可能再一次进入可运行状态。下面描述了针对每个进入不可运行状态的出口。

- If a thread has been put to sleep, the specified number of milliseconds must elapse.
- If a thread is waiting for a condition, then another object must notify the waiting thread of a change in condition by calling notify or notifyAll; more information on this is available in the Synchronizing Threads section.
- If a thread is blocked on I/O, the I/O must complete.

> 如果一个线程在等待一个条件，另一个对象必须通过调用 notify 或 notifyall 方法通报等待线程条件的改变，更多信息可在线程同步部分找到。
> 如果线程在 I/O 受阻，则 I/O 必须完成。

11.3.4 Stopping a Thread

Although the Thread class does contain a stop method, this method is deprecated and should not be used to stop a thread because it is unsafe. Rather, a thread should arrange for its own death by having a run method that terminates naturally.

> 尽管线程类具有 stop 方法，但这个方法不建议用来终止线程，因为它是不安全的。线程应在 run 方法中安排自己死亡而自然终止线程。

For example, the while loop in this run method is a finite loop: It will iterate 100 times and then exit.

```
public void run() {
    int i = 0;
    while (i < 100) {
        i++;
        System.out.format("i = %d%n", i);
    }
}
```

A thread with this run method dies naturally when the loop completes and the run method exits.

> 当循环结束及 run 方法退出时，具有这种 run 方法的线程将自然死亡。

Let's look at how the Clock applet thread arranges for its own death. (You might want to use this technique with your applet threads.) Recall the code for the Clock's run method.

```
public void run() {
    Thread myThread = Thread.currentThread();
    while (clockThread == myThread) {
        repaint();
        try {
            Thread.sleep(1000);
        } catch (InterruptedException e) {
                //虚拟机要求不能再休息更多时间了，因此必须返回去工作
        }
    }
}
```

The exit condition for this run method is the exit condition for the while loop because there is no code after the while loop.

```
while (clockThread == myThread) {
```

This condition indicates that the loop will exit when the currently executing thread is not equal to clockThread. When would this ever be the case?

When you leave the page, the application in which the applet is running calls the applet's stop method. This method then sets the clockThread to null, thereby telling the main loop in the run method to terminate.

```
public void stop() {    //applet 的 stop 方法
    clockThread = null;
}
```

If you revisit the page, the start method is called again and the clock starts up again with a new thread. Even if you stop and start the applet faster than one iteration of the loop, clockThread will be a different thread from myThread and the loop will still terminate.

11.3.5　Testing Thread State

A final word about Thread state: Release 1.5.0 introduced the Thread.getState method. When called on a thread, one of the following Thread.State values is returned:

> 关于线程状态的结束语：1.5.0 版引入了 Thread.getState 方法，当一个类调用此方法时，下列状态之一的值将返回：

- NEW
- RUNNABLE
- BLOCKED
- WAITING
- TIMED_WAITING

- **TERMINATED**

The API for the Thread class also includes a method called isAlive. The isAlive method returns true if the thread has been started and not stopped. If the isAlive method returns false, you know that the thread either is a New Thread or is Dead. If the isAlive method returns true, you know that the thread is either in a Runnable or Not Runnable state.

> 在线程类的API中还包括一个叫isAlive的方法,如果线程已开始并没有结束时,isAlive方法返回true。如果isAlive方法返回false,则线程要么是一个新的线程,要么就是处于死亡状态的线程。如isAlive方法返回true,则线程要么处于可运行状态,要么处于不可运行状态。

Prior to release 5.0, you couldn't differentiate between a New Thread or a Dead thread; nor could you differentiate between a Runnable thread and a Not Runnable thread.

isAlive( )

The thread class provides a method which can be called on an instance of a Thread to determine if it is currently executing. If isAlive( ) returns false the thread is either a new thread that is not currently executing or it is dead.

```
public boolean isAlive();
```

The following code creates a thread and executes it by calling it's start( ) method. Before the thread is started a test is performed to determine if the thread is currently executing. Since the thread is new, isAlive( ) returns false. You will also notice within the run( ) method a reference to the current thread is grabbed using the following method.

Code:

```
static Thread currentThread()
```

Since isAlive( ) is called within the run( ) method isAlive( ) returns true because the thread is executing at that time.

Code:

```java
public class ThreadTest{
    public static void main(String[] args){
        T t = new T("T Thread");
        threadStatus(t.isAlive(),t.getName());
        t.start();
    }
    public static void threadStatus(boolean isalive, String tname){
        String status = isalive ? "is alive" : " is not alive";
        System.out.println(tname +""+ status);
    }
}
class T extends Thread{
    public T(String tname){
        super(tname);
    }
    public void run(){
        Thread t = Thread.currentThread();
        String tname = t.getName();
```

```
        boolean isalive = t.isAlive();
        for(int i = 0; i < 9; ++i){
            System.out.println(tname + " is alive ?" + " : " + isalive);
        }
    }//run 方法定义结束
}//类 T 定义结束
join
public final void join()
        throws InterruptedException
```

Waits for this thread to die.

Throws:

InterruptedException-if another thread has interrupted the current thread. The *interrupted status* of the current thread is cleared when this exception is thrown.

> InterruptedException: 如果另一线程中断了当前线程，这个异常会被抛出，当前线程被中断的状态被清空。

11.3.6 Why Pause and Resume Processes

Anyone using the Windows NT product line (Windows 2000 and Windows XP) must have used the task manager utility. This utility, activated by pressing CTRL+SHIFT+ESC, brings up a list of the active processes and allows you several actions for controlling them: starting new processes, stopping processes and setting their priority. When you have some process that is taking a lot of resources (normally CPU time), you can easily assign it the lowest priority and the system will take care of assigning only the remaining resources or the "idle time" on the machine.

> 使用 Windows NT 产品(Windows 2000 和 Windows XP)的用户，都熟悉任务管理器组件。这个组件可通过同时按下 Ctrl+Shift+Esc 激活，弹出一个活动进程的列表，并允许进行一些控制：开始新的进程，停止进程以及设置它们的优先级。但有些占用多量系统资源(通常是 CPU 时间)的进程，很容易为其设置较低的优先级，系统将只用保留的系统资源或"空闲"的机器时间来管理它。

- You have a time-consuming operation, e.g. a big build, and want to pause it for doing something quickly and resuming it after doing this.
- You have some P2P software or download running and want to pause and resume it without reconnecting and want to browse some pages quickly.
- A program starts a disk trashing operation and you want to send and e-mail.
- A program starts working in a way it shouldn't for just a moment and you want to attach a debugger to it.
- You have a buggy process running and want to kill it fast.

> 很耗时的操作，如大规模构建操作，想暂停以便其他进程工作更快一点，等操作完成后再恢复这个进程。
> 在运行 P2P 软件或下载程序在运行，想暂停并无差错地恢复，并想浏览页面快一点。
> 程序开始进行磁盘扫描操作时，想发送一邮件。
> 有一些臭虫进程在运行，想尽快杀死它。

How it is done

The main problem is: there is no SuspendProcess API function. And there is no documented or safe way of doing this.

> 主要问题在于：没有 SuspendProcessAPI 函数，且没有完成这个操作的安全或被推荐的方式。

The only simple way of doing this is via SuspendThread/ResumeThread. This pair of API functions allows you to suspend and resume a thread. More than that, for the sake of safety, they maintain an internal "suspend count". Each time you call SuspendThread, it increments this counter. ResumeThread, on the other hand, decreases this counter. If this was not done this way, the caller of SuspendThread would have no way of knowing how to restore the original state of the thread. Calling ResumeThread after calling SuspendThread effectively restores the original thread's state.

> 简单处理这种操作的方式是通过 SuspendThread/ResumeThread。这一对 API 函数允许挂起及恢复一个线程。更重要的是，从安全角度来说，它们保持了一种内在的"挂起账目"。每次调用 SuspendThread 时，它将在这个计算器中加 1。而 ResumeThread，反过来，则在这个计数器中减 1。如果不通过这种方式，则 SuspendThread 调用无法知道如何存储线程的起始位置。在调用 SuspendThread 之后调用 ResumeThread，能有效存储线程的起始状态。

Knowing this, it is very straightforward suspending a process: it is just a matter of listing all the threads on a process, opening a handle for each of them and calling SuspendThread. The resuming is done the same way.

> 在了解上述原理后，挂起一个进程很简单：如进程中所有线程列表一样，打开属于它们的名柄并调用 SuspendThread，同样的方法也用于恢复操作。

11.4 Sample Examples

Example 11.1

```
//本例通过 Runnable 创建多线程
class newThread implements Runnable
{
    String name;
    Thread t;
    newThread(String threadName)
    {
        name = threadNa,e;
        t = new Thread(this,name);
        Syste.out.println("newThread" + t);
        t.start();
    }
    public void run()
    {
        try
        {
```

```
                for(int i=5; i>0; --i)
                {
                    System.out.println(i);
                }
            }
            catch(InterruptedException e)
            {
                System.out.println(name + "interrupted");
            }
            System.out.println("Thread existing");
        }
    }
    class multiThred
    {
        public static void main(String args[])
        {
            new newThread("one");
            new newThread("two");
            new newThread("three");
            try
            {
                Thread.sleep(1000);
            }
            catch(InterruptedException e)
            {
                System.out.println("Main thread interrupted");
            }
            System.out.println("Main thread existing");
        }
    }
```

Example 11.2

```
    //本例创建一个线程
    class newThread implements Runnable
    {
        Thread t;
        newThread()
        {
            t = new Thread(this.name + "Demo thread");
            System.out.println(t);
            t.start();
        }
        public void run()
        {
            try
            {
                for(int i=5; i>0; --i)
                {
                    System.out.println("New thread " + i);
                    Thread.sleep(1000);
                }
            }
            catch(InterruptedException e)
```

```java
            {
                System.out.println("Thread interrupted");
            }
        }
        System.out.println("Existing the thread");
    }
}

class newThread
{
    public static void main(String args[])
    {
        try
        {
            for(int i=5; i>0; --i)
            {
                System.out.println("Main thread " + i);
                Thread.sleep(1000);
            }
        }
        catch(InterruptedException e)
        {
            System.out.println("Main thread interrupted");
        }
    }
}
```

Example 11.3

```java
//从 Thread 类继承创建一个线程
public class PingPong extends Thread
{
    private String word;
    private int delay;
    public PingPong(String whatTosay, int delaytime)
    {
        word = whatTosay;
        delay = delaytime;
    }
    public void run()
    {
        try
        {
            for(int i=0; i<10; ++i)
            {
                System.out.print(word + " ");
                Thread.sleep(delay);
            }
        }
        catch(InterruptedException e)
        {
            return;
        }
    }
```

```java
    public static void main(String args[])
    {
        new PingPong("ping",1000).start();
        new PingPong("pong",1000).start();
    }
}
```

Example 11.4

```java
//通过实现 Runnable 接口创建一个线程
class PingPong2 implements Runnable
{
    private String word;
    private int delay;
    PingPong2(String whatTosay, int delaytime)
    {
        word = whatTosay;
        delay = delaytime;
    }
    public void run()
    {
        try
        {
            for(int i=0; i<10; ++i)
            {
                System.out.print(word + " ");
                Thread.sleep(delay);
            }
        }
        catch(InterruptedException e)
        {
            return;
        }
    }
    public static void main(String args[])
    {
        Runnable ping = new PingPong2("ping",1000);
        Runnable pong = new PingPong2("pong",1000);
        new Thread(ping).start();
        new Thread(pong).start();
    }
}
```

Example 11.5

```java
//使用 isAlive 和 join 方法
class newThread implements Runnable
{
    Thread t;
    String name;
    newThread(String threadname)
    {
        name = threadname;
        t = new Thread(this,name);
        System.out.println("New Thread " + this.name);
```

```java
            t.start();
    }
    public void run()
    {
        try
        {
            for(int i=5; i>0; --i)
            {
                System.out.println(name + " " + i);
                Thread.sleep(1000);
            }
        }catch(InterruptedException e)
        {
            System.out.println(name + "Interrupted");
        }
        System.out.println(name + "Existing");
    }
}
class threadDemo
{
    public static void main(String args[])
    {
        newThread ob1 = new newThread("one");
        newThread ob2 = new newThread("two");
        newThread ob3 = new newThread("three");
        System.out.println(ob1.t.isAlive());
        System.out.println(ob2.t.isAlive());
        System.out.println(ob3.t.isAlive());
        try
        {
            ob1.t.join();
            ob2.t.join();
            ob3.t.join();
        }
        catch(InterruptedException e)
        {
            System.out.println("Main thread");
        }
        System.out.println(ob1.t.isAlive());
        System.out.println(ob2.t.isAlive());
        System.out.println(ob2.t.isAlive());
    }
}
```

11.5　Exercise for you

1. What is a thread priority and how to set those? Give one example.

2. Give an example to extend a thread. Can you extend the same thread more than 1 time? If yes, how? If no, why?

3. What is a synchronization method and how to make it?

Chapter 12   Input and Output

This chapter explains one of the most important packages IO, Which make the Input and Output operation possible in the program. Most of the cases, the program needs the external data to accomplish the task. Be it a file or string or anything, they are broadly defined in to two parts, the source and the destination. Opening, reading, writing, updating, creating and deleting files are very common approach when we work on the network. Those all are supported by the Java IO package.

> 本章介绍的 IO 包是 Java 中最重要的包之一，该包主要负责程序中输入和输出操作。在多大数情况下，程序需要外部数据来完成任务，如一个文件、字符串或其他的相关数据。文件的打开、读取、写入、更新、创建和删除等都是网络上的常见操作。Java IO 包均支持以上操作。

12.1   Basic Java I/O

Unlike most of the previous discussion of Java, file I/O will be discussed in the context of a Java application rather than an applet. The reason for this is that browsers, such as Netscape, block access to the local file system as a security precaution. Consequently, you will not be able to execute the example programs discussed below from this.

> 与前面讨论的大多数 Java 内容不同，文件输入/输出(I/O)是在 Java 应用程序上讨论，而不是在 applet 程序中。主要原因是如 Netscape 等浏览器，鉴于安全原因会禁止操作本地文件系统。因此，讨论完下列内容后再执行示例程序。

An important concept to understand is that in Java files and their data are objects. Thus, the basic strategy you will follow is to first create a File object. It identifies the file in the local file system through a filename and, optionally, a path. When you instantiate a variable of type File, you get an object that you can then pass to various methods. After you create a file object, you will create a data object that you associate with the file object. Thus, you should think of the data in a file as an object whose class provides various methods that allow you to access its contents.

> Java 中文件和数据都是对象，这是一个很重要的概念。因此，操作文件的首要步骤就是创建 File 对象。File 对象通过文件名或路径识别本地文件系统中的文件文件。当实例化一个文件类型的变量后将产生一个文件对象，并可以传递给不同的方法。在新建一个文件对象之后，就创建了与文件对象相联系的数据对象。因此应考虑将文件中的数据看做一个提供了访问这些数据本身的不同方法的类对象。

I/O is one of the areas that has been changed significantly in Java 1.1. In the earlier 1.0 version, I/O for both byte data and character data was handled using classes that derived from InputStream and OutputStream. Frequently used subclasses were FilteredInputStream and FilteredOutputStream.

Java 1.1 still includes the InputStream and OutputStream classes and their subclasses, but they have been assigned primary responsibility for byte I/O. A new set of classes has been added that are intended to be used with character data. The abstract classes that all these character classes derive from are Reader and Writer, with BufferedReader and BufferedWriter likely to be the most frequently used subclasses for basic character I/O. For example, they include convenient methods for reading and writing a line of data and, for input, noting the end_of_file condition.

> 在 Java 1.1 中仍包括 InputStream 和 OutputStream 类及其子类，但其功能仅限于负责字节 I/O。在现有的 Java 中加入了一组用于字符数据操作的新类。所有这些类的抽象类都是从 Reader 和 Writer 类派生而来。而 BufferedReader 和 BufferedWriter 是使用最频繁的字符输入/输出基类的子类。例如，它们包含了一组方便地用于读写数据行和根据文件结束条件输入数据的方法。

In the discussion below, we will focus on the File class in order to view and test file attributes and on the BufferedReader and BufferedWriter classes for basic I/O.

> 下面集中讨论可查看和测试文件属性的文件类，以及用于基本输入/输出的 BufferedReader 和 BufferedWriter 类。

12.1.1 Background

Before looking at these I/O classes, we need to consider another Java concept——Exceptions——since many of the I/O classes incorporate Java exception handling facilities.

12.1.2 Exceptions

The concept of Exception in basic to Java and is used in a number of contexts in addition to I/O. In an effort to help the programmer write code that is reliable, Java provides a general mechanism that enables the programmer to look for possible run time errors and to take appropriate action when they occur, rather than letting the program continue until possibly crashing or producing erroneous results.

> 异常是 Java 中的基本概念，除了 I/O 方面之外，还广泛用于其他内容。为确保程序代码可靠，Java 提供了通用机制使程序员能查找可能的运行时错误，以及当错误发生时采取适当的措施处理，而不是让程序继续运行直到程序可能崩溃或产生错误的结果。

12.1.3 Applications

In this section, we will look at four applications. The first checks attributes for files and directories. The next two illustrate output and input at the "Hello, World!" level. The last reads from the keyboard and writes the lines read to a file. It should be an easy extrapolation from these example programs to handling more substantial I/O for actual data.

> 这部分将介绍四个应用程序。第一个是检测文件和目录的属性；其后两个通过"hello,world"程序演示输出和输入；最后一个从键盘读取数据并逐行写入一个文件。通过这些简单的程序掌握大多数实际的 I/O 操作。

12.1.4 File Attributes

This application illustrates several of the methods included in the File class for accessing the various file attributes, such as length, date of creation, etc.

Output

This application illustrates the use of the PrintWriter class for writing a stream of character data to a file.

Input

This application illustrates the use of the BufferedReader class for reading a stream of character data from a file.

> 这个应用程序阐明了 BufferedReader 类从文件中读字符数据类的使用方法。

Keyboard

This application illustrates the use of both the BufferedReader and the BufferedWriter classes for reading lines of data from the keyboard and writing them to a file.

> 这个应用程序阐明了用 BufferedReader 和 BufferedWriter 两个类完成逐行从键盘获取数据并写入文件的方法。

Object I/O

This application illustrates the writing and reading of a Java Object to and from a file.

Socket I/O

This discussion describes the process of building up a basic set of I/O tools——writeUTF and readUTF——to support socket communication over a network between Java programs running on different machines.

> 这些讨论描述建立基本 I/O 工具集即 writeUTF 和 readUTF 的过程，它为网络中运行在不同计算机的 Java 程序之间提供了 socket 通信功能。

Most of the programs work with external data stored either in local files or coming from other computers on the network. Java has a concept of working with so-called streams of data. After a physical data storage is mapped to a logical stream, a Java program reads data from this stream serially - byte after byte, character after character, etc. Some of the types of streams are byte streams (InputStream, OutputStream) and character streams (Reader and Writer). The same physical file could be read using different types of streams, for example, FileInputStream, or FileReader.

> 大多数程序的运行需要访问存储在本地文件或网络中其他计算机上文件的数据。Java 有一个数据流的概念。通过将存储的物理数据映射成逻辑数据流后，Java 程序能逐字节、逐字符连续地从这个流中读取数据。一部分数据流称为字节流(InputStream, OutputStream)，另一部分称为字符流(Reader 和 Writer)。同一个物理文件可以被不同类型的数据流读取，例如 FileInputStream 或 FileReader。

12.2 Streams

Classes that work with streams are located in the package java.io. Java 1.4 has introduced the new package java.nio with improved performance, which is not covered in this lesson.

> 与流操作相关的类放在 java.io 包中。Java 1.4 引入了性能有所提高的新的包 java.nio，该包不在本课程内。

There are different types of data, and hence different types of streams.

Here's the sequence of steps needed to work with a stream:

(1) Open a stream that points at a specific data source: a file, a socket, URL, etc; (2) Read or write data from/to this stream; (3) Close the stream.

> 流的操作步骤：打开一个指向特定数据源的流，例如一个文件、socket、URL 等；从(向)流中读出(写入)数据；关闭流。

Let's have a closer look at some of the Java streams.

12.2.1　Byte Streams

If a program needs to read/write bytes (8-bit data), it could use one of the subclasses of the InputStream or OutputStream respectively. The example below shows how to use the class FileInputStream to read a file named abc.dat. This code snippet prints each byte's value separated with white spaces. Byte values are represented by integers from 0 to 255, and if the read() method returns -1, this indicates the end of the stream.

> 当程序需要读/写字节数据(8 比特数据)时，应分别使用 InputStream 或 OutputStream 类中的子类。下面的例子显示如何使用 FileInputSteam 类来读取 abc.dat 文件。代码段打印输出用空白符隔开的每个字节的值。字节的值用 0～255 的整数来表示。如果 read()方法返回 -1，则表示到达了字节流的结束。

```java
        FileInputStream myFile = null;
        try {
            myFile = new  FileInputStream("abc.dat");       //打开流
            boolean eof = false;
            while (!eof) {
                int byteValue = myFile.read();              //读取流
                System.out.print(byteValue + " ");
                if (byteValue  == -1)
                    eof = true;
            }
        //myFile.close();                                   //不在这里执行
        }
        catch (IOException e) {
            System.out.println("Could not read file: " + e.toString());
        }
        finally{
            try{
                myFile.close();                             //关闭流
            }
            catch (Exception e1){
                e1.printStackTrace();
            }
        }
```

Please note that the stream is closed in the clause finally. Do not call the method close() inside of the try/catch block right after the file reading is done. In case of exception during the file read, the program would jump over the close() statement and the stream would never be closed!

> 请注意字节流在 finally 子句中关闭。try/catch 块中不要在文件读操作完成之后立即调用 close( )方法。主要是由于在读文件的过程中发生异常后,程序将跳过 close( )语句,则该流将永远不被关闭。

Let's modify the example that reads the file abc.dat to introduce the buffering:

```
FileInputStream myFile = null;
BufferedInputStream buff =null
try {
    myFile = new  FileInputStream("abc.dat");
    BufferedInputStream buff = new BufferedInputStream(myFile);
    boolean eof = false;
    while (!eof) {
        int byteValue = buff.read();
        System.out.print(byteValue + " ");
        if (byteValue  == -1)
            eof = true;
    }
}
catch (IOException e) {
    e.printStackTrace();
}
finally{
    buff.close();
    myFile.close();
}
```

It's a good practice to call the method flush( ) when the writing into a BufferedOutputStream is done - this forces any buffered data to be written out to the underlying output stream. While the default buffer size varies depending on the OS, it could be controlled.

> 在进行带缓冲的输出流写操作时最好调用 flush( )方法,将使任何缓冲的数据写入特定的输出流。缺省缓冲区大小依赖并受控于操作系统。

For example, to set the buffer size to 5000 bytes do this:

```
BufferedInputStream buff = new BufferedInputStream(myFile, 5000);
```

12.2.2 Character Streams

Java uses two-byte characters to represent text data, and the classes FileReader and FileWriter work with text files. These classes allow you to read files either one character at a time with read( ), or one line at a time with readLine( ).

> Java 使用 2 字节字符来存储文本数据,并使用 FileReader 和 FileWriter 类进行文本文件操作。这些类允许使用 read( )方法进行逐字符的读文件操作,或用 readLine( )方法进行逐行读文件操作。

The classes FileReader and FileWriter also support have buffering with the help of BufferedReader and BufferedWriter.
The following example reads text one line at a time:

```
        FileReader myFile = null;
        BufferedReader buff = null;
        try {
            myFile = new FileReader("abc.txt");
            buff = new BufferedReader(myFile);
            boolean eof = false;
            while (!eof) {
                String line = buff.readLine();
                if (line == null)
                    eof = true;
                else
                    System.out.println(line);
            }
            ...
        }
```

For the text output, there are several overloaded methods write( ) that allow you to write one character, one String or an array of characters at a time.

对文本输出，通过重载write()方法可实现某字符、字符串或字符数组的写操作。

To append data to an existing file while writing, use the 2-arguments constructor (the second argument toggles the append mode):

为向已存在的文件中添加数据，可使用两参数的构造函数(第2个参数设置为追加模式)。

```
        FileWriter fOut = new FileWriter("xyz.txt", true);
        Below is yet another version of the tax calculation program
        (See the lesson Intro to Object-Oriented Programming with Java).
        This is a Swing version of the program and it populates the dropdown box chState
with the data from the text file states.txt.
        import java.awt.event.*;
        import java.awt.*;
        import java.io.FileReader;
        import java.io.BufferedReader;
        import java.io.IOException;
        public class TaxFrameFile extends java.awt.Frame implements ActionListener {
            Label lblGrIncome;
            TextField txtGrossIncome = new TextField(15);
            Label lblDependents=new Label("Number of Dependents:");
            TextField txtDependents = new TextField(2);
            Label lblState = new Label("State: ");
            Choice chState = new Choice();
            Label lblTax = new Label("State Tax: ");
            TextField txtStateTax = new TextField(10);
            Button bGo = new Button("Go");
            Button bReset = new Button("Reset");
            TaxFrameFile() {
                lblGrIncome = new Label("Gross Income: ");
                GridLayout gr = new GridLayout(5,2,1,1);
                setLayout(gr);
                add(lblGrIncome);
                add(txtGrossIncome);
                add(lblDependents);
```

```java
            add(txtDependents);
            add(lblState);
            add(chState);
            add(lblTax);
            add(txtStateTax);
            add(bGo);
            add(bReset);
            //从一个文件填充语句
            populateStates();
            txtStateTax.setEditable(false);
            bGo.addActionListener(this);
            bReset.addActionListener(this);
            //定义且实例化并注册一个 WindowAdapter
            //处理本窗体的 windowClosing 事件
            this.addWindowListener(new WindowAdapter() {
                public void windowClosing(WindowEvent e) {
                    System.out.println("Good bye!");
                    System.exit(0);
                }
            });
    }
    public void actionPerformed(ActionEvent evt) {
        Object source = evt.getSource();
        if(source == bGo ){
            //如果是 Go 按钮，则进行处理
            try{
                int grossInc =
                Integer.parseInt(txtGrossIncome.getText());
                int dependents  =
                Integer.parseInt(txtDependents.getText());
                String state = chState.getSelectedItem();
                Tax tax=new Tax(dependents,state,grossInc);
                String sTax =Double.toString(tax.calcStateTax());
                txtStateTax.setText(sTax);
            }
            catch(NumberFormatException e){
                txtStateTax.setText("Non-Numeric Data");
            }
            catch (Exception e){
                txtStateTax.setText(e.getMessage());
            }
        }
        else if (source == bReset ){
            //如果是 Reset 按钮，则执行下面处理
            txtGrossIncome.setText("");
            txtDependents.setText("");
            chState.select(" ");
            txtStateTax.setText("");
        }
    }
    //本方法将读 states.txt 文件，并填充下拉的 chStates
    private void populateStates(){
```

```java
            FileReader myFile = null;
            BufferedReader buff = null;
            try {
                myFile = new FileReader("states.txt");
                buff = new BufferedReader(myFile);
                boolean eof = false;
                while (!eof) {
                    String line = buff.readLine();
                    if (line == null)
                        eof = true;
                    else
                        chState.add(line);
                }
            }
            catch (IOException e){
                txtStateTax.setText("Can't read states.txt");
            }
            finally{
            //关闭流
                try{
                    buff.close();
                    myFile.close();
                }
                catch(IOException e){
                    e.printStackTrace();
                }
            }
        }
        public static void main(String args[]){
            TaxFrameFile taxFrame = new TaxFrameFile();
            taxFrame.setSize(400,150);
            taxFrame.setVisible(true);
        }
}
```

12.2.3 Buffered Streams

So far we were reading and writing one byte at a time. Disk access is much slower than the processing performed in memory. That's why it's not a good idea to access disk 1000 times for reading a file of 1000 bytes. To minimize the number of time the disk is accessed, Java provides buffers, which are sort of "reservoirs of data". For example, the class BufferedInputStream works as a middleman between the FileInputStream and the file itself. It reads a big chunk of bytes from a file in one shot into memory, and, then the FileInputStream will read single bytes from there. The BufferedOutputStream works in a similar manner with the class FileOutputStream. Buffered streams just make reading more efficient.

> 到目前为止，介绍了按字节模式进行文件读写。磁盘访问远比内存中的处理要慢。因此，为读一个1000字节的文件而访问磁盘1000次肯定不是一个好办法。为减少磁盘访问的字符次数，Java提供了"数据池"的缓冲区操作模式。例如，BufferedInputStream类为FileInputStream类和操作文件充当中间人的角色。它一次从文件中读一大块字符放入内存，然后，FileInputStream再从此内存中逐字节读出。BufferedOutputStream类也以相同的模式与FileOutputStream类协同工作，缓冲流使读操作变得更高效。

You can use stream chaining (or stream piping) to connect streams - think of connecting two pipes in plumbing. Let's modify the example that reads the file abc.dat to introduce the buffering:

```java
FileInputStream myFile = null;
BufferedInputStream buff =null
try {
    myFile = new FileInputStream("abc.dat");
    BufferedInputStream buff = new BufferedInputStream(myFile);
    boolean eof = false;
    while (!eof) {
        int byteValue = buff.read();
        System.out.print(byteValue + " ");
        if (byteValue == -1)
            eof = true;
    }
}
catch (IOException e) {
    e.printStackTrace();
}
finally{
    buff.close();
    myFile.close();
}
```

It's a good practice to call the method flush( ) when the writing into a BufferedOutputStream is done - this forces any buffered data to be written out to the underlying output stream.

While the default buffer size varies depending on the OS, it could be controlled. For example, to set the buffer size to 5000 bytes:

```java
BufferedInputStream buff = new BufferedInputStream(myFile, 5000);
```

12.2.4 Data Streams

If you are expecting to work with a stream of a known data structure, i.e. two integers, three floats and a double, use either the DataInputStream or the DataOutputStream. A method call readInt( ) will read the whole integer number (4 bytes) at once, and the readLong( ) will get you a long number (8 bytes).

> 如果对一个已知数据结构的流操作，例如 2 个整数，3 个单精度和 1 个双精度数据，可使用 DataInputStream 或 DataOutputStream 类实现。readInt( )方法以每次整型数据(4 字节)方式读取数据，而 readLong( )方法将以长整型(8 字节)方式读取。

The DataInput stream is just a filter. We are building a "pipe" from the following fragments:
FileInputStream --> BufferedInputStream --> DataInputStream

```java
FileInputStream myFile = new FileInputStream("myData.dat");
BufferedInputStream buff = new BufferedInputStream(myFile);
DataInputStream data = new DataInputStream(buff);
try {
    int num1 = data.readInt();
    int num2 = data.readInt();
    float num2 = data.readFloat();
    float num3 = data.readFloat();
```

```
        float num4 = data.readFloat();
        double num5 = data.readDouble();
} catch (EOFException eof) {...}
```

12.2.5 Class StreamTokenizer

Sometimes you need to parse a stream without knowing in advance what data types you are getting. In this case you want to get each "piece of data" (token) based on the fact that a delimiter such as a space, comma, etc separates the data elements.

The class java.io.StreamTokenizer reads tokens one at a time. It can recognize identifiers, numbers, quoted strings, etc. Typically an application creates an instance of this class, sets up the rules for parsing, and then repeatedly calls the method nextToken( ) until it returns the value TT_EOF (end of file).

> 有时需要解析一个预先不知道数据类型的流,且希望通过如空格、逗号等数据分隔符来获得每个数据块(标记)。java.io.StreamTokenizer 类每次读取一个标记。它能识别标识符、数字、引用字符串等。通常一个应用程序创建这个类的一个实例,设置解析规则,然后重复调用 nextToken( )方法直到返回 TT_EOF(文件结束)。

Let's write a program that will read and parse the file customers.txt distinguishing strings from numbers. Suppose we have a file customers.txt with the following content:

Hello Everyone   50.24

ECIT Good   234.29

Nanchang Jinagxi   330013

Here is the program that parses it:

```
//本例展示StreamTokenizer的用法
import java.io.StreamTokenizer;
import java.io.FileReader;
public class CustomerTokenizer{
    public static void main(String args[]){
    StreamTokenizer stream =null;
    try{
        stream = new StreamTokenizer(new FileReader("customers.txt"));
        while (true) {
            int token = stream.nextToken();
            if (token == StreamTokenizer.TT_EOF)
                break;
            if (token == StreamTokenizer.TT_WORD) {
                System.out.println("Got the string: " +
                                    stream.sval);
            }
            if (token == StreamTokenizer.TT_NUMBER) {
                System.out.println("Got the number: " +
                                    stream.nval);
            }
        }
    }
    catch (Exception e){
        System.out.println("Can't read Customers.txt: " +
                            e.toString());
```

```
                }
            finally{
                try{
                    stream.close();
                }
                catch(Exception e){
                    e.printStackTrace();
                }
            }
        }
    }
```

After compiling and running the program CustomerTokenizer, the system console will look like this:

javac CustomerTokenizer.java

java CustomerTokenizer

Got the string:Hello

Got the string: Everyone

Got the number: 50.24

Got the string: ECIT

Got the string: Good

Got the number: 234.29

Got the string: Nanchang

Got the string: Jiangxi

Got the number: 330013

When a StreamTokenizer finds a word, it places the value into the sval member variable, and the numbers are placed into the variable nval.

> 当 StreamTokenizer 找到一个单词,就把它放到 sval 成员变量,而数字被放到 nval 成员变量中。

You can specify characters that should be treated as delimiters by calling the method whitespaceChars( ). The characters that represent quotes in the stream are set by calling the method quoteChar( ).

> 可通过调用 whitespaceChars( )方法定制分隔字符,通过调用 quoteChar( )方法可以设置流中的分隔符。

To make sure that certain characters are not misinterpreted, call a method ordinaryChar( ), for example ordinaryChar('/');

12.2.6 Class StringTokenizer

The class java.util.StringTokenizer is a simpler version of a class StreamTokenizer, but it works only with strings. The set of delimiters could be specified at the creation time, i.e. command angle brackets:

```
        StringTokenizer st = new StringTokenizer(
            "Heyueshun, President Ruanjianxueyuan, China", ",()");
        while (st.hasMoreTokens()) {
```

```
            System.out.println(st.nextToken());
    }
```

The above code fragment would print the following:

    Heyueshun

    President Ruanjianxueyuan

    China

The previous sample would not return the value of a delimiter - it just returned the tokens. But sometimes, in case of multiple delimiters, you may want to know what the current delimiter is. The 3-argument constructor will provide this information. The following example defines 4 delimiters: greater and less then signs, comma and a white space:

```
StringTokenizer st=new StringTokenizer("...IBM...price<...>86.3", "<>, ", true);
```

If the third argument is true, delimiter characters are also considered to be tokens and will be returned to the calling method, so a program may apply different logic based on the delimiter. If you decide to parse HTML file, you'll need to know what the current delimiter is to apply the proper logic.

> 注：StringTokenizer 类的第三个参数有特定作用，当希望将界定符也作为标记返回时可设置构造方法的第三个参数。

12.3 Class File

This class has a number of useful file maintenance methods that allow rename, delete, perform existence check, etc.

> File 类拥有大量非常有用的文件操作的方法，包括重命名、删除、文件检查等。

First you have to create an instance(实例)of this class:

```
File myFile = new File("abc.txt");
```

The line above does not actually create a file - it just creates an instance of the class File that is ready to perform some manipulations with the file abc.txt. The method createNewFile( ) should be used for the actual creation of the file.

> 上述语句并不真正创建一个文件，它仅创建一个 File 类的实例，可以执行对 abc.txt 文件的一些操作。而 createNewFile()方法则用于真正的文件创建。

Below are some useful methods of the class File:

createNewFile( )：creates a new, empty file named according to the file name used during the File instantiation. It creates a new file only if a file with this name does not exist.

delete( )：deletes file or directory

renameTo( )：renames a file

length( )：returns the length of the file in bytes

exists( )：tests whether the file with specified name exists

list( )：returns an array of strings naming the files and directories in the specified   directory

lastModified( )：returns the time that the file was last modified

mkDir( ): creates a directory

The code below creates a renames a file customers.txt to customers.txt.bak. If a file with such name already exists, it will be overwritten.

```
File file = new File("customers.txt");
File backup = new File("customers.txt.bak");
if (backup.exists()){
    backup.delete();
}
file.renameTo(backup);
```

12.3.1　The PrintWriter Class

Using FileWriter for saving into a file is pretty inconvenient: The FileWriter methods take arrays of characters as parameters, but generally we just want to print a value into the file. The java.io library provides class called PrintWriter that we can layer on top of a FileWriter, to work with the file the way we actually want to work.

> FileWriter 类可以方便地进行文件写操作。FileWriter 方法将字符数组作为参数，但通常可能只将一个值写入文件。java.io 库中提供了 PrintWriter 类，通过此类可以根据实际需要操作文件。

12.3.2　Constructor Method

The constructor method for a PrintWriter takes a FileWriter as a parameter.

```
PrintWriter(FileWriter writer)
```

Creates a PrintWriter object that uses writer whenever it wants to actually write something into a file. A PrintWriter object will build up the character arrays for you, and it pass them along to the FileWriter that you give it in the constructor. (Technically, the PrintWriter constructor actually takes a Writer object as a parameter, an abstract class which FileWriter extends. There are other classes that also extend the Writer class, and PrintWriter can be layered on top of any of these.)

> 技术上来说，PrintWriter 构造方法将 Writer 对象作为参数，而 FileWriter 类从 Writer 这个抽象类扩展而来。当然，还有其他的类从 Writer 类扩展，PrintWriter 类位于这些类层次结构中的顶层。

Instance methods

The PrintWriter instance methods are as follows.

```
void print(char c):Prints the single character c.
void print(int i):Prints the integer i in decimal.
void print(double d):Prints the floating-point value d.
void print(String s):Prints the string s.
void print(Object obj):Prints the string returned by obj.toString().
void close():Closes the file.
```

Additionally, the PrintWriter instance methods have a println( ) method corresponding to each of the print( ) methods, which prints the data and then terminates the current line of input.

> 此外,PrintWriter 实例方法有一个 println( )方法与每个 print( )方法相对应,实现数据的换行打印输出。

Example

This allows us to save into a file just as conveniently as we can print to the screen. As an example, the following program would create a table of numbers and their square roots.

```java
//本例展示就像把数据写入文件一样,可以方便地将数据输出在屏幕上
import java.io.*;
public class SqrtTable {
    public static void main(String[] args) {
        PrintWriter file;
        try {
            if(args.length != 1) {
                System.err.println("usage: java SqrtTable filename");
                return;
            }
            file = new PrintWriter(new FileWriter(new File(args[0])));
        }
        catch(IOException e) {
            System.err.println("Error opening file " + args[0]
                                + ": " + e.getMessage());
            return;
        }
        for(int i = 1; i < 10; i++) {
            file.print(i);
            file.print(" ");
            file.print(Math.sqrt((double) i));
        }
        file.close();
    }
}
```

Notice how the PrintWriter methods don't throw exceptions, so that the only the process of opening the file needs to go into the try block.

> 注意,PrintWriter 方法不抛出异常,因此打开文件操作应在 try 语句块中执行。

12.3.3 File Handling and Input/Output

java.io package

Classes related to input and output are present in the Java language package java.io . Java technology uses "streams" as a general mechanism of handling data. Input streams act as a source of data. Output streams act as a destination of data.

Chapter 12　Input and Output

Java 语言中与输入和输出相关的类都保存在 java.io 包中。Java 技术上采用"流"作为数据处理的通用机制。输入流作为数据源，而输出流将作为数据终点。

- **File Class**

The file class is used to store the path and name of a directory or file. The file object can be used to create, rename, or delete the file or directory it represents.

文件类用于保存目录或文件的路径和名称。文件对象能用于创建、重命名或删除已有的文件或目录。

The File class has the following constructors:

```
File(String pathname);  //路径名可以是文件或目录名
File(String dirPathname, String filename);
File(File directory, String filename);
```

The File class provides the getName( ) method which returns the name of the file excluding the directory name.

```
String getName();
```

- **Byte Streams**

The package java.io provides two set of class hierarchies——one for handling reading and writing of bytes, and another for handling reading and writing of characters. The abstract classes InputStream and OutputStream are the root of inheritance hierarchies handling reading and writing of bytes respectively.

java.io 包提供了两个类层次结构：一个处理字节的读写；另一个处理字符的读写。InputStream 和 OutputStream 抽象类是分别处理字节读写操作的顶层基类。

- **Read and Write Methods**

InputStream class defines the following methods for reading bytes:

```
int read() throws IOException
int read(byte b[]) throws IOException
int read(byte b[], int offset, int length) throws IOException
```

Subclasses of InputStream implement the above mentioned methods.
OutputStream class defines the following methods for writing bytes：

```
void write(int b) throws IOException
void write(byte b[]) throws IOException
void write(byte b[], int offset, int length) throws IOException
```

Subclasses of OutputStream implement the above mentioned methods.
The example below illustrates code to read a character：

```
//首先创建一个 FileInputStream 类型的一个对象
FileInputStream inp = new FileInputStream("filename.ext");
//创建一个 DataInputStream 的对象
```

```
DataInputStream dataInp = new DataInputStream(inp);
int i = dataInp.readInt();
```

- **Reader and Writer Classes**

Similar to the InputStream and OutputStream class hierarchies for reading and writing bytes, Java technology provides class hierarchies rooted at Reader and Writer classes for reading and writing characters.

> 与InputStream和OutputStream类的字节类层次结构类似,Java技术也提供了基于Reader和Writer类的字符读写的类层次结构。

A character encoding is a scheme for internal representation of characters. Java programs use 16 bit Unicode character encoding to represent characters internally. Other platforms may use a different character set (for example ASCII) to represent characters. The reader classes support conversions of Unicode characters to internal character shortage. Every platform has a default character encoding. Besides using default encoding, Reader and Writer classes can also specify which encoding scheme to use.

> 字符编码是字符内部表示方式。Java程序采用16位统一字符编码标准进行字符表示。其他操作平台可能采用不同的字符集(如ASCII)来表示字符。Reader类支持统一字符编码标准与内部字符存储方式间的转换。每个平台都有一缺省的字符编码。Reader和Writer类不仅可以使用缺省编码,也可指定特定的编码方式。

Table 12.1 gives a brief overview of key Reader classes.

Table 12.1 Reader Classes

CharArrayReader	The class supports reading of characters from a character array
InputStreamReader	The class supports reading of characters from a byte input stream. A character encoding may also be specified
FileReader	The class supports reading of characters from a file using default character encoding

Table 12.2 gives a brief overview of key Writer classes.

Table 12.2 Writer Classes

CharArrayWriter	The class supports writing of characters from a character array
OutputStreamReader	The class supports writing of characters from a byte output stream. A character encoding may also be specified
FileWriter	The class supports writing of characters from a file using default character encoding

The example below illustrates reading of characters using the FileReader class.

```
//创建一个FileReader类的对象
FileReader fr = new FileReader("filename.txt");
int i = fr.read(); //读取一个字符
```

When Reading and Writing Text Files:

It is almost always a good idea to use buffering (default size is 8K), it is often possible to use generic interface references, instead of references to specific classes ,there is always a need to pay attention to exceptions (in particular, IOException and FileNotFoundException).

The close method always needs to be called, or else resources will leak will automatically flush the stream, if necessary calling close on a "wrapper" stream will automatically call close on its underlying stream closing a stream a second time has no consequence.

> 读写文本时，在绝大多数情况下使用缓冲都是不错的选择（缺省缓冲大小一般是 8 K）；通常情况下尽量使用标准接口，而不是特定类；并应能注意常见异常的处理；尽可能调用 close 方法以防止资源泄漏，该方法可以自动释放缓冲区中的流。

Commonly used methods:
BufferedReader- readLine
BufferedWriter- write + newLine

The FileReader and FileWriter classes always use the system's default character encoding. If this default is not appropriate (for example, when reading an XML file which specifies its own encoding), the recommended alternatives are:

```
FileInputStream fis = new FileInputStream("test.txt");
InputStreamReader in = new InputStreamReader(fis, "UTF-8");
FileOutputStream fos = new FileOutputStream("test.txt");
OutputStreamWriter out = new OutputStreamWriter(fos, "UTF-8");
```

Example :

```
import java.io.*;
public class ReadWriteTextFile {
  /**
   * 提取一个文本文件的所有内容，并将其作为一个字符串返回。
   * 这种实现方式不对调用者抛出异常。
   * 参数 aFile 是一个已经存在且能够被读的文件。
   */
  static public String getContents(File aFile) {
     StringBuffer contents = new StringBuffer();
     //在本地定义，主要是使 finally 部分可见
     BufferedReader input = null;
     try {
     //使用缓冲
     //这种实现一次可读取一行
     //FileReader 总是假设默认编码可以正确工作
        input = new BufferedReader( new FileReader(aFile) );
        String line = null; //不要在 while 循环中定义
        while (( line = input.readLine()) != null){
            contents.append(line);
            contents.append(System.getProperty("line.separator"));
        }
     }
     catch (FileNotFoundException ex) {
         ex.printStackTrace();
     }
     catch (IOException ex){
         ex.printStackTrace();
     }
     finally {
```

```java
            try {
                if (input!= null) {
//必须对 input 执行 flush 和 close, 它会在底层作用于 FileReaderinput.close();
                }
            }
            catch (IOException ex) {
                ex.printStackTrace();
            }
        }
        return contents.toString();
    }
    /**
    * 更改文本文件的所有内容, 覆盖原有的文本。
    * 这种实现方式, 将所有异常抛给调用者
    * 参数 aFile 是一个已经存在能够被写入的文件。
    * 如果参数不符合要求, 则抛出 IllegalArgumentException
    * 如果文件不存在, 则抛出 FileNotFoundException
    * 如果在写过程出现问题, 则抛出 IOException 异常
    */
    static public void setContents(File aFile, String aContents)
        throws FileNotFoundException, IOException {
      if (aFile == null) {
        throw new IllegalArgumentException("File should not be null.");
      }
      if (!aFile.exists()) {
        throw new FileNotFoundException ("File does not exist: " + aFile);
      }
      if (!aFile.isFile()) {
        throw new IllegalArgumentException("Should not be a directory: " + aFile);
      }
      if (!aFile.canWrite()) {
        throw new IllegalArgumentException("File cannot be written: " + aFile);
      }
      //在此定义, 主要使得其对 finally 块可见
      Writer output = null;
      try {
        //使用缓冲
        //FileWriter 总是假设默认编码可以正常工作
        output = new BufferedWriter( new FileWriter(aFile) );
        output.write( aContents );
      }
      finally {
        //对 output 必须执行 flush 和 close, 会影响底层的 FileWriter
        if (output != null) output.close();
      }
    }
    public static void main ( String[] aArguments ) throws IOException {
      File testFile = new File("C:\\Temp\\blah.txt");
      setContents(testFile, "blah blah blah");
      System.out.println( "File contents: " + getContents(testFile) );
    }
}
```

Writing to a File

If the file does not already exist, it is automatically created.

```
try {
    BufferedWriter out = new BufferedWriter(new FileWriter("outfilename"));
    out.write("aString");
    out.close();
}
catch (IOException e) {
}
```

Reading Text from a File

```
try {
    BufferedReader in = new BufferedReader(new FileReader("infilename"));
    String str;
    while ((str = in.readLine()) != null) {
        process(str);
    }
    in.close();
}
catch (IOException e) {
}
```

Reading a File into a Byte Array

This example implements a method that reads the entire contents of a file into a byte array.

```
// 将文件的内容以一个字节数组的形式返回
public static byte[] getBytesFromFile(File file) throws IOException {
    InputStream is = new FileInputStream(file);
    //获取文件的大小
    long length = file.length();
    // 不能使用 long 类型创建一个数组
    // 必须使用 int 类型的数组
    // 在转变一个 int 类型之前，必须确保文件的大小不超过 Integer.MAX_VALUE
    if (length > Integer.MAX_VALUE) {
        // 文件太大
    }
    // 创建字节数组用来存储数据
    byte[] bytes = new byte[(int)length];
    // 以字节形式读入
    int offset = 0;
    int numRead = 0;
    while (offset < bytes.length
      && (numRead=is.read(bytes, offset, bytes.length-offset)) >= 0) {
        offset += numRead;
    }
    // 确保所有字节已经读出
    if (offset < bytes.length) {
        throw new IOException("Could not completely read file "+file.getName());
    }
    // 关闭 input 流，并返回字节数
    is.close();
    return bytes;
}
```

Appending to a File

```
try {
    BufferedWriter out = new BufferedWriter(new FileWriter("filename", true));
    out.write("aString");
    out.close();
}
catch (IOException e) {
}
```

Using a Random Access File

```
try {
    File f = new File("filename");
    RandomAccessFile raf = new RandomAccessFile(f, "rw");
    // 读一个字符
    char ch = raf.readChar();
    // 跳到文件末尾
    raf.seek(f.length());
    // 在末尾添加
    raf.writeChars("aString");
    raf.close();
}
catch (IOException e) {
}
```

Strictfp

Floating point hardware calculates with more precision, and with a greater range of values than the Java specification requires. It would be confusing if some platforms gave more precision than others. When you use the strictfp modifier on a method or class, the compiler generates code that adheres strictly to the Java spec for identical results on all platforms. Without strictfp, is it is slightly laxer, but not so lax as to use the guard bits in the Pentium to give 80 bits of precision. With Jet, there is an option to use the full 80 bits in calculation.

> 精确性考虑
> 机器浮点运算比 Java 规范要求的精度更高、计算值的范围更大。不同平台对浮点运算精度要求不一致。当在一个方法或类上使用 strictfp 时，浮点运算在不同平台的计算机上将会得到相同的结果。

12.3.4 The Basic Input Output

Consists of Following Basic Classes:
Stream Classes
Processing External Files
Data Streams
Print Streams
Buffered Streams
Use JFileChooser
Text Input and Output on the Console
Object Streams

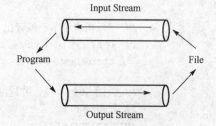

Figure 12.1 Stream's Work

Random Access Files
Parsing Text Files
Input Stream and Output Stream's work in program is shown in figure 12.1.
Basic classes of stream is shown in figure 12.2.
A stream is an abstraction of the continuous one-way flow of data.

流是指某一个方向持续数据流动的抽象。

(1) Stream Classes

The stream classes can be categorized into two types: byte streams and character streams.

The InputStream/OutputStream class is the root of all byte stream classes, and the Reader/Writer class is the root of all character stream classes. The subclasses of InputStream/OutputStream are analogous to the subclasses of Reader/Writer.

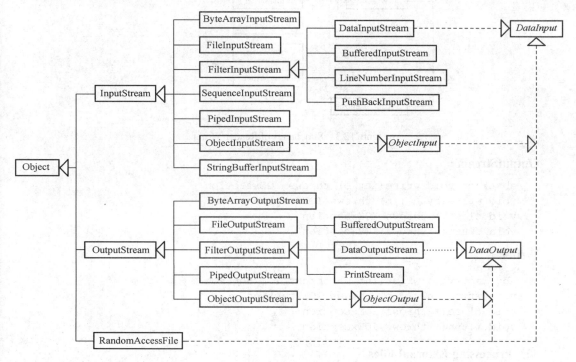

Figure 12.2　Stream basic classes

The subclasses of Reader/Writer is shown in figure 12.3.

InputStream
```
abstract int read() throws IOException
int read(byte[] b) throws IOException
void close() throws IOException
int available() throws IOException
long skip(long n) throws IOException
```

Reader

The Reader class is similar to the InputStream class. The methods in Reader are subject to character interpretation.

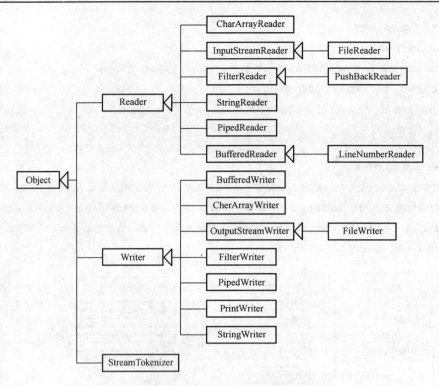

Figure 12.3 Subclasses of Reader/Writer

OutputStream

```
abstract void write(int b) throws IOException
void write(byte[] b) throws IOException
void close() throws IOException
void flush() throws IOException
```

Writer

```
abstract void write(int b) throws IOException
void write(char[] b) throws IOException
void close() throws IOException
void flush() throws IOException
```

(2) **Processing External Files**

You must use file streams to read from or write to a disk file. You can use FileInputStream or FileOutputStream for byte streams, and you can use FileReader or FileWriter for character streams.

File I/O Stream Constructors

Constructing instances of FileInputStream, FileOutputStream, FileReader, and FileWriter from file names.

(3) **Data Streams**

The data streams (DataInputStream and DataOutputStream) read and write Java primitive types in a machine-independent fashion, which enables you to write a data file in one machine and read it on another machine that has a different operating system or file structure.

DataInputStream Methods

```
int readByte() throws IOException
```

```
int readShort() throws IOException
int readInt() throws IOException
int readLong() throws IOException
float readFloat() throws IOException
double readDouble() throws IOException
char readChar() throws IOException
boolean readBoolean() throwsIOException
String readUTF() throws IOException
```

DataOutputStream Methods

```
void writeByte(byte b) throws IOException
void writeShort(short s) throws IOException
void writeInt(int i) throws IOException
void writeLong(long l) throws IOException
void writeFloat(float f) throws IOException
void writeDouble(double d) throws IOException
void writeChar(char c) throws IOException
void writeBoolean(boolean b) throws IOException
void writeBytes(String l) throws IOException
void writeChars(String l) throws IOException
void writeUTF(String l) throws IOException
```

Data I/O Stream Constructors

```
DataInputStream in = new DataInputStream(inputstream);
DataInputStream infile = new DataInputStream(new FileInputStream("in.dat"));
Creates a data input stream for file in.dat.
DataOutputStream out = new DataOutputStream(outputstream);
DataOutputStream outfile = new DataOutputStream(new FileOutputStream("out.dat"));
Creates a data output stream for file out.dat.
```

(4) Print Streams

The data output stream outputs a binary representation of data, so you cannot view its contents as text. In Java, you can use print streams to output data into files. These files can be viewed as text.

The PrintStream and PrintWriter classes provide this functionality.

PrintWriter Constructors

```
PrintWriter(Writer out)
PrintWriter(Writer out, boolean autoFlush)
PrintWriter(OutputStream out)
PrintWriter(OutputStream out, boolean autoFlush)
```

PrintWriter Methods

```
void print(Object o)
void print(String s)
void println(String s)
void print(char c)
void print(char[] cArray)
void print(int i)
void print(long l)
void print(float f)
void print(double d)
void print(boolean b)
```

(5) Buffered Streams

Java introduces buffered streams that speed up input and output by reducing the number of reads and writes. In the case of input, a bunch of data is read all at once instead of one byte at a time. In the case of output, data are first cached into a buffer, then written all together to the file.

Using buffered streams is highly recommended.

Buffered Stream Constructors

```
BufferedInputStream (InputStream in)
BufferedInputStream (InputStream in, int bufferSize)
BufferedOutputStream (OutputStream in)
BufferedOutputStream (OutputStream in, int bufferSize)
BufferedReader(Reader in)
BufferedReader(Reader in, int bufferSize)
BufferedWriter(Writer out)
BufferedWriter(Writer out, int bufferSize)
```

Displaying a File in a Text Area

Objective: View a file in a text area. The user enters a filename in a text field and clicks the View button; the file is then displayed in a text area.

(6) Using JFileChooser

Objective: Create a simple notepad using JFileChooser to open and save files. The notepad enables the user to open an existing file, edit the file, and save the note into the current file or to a specified file. You can display and edit the file in a text area.

> 目的：通过 JFileChooser 来打开和保存文件创建一个简单的记事本。这个记事本能让用户打开一个已有的文件，编辑文件并保存记录到当前文件或特定文件。

(7) Text Input and Output on the Consoles

There are two types of interactive I/O. One involves simple input from the keyboard and simple output in a pure text form. The other involves input from various input devices and output to a graphical environment on frames and applets. The former is referred to as text interactive I/O, and the latter is known as graphical interactive I/O.

> 控制台下的文本输入和输出：
> 有两种交互 I/O：一种是指通过键盘的简单输入和通过纯字符界面的简单输出；另一种是包括不同输入设备输入并输出到框架或 Java 小应用程序的图形化环境。前者是文本交互 I/O，后者则称为图形化交互 I/O。

Console Output/Input

To perform console output, you can use any of the methods for PrintStream in System.out. However, keyboard input is not directly supported in Java. In order to get input from the keyboard, you first use the following statements to read a string from the keyboard.

> 为执行控制台输出，可使用 System.out 包中的 PrintStream 类的任何方法。然而，Java 中并不直接支持键盘输入，为从键盘获取输入数据，应先采用下面讨论的语句实现键盘读取字符串。

(8) Object Streams

Object streams enable you to perform input and output at the object level. To enable an object to be read or write, the object's defining class has to implement the java.io.Serializable interface or the java.io.Externalizable interface.

> 对象流能在对象级上执行输入和输出。为使对象能被读或写，对象定义类必须实现 java.io.Serializable 接口或 java.io.Externalizable 接口。

The Serializable Interface

The Serializable interface is a marker interface. It has no methods, so you don't need to add additional code in your class that implements Serializable. The Serializable interface Relations is shown in figure 12.4.

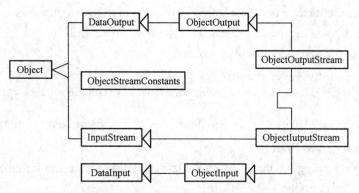

Figure 12.4　The Serializable interface Relations

Implementing this interface enables the Java serialization mechanism to automate the process of storing the objects and arrays.

The Object Streams

You need to use the ObjectOutputStream class for storing objects and the ObjectInputStream class for restoring objects.

These two classes are built upon several other classes.

Testing Object Streams

Objective: Stores objects of MessagePanel and Date, and Restores these objects.

(9) Random Access Files（随机访问文件）

Java provides the RandomAccessFile class to allow a file to be read and updated at the same time. The RandomAccessFile class extends Object and implements DataInput and DataOutput interfaces.

> Java 提供 RandomAccessFile 类允许文件能同时被读或更新。RandomAccessFile 类扩展了 Object 类，并实现了 DataInput 和 DataOutput 接口。

RandomAccessFile Methods

Many methods in RandomAccessFile are the same as those in DataInputStream and DataOutputStream. For example, readInt( ), readLong( ), writeDouble( ), readLine( ), writeInt( ), and write-Long( ) can be used in data input stream or data output stream as well as in RandomAccessFile streams.

RandomAccessFile Methods, cont.

```
void seek(long pos) throws IOException;
```
sets the offset from the beginning of the RandomAccessFile stream to where the next read or write occurs.

```
long getFilePointer() IOException;
```
returns the current offset, in bytes, from the beginning of the file to where the next read or write occurs.

RandomAccessFile Methods, cont.

```
long length()IOException
```
returns the length of the file.

```
final void writeChar(int v)   throws IOException
```
writes a character to the file as a two-byte Unicode, with the high byte written first.

```
final void writeChars(String s)   throws IOException
```
writes a string to the file as a sequence of characters.

```
RandomAccessFile Constructor
RandomAccessFile raf =new RandomAccessFile("test.dat", "rw"); //allows read and write
RandomAccessFile raf =new RandomAccessFile("test.dat", "r"); //read only
```

Using Random Access Files

Objective: Create a program that registers students and displays student information.

Parsing Text Files (Optional)

The StreamTokenizer class lets you take an input stream and parse it into words, which are known as tokens. The tokens are read one at a time. The following is the StreamTokenizer constructor:

```
StreamTokenizer st=StreamTokenizer(Reader is)
```

StreamTokenizer Constants:

TT_WORD: The token is a word.

TT_NUMBER: The token is a number.

TT_EOL: The end of the line has been read.

TT_EOF: The end of the file has been read.

StreamTokenizer Variables

int ttype: Contains the current token type, which matches one of the constants listed on the preceding slide.

double nval: Contains the value of the current token if that token is a number.

String sval: Contains a string that gives the characters of the current token if that token is a word.

StreamTokenizer Methods

```
public int nextToken() throws IOException
```
Parses the next token from the input stream of this StreamTokenizer.

The type of the next token is returned in the ttype field. If ttype == TT_WORD, the token is stored in sval; if ttype == TT_NUMBER, the token is stored in nval.

12.4 Sample Examples

Example 12.1

```java
//简单的 Java 读程序
import java.io.*;
class Input
{
    public static void main(String args[]) throws IOException
    {
        int a;
        String s;
        BufferedReader bf = new BufferedReader(new InputStreamReader(System.in));
        s = bf.readLine();
        a = Integer.parseInt(s);
        System.out.println(a);
    }
}
```

Example 12.2

```java
//简单的字符串转浮点型程序
import java.io.*;
class xhc
{
    public static void main(String args[]) throws IOException
    {
        String s;
        double d;
        d = Double.valueOf(s).doubleValue();
        System.out.println(d);
    }
}
```

Example 12.3

```java
//从键盘读入一个字符串,并转变为对应的数据类型输出,主程序中输出 Double 类型的数据
import java.io.*;
class Input3
{
    public static void main(String args[]) throws IOException
    {
        double value;
        value = getDouble();
        System.out.println(value);
    }
    public static String getString() throws IOException
```

```
        {
            BufferedReader bf = new BufferedReader(new InputStreamReader
(System.in));
            return bf.readLine();
        }
        public static int getInt() throws IOException
        {
            String s = getString();
            return Integer.parseInt(s);
        }
        public static char getChar() throws IOException
        {
            String s = getString();
            return s.charAt(0);
        }
        public static double getDouble() throws IOException
        {
            String s = getString();
            return Double.parseDouble(s);
        }
        public static long getLong() throws IOException
        {
            String s = getString();
            return Long.parseLong(s);
        }
    }
```

Example 12.4

```
//通过字符流类实现对文件的读和写
import java.io.*;
class fileDemo4
{
    public static void main(String args[]) throws IOException
    {
        BufferedReader br1 = new BufferedReader(new InputStreamReader
(System.in));
        System.out.print("Enter the file name:");
        String fname = br1.readLine();
        FileWriter fo = new FileWriter(fname);
        BufferedReader br2 = new BufferedReader(new InputStreamReader
(System.in));
        String s;
        System.out.print("Enter the content of the file:");
        System.out.flush();
        s = br2.readLine();
        for(int i=0; i<s.length(); ++i)
            fo.write(s.charAt(i));
        fo.close();
        FileReader fi = new FileReader(fname);
        int read;
```

```
            System.out.print("\nIn the " + fname + " file: ");
            while((read=fi.read()) != -1)
                System.out.print((char)read);
            fi.close();
        }
    }
```

Example 12.5

```
//从键盘读入一些特定类型的数据
import java.io.*;
import java.util.*;
public class input_reader {
    public static void main(String args[]) throws IOException {
        Scanner kb = new Scanner(System.in);
        System.out.print("Enter an integer: ");
        int a = kb.nextInt();      //读入一个整数
        System.out.println(a);
        System.out.print("Enter a double: ");
        double d = kb.nextDouble(); //读入一个Double型数据
        System.out.println(d);
        System.out.print("Enter a word: ");
        String s = kb.next();      //读入一个字符串
        System.out.println(s);
    }
}
```

12.5 Exercise for you

1. Create a text file ECIT.bat and write your name in it using I/O.
2. How you can make a file input and file output stream together? Give one example.
3. Give examples of buffer reader and buffer write.

Chapter 13   String Handling

In Java string is a sequence of characters. Here the difference is, Java considers strings as object of type string, while other languages consider string as character array. Implementing strings as built in objects allows Java to provide full features that makes string handling convenient. For example, Java has methods like comparing two strings, searching for a sub string, concatenating two strings and changing the case of letters within a string. Also, string objects can be constructed through a number of ways, making it easy to obtain a string when needed.

> 在 Java 中，字符串是指一组字符序列，Java 将字符串作为字符串类型对象看待，将字符串作为内置对象使得 Java 能提供全面的功能，方便完成字符串的相关操作。

13.1   The String Class

The String Class represents character strings. All string literals in Java programs, such as "abc", are considered as instances of this class.

Strings are constant, their values cannot be changed after they are created. String buffers support mutable strings. For String objects are immutable, they can be shared.

> 字符串是常量，它们的值在类创建后就不能被更改。字符串缓冲支持可变字符串。由于字符串对象是不变的，因此它们能被共享。

For example:
```
String str = "abc";
```
is equivalent to
```
char data[] = {'a', 'b', 'c'};
String str = new String(data);
```
Here are some more examples of how strings can be used:
```
System.out.println("abc");
String cde = "cde";
System.out.println("abc" + cde);
String c = "abc".substring(2,3);
String d = cde.substring(1, 2);
```

The class String includes methods for examining individual characters of the sequence and comparing strings, searching strings, extracting substrings, and creating a copy of a string with all characters translated to uppercase or to lowercase. Case mapping relies heavily on the information provided by the Unicode Consortium's Unicode 3.0 specification. The specification's UnicodeData.txt and SpecialCasing.txt files are used extensively to provide case mapping.

> 字符串类中包括了检测字符序列中特定字符的方法，包括比较字符串、查询字符串、提取子字符串，以及创建一个完成字符大小写转换后的原字符串副本。字符映射很大程度依赖于统一字符编码组织制定的 Unicode 3.0 编码规范，并由该规范中的 UnicodeData.txt 和 SpecialCasing.txt 这两个文件提供映射支持。

The Java language provides special support for the string concatenation operator (+), and conversion of other objects to strings. String conversions are implemented through the method toString, defined by Object and inherited by all classes in Java. For additional information on string concatenation and conversion, see Gosling, Joy, and Steele, The Java Language Specification. Unless noted, passing a null argument to a constructor or method in this class will cause a NullPointerException to be thrown.

> Java 语言为字符串连接操作(+)和字符串转换操作提供特殊支持。字符串转换通过定义在 Object 对象中并被其他 Java 类继承的 toString 方法执行。传递一个空参数到一个字符类的构造方法或成员方法将引发一个 NullPointExcetion 的异常。

13.2 Strings in Java

Introducing strings in the Java programming language, focusing on a selection of useful operations.

Important Classes
- Java.lang.String
- Java.lang.StringBuffer

Contents:
- String Basics
- Comparing Strings
- Other Operations
- StringBuffer Objects

13.2.1 String Basics

In the terminology of computer science, a string is a sequence of simple characters. (A character is, most frequently, something that you can type on the keyboard, such as a letter, number, or piece of punctuation.) Almost every modern programming language permits programmers to use variables of the string data type. Languages typically permit a variety of operations on strings, including input, output, and to extract individual characters.

In Java, strings are objects that belong to class Java.lang.String. Because that is such a common class, Java permits you to skip the package name, even without a corresponding import statement.

> 在 Java 中，字符串属于 Java.lang.String 类对象。鉴于它是一个公共类，Java 允许省略类库名称，甚至可不包括相应的导入语句。

For example:
```
String username;
```
is equivalent to
```
Java.lang.String username;
```
String constants are represented by placing the sequence of characters between a pair of double-quotation marks.

For example:
```
username = "Michael";
```
Java strings can contain essentially any character you can type or otherwise generate. They can contain numbers, spaces, punctuation, and so on and so forth.

For example:
```
String address;
address = "Apt. 302, Foreign Appartment, ECIT.";
```

13.2.2 Creating a String

Internally, a String encapsulates an array of characters. The reason for this encapsulation is that the String object ensures that the string will be maintained properly. It also encapsulates a number of useful methods that can be performed on a string.

The most important thing to remember is that a String is immutable, which means that after you assigned your String object a value, it cannot change anymore during its lifetime. If you want to manipulate the characters, consider using the StringBuffer class!

> 一个字符串封装了一组字符，String 对象能确保正确地管理这一组字符，同时还封装一组字符串操作方法。字符串值不能被改变，一旦给字符串赋值，在其生命周期中就不能被更改。如果需要操作字符串中字符，可考虑使用 StringBuffer 类。

There are two ways to create a String object:
- implicitly: when you use a string literal, like "Hello, world!", Java automatically creates a String object for you.

For example:
```
String s = "Hello, world!";
```
You can use all String methods on a string literal, eg.
```
s = "Hello, World!".substring(7);   //给 s 赋 "world!"
```
- explicitly: when you use the new operator to instantiate a String object. For example:
```
String s = new String("Hello, world!");
```
This will call the String class constructor that takes another String as parameter. (remember, "Hello, world!" is represented internally as a String object!).The String object has other constructors to create Strings as shown in the following test program:

> 创建字符串对象有两种方式：
> 隐含方式：当使用任一字符串文字时，像 "hello,world"，Java 会自动创建一个字符串对象；
> 直接方式：通过 new 创建一个新的字符串对象实例。

```java
public class TestProg
{
    public static void main(String args[])
    {
        try {
            String s;
            StringBuffer sb = new StringBuffer("A StringBuffer");
            //创建一个字符数组
```

```
            char ca[] = "A character array!".toCharArray();
            //创建一个默认的编码的字节数组，默认编码是 ISO-Latin-1
            byte ba[] = "A byte array!".getBytes();
            //创建一个字节数组，包含 Unicode 编码
            byte bau[] = "A Unicode byte array!".getBytes("Unicode");
            //创建一个空字符串
            s = new String();
            System.out.println(s);          //输出是空
            //从一个字节数组创建一个字符串
            s = new String(ba);
            System.out.println(s);          //输出 "A byte array!"
            //从一个字节数组索引 2（包含）开始总共 1 字节的字符串
            s = new String(ba, 2, 1);
            System.out.println(s);          //输出 "b"
            //从一个包含 Unicode 的字符子数组中，创建一个字符串
            s = new String(bau, "Unicode");
            System.out.println(s);          //输出 "A Unicode byte array!"
            //从一个包含 Unicode 的字符子数组中，创建一个字符串
            s = new String(bau, 0, 20, "Unicode");
            System.out.println(s);          //输出 "A Unicode"
            //从一个 StringBuffer 创建一个字符串对象
            s = new String(sb);
            System.out.println(s);          //输出 "A StringBuffer"
            //从一个子字符数组创建一个字符串
            s = new String(ca, 2, 4);
            System.out.println(s);          //输出 "char"
            //从一个字符数组创建一个字符串
            s = new String(ca);
            System.out.println(s);          //输出 "A character array!"
        }
        catch(Java.io.UnsupportedEncodingException e) {
            System.out.println(e);
        }
    }
}
```

What is the difference between String a="a" and String a=new("a") ?

I'm assuming you mean "what's the difference between String a = "a"; and String a = new String("a");"

The simple answer is that in the 2nd case an extra, unnecessary String object is created. The reason for that is that even string literals in Java are instances of class String, so what happens in the 2nd case is that a string literal "a" is created and then another String object with a copy of the same value is created. The complicated answer is that there is another difference. All Java string literals are saved by the compiler directly into the class file, the JVM then reads the class file and creates instances of class String for all the string literals in the class file. All these instances are saved in a special internal pool of Strings inside the String class.

What does this mean? It means that if you do:

```
String a = "a";
String b = "a";
```
then both a and b will reference the SAME object (so a==b is true). But if you do this:

```
String a = "a";
```

String b Then a and b will reference different String objects (both strings will have the same value, but they will be 2 strings, not one like in the first case). You can add a String into the internal pool by invoking the intern( ) method on a String. That method creates a new string with the same value (if none exists in the pool already), adds it to the pool and returns = new String("a");

13.2.3 Comparing Strings

Strings are also comparable. That means we can determine whether two strings are the same and whether one string naturally precedes another. (Java typically uses a local configuration file for determining order so that it matches the local language and its ordering rules.) Once you can compare strings, you have the ability to develop more sophisticated operations on collections of strings, such as placing them in order or searching for a particular string.

> 字符串是可比较的，可以判断两个字符串是否相等或一个字符串是否大于另一个字符串。一旦能比较字符串，就能开发更复杂的字符串操作，如排序或特定字符查找等。

Java provides two basic operations for comparing strings:

equals(String other) determines whether this string is the same as another string. Using the method Java.lang.String.equals(Object) will return true if it matches and false if it doesn't.

```
String a = "some text";
 String b = "some text";
if (a.equals(b)){
     System.out.println("String a is equal to String b");
}
else{
     System.out.println("String a is NOT equal to String b");
}
```

> equals(String other)判断这个字符串是否与另一个字符串相等。字符串比较可通过使用 Java.lang.String.equals(Object)方法实现，如果匹配返回 true，否则返回 false。

compareTo(String other) determines the relative order for two strings. It returns zero if they are equal, a negative number returns if the string executing compareTo naturally precedes another, and a positive number returns if the string executing compareTo naturally follows another.

> compareTo(String other)判断两个字符串的相对顺序。如果相等返回零，如果本字符串大于另一个字符串，返回一个负数；如果本字符串小于另一字符串之后，则返回一个正数。

How to determine whether two Strings are equal.

Lots of beginner Java programmers make the mistake of using the equality operator (==) to compare whether two Strings contain the same characters. The == operator just checks whether two object handles point to the same memory location. Sometimes this might work, ie if the Strings that are being compared point to the same space in memory. Other times, even when the contents of the two String are the same, the == operator return false. Instead of == you should use the equals method.

> 很多 Java 编程初学者都错误地使用等于运算符(==)来比较包括相同字符的两个字符串。实际上这种方式只是检查两个对象句柄是否指向相同的内存位置。有时这也可能得出正确结果(如果比较的字符串指向相同的内存位置)。更多的时候，即使两个字符串的内容是相同的，==操作符也返回 false。在 Java 中，正确的两个字符串相等的比较应该使用 equals 方法。

An example should clear this up:

```java
public class TestProg
{
    public static void main(String args[])
    {
        String s1 = "Hello, world!";
        String s2 = s1;                  //s1 的别名
        String s3 = new String(s1);      //创建一个新字符串对象，并使用 s1 初始化
        System.out.println(s1 == s2);           //输出 "true"
        System.out.println(s1.equals(s2));      //输出 "true"
        System.out.println(s1 == s3);           //输出 "false"
        System.out.println(s1.equals(s3));      //输出 "true"
    }
}
```

Conclusion: always use the equals method unless you want to compare whether two objects share the same memory. If you don't care about the case when comparing two Strings, the String class provides the method equalsIgnoreCase.

> 结论：尽量使用 equals 方法，除非想比较两个对象是否共享相同的内存。如果比较两个字符串时不关心大小写，字符串类也提供了 equalsIgnoreCase 方法。

13.2.4 Other Operations

We can do many other things with strings. Java's primary string class, Java.io.String gives us access not just to the sequence of characters as a whole, but also to subsequences and to operations on those subsequences. Interestingly enough, Java takes advantage of the encapsulation for the String class and makes strings immutable: Operations do not change strings, but only return new strings based on the original string executing an operation.

Some of the string operations you may find useful include:

`toLowerCase( )`: Build a new string, similar to the original, but with all letters in the string converted to lower-case (using the local installations definition of lower-case and upper-case to lower-case conversion).

`toUpperCase( )`: Build a new string, similar to the original, but with all letters in the string converted to upper-case.

`trim( )`: Build a new string, similar to the original, but without leading and trailing whitespace. This command is often useful when you're processing unstructured user input.

`substring(startindex)`: Extract the portion of the string that begins at startindex. Note that indices start at 0.

`substring(startindex,endindex)`: Extract the portion of the string that begins at startindex and ends directly before endindex.

`replace(target,replacement)`: Build a new string, similar to the original, but with all substrings that match target replaced by replacement.

`indexOf(target)`: Determine the first index of target within the string. (This method is often useful in conjunction with substring.)

`indexOf(target, startHere)`: A variant that starts looking at the specified index.

`concat(target)`: Build a new string by joining the specified target to the end of the current string.

`length( )`: Determine how many characters are in the string.

- **Determining the length of a String.**

Use the length( ) method from the String class.

```
String s = "Hello, world!";
System.out.println(s.length());   // 输出 13
```

- **Converting a character array to a String.**

The String class has a useful contructor especially for this, as you can see in the following example:

```
public class TestProg
{
    public static void main(String args[])
    {
        char [] array = {'H','e','l','l','o',',',' ','w','o','r','l','d','!' };
        String s = new String(array);
        System.out.println(s);   // 输出 Hello, world!
    }
}
```

- **Determining if a String contains another String or not.**

Use Java.lang.String.indexOf(String) to check if one string contains another. indexOf( ) will return -1 if the string is not found and the starting index of the string if it is found.

```
String a = "Some text";
String b = "text";
if (a.indexOf(b) != -1) {
    System.out.println("String a contains String b, starting at position " + a.indexOf(b));
} else {
    System.out.println("String a does not contain String b");
}
```

- **Getting a particular portion of a String.**

Java's String class provides an instant method called substring( ) for it's String object. The usage is as follows :

```
String str = new String("ABCD");
String newStr = str.substring(0, 3);
```

- **Determine whether parts of two Strings are equal?**

part of the a-String : efg

String "efg" was found in "defghijk", starts at position 1

String "efg" not found in "eghijklm"

```
public class StringCompare {
```

```java
        public static void compare(String tmp, String b) {
            // 在b中查找tmp
            // 使用 indexOf 方法
            // 如果tmp不存在于b中，本方法将返回-1
            int pos;
            if ((pos = b.indexOf(tmp)) != -1) {
                System.out.println("String \"" + tmp + "\" was found in \"" +
                                        b +"\", starts at position " + pos);
            }
            else {
                System.out.println("String \"" + tmp + "\" not found in \"" +
                                                            b + "\"");
            }
        }
        public static void main(String args[]) {
            String a = "abcdefgh";
            String b = "defghijk";
            String c = "eghijklm";
            //要在b中找的efg, 也存在于a中
            System.out.println("part of the a-String : " + a.substring(4,7));
            compare(a.substring(4,7), b);
            compare(a.substring(4,7), c);
        }
    }
```

- **Find the first occurrence of a character or String in another String or not.**

Use the method indexOf which will return the index as an int. If there is more than one occurrence in the target String, indexOf will return the index of the first occurrence.

The function will return -1 if the String you are looking for does not exist within the String. Remember, Java Strings are zero based!

```java
    public class TestProg
    {
        public static void main(String args[])
        {
            String s = "Hello, World!";
            System.out.println(s.indexOf("Hello"));    //输出 0
            System.out.println(s.indexOf("o"));        //输出 4
            System.out.println(s.indexOf('h'));        //输出 -1
        }
    }
```

- **Converting a String to lowercase or uppercase.**

```java
    public class TestProg
    {
        public static void main(String args[])
        {
            String s = "Hello, World!";
            String supper = s.toUpperCase();
            String slower = s.toLowerCase();

            System.out.println(supper);         //输出 HELLO, WORLD!
```

```
        System.out.println(slower);          //输出 hello, world!
    }
}
```

For internationalization purposes, the methods toUpperCase( ) and toLowerCase( ) perform some checking for languages as German and Turkish to do the correct conversion. If you need these functions in time-critical applications, you might want to consider rewriting them without these checkings.

- **Concatenation of two Strings.**

There are two ways to concatenate one string to another. The most common is to use the (rather ambiguous) + operator:

```
String s1 = "Hello, ";
String s2 = "world!";
String s3 = s1 + s2;
System.out.println(s3);           //输出 Hello, world!
```

The String class also provides a method concat that does just the same:

```
String s1 = "Hello, ";
String s2 = "world!";
String s3 = s1.concat(s2);
System.out.println(s3);           //输出 Hello, world!
```

In the previous example, the method concat is invoked on s1. s1 remains unchanged and a completely new String will be created. A String is immutable! Once you assigned it a value, it cannot change anymore during its lifetime. You will always have to create a new String object.

> 在 Java 中，字符串的串联可通过两种方式实现，最常用的是 "+" 运算符；同时字符串类也提供了一个 concat 方法来实现同样的操作。

- **Getting a single character from a String.**

Use the method charAt in the String class. It will return the character at that position or throw the Exception StringIndexOutOfBoundsException if index < 0 or > than the length of the string.

> 从字符串中获取单个字符，可使用 charAt( )方法来实现。它将返回指定位置的字符，当 index<0 或 index 大于字符串长度时，抛出 StringIndexOutOfBoundsException 异常。

Example:

```
System.out.println("Hello, world!".charAt(1));    // 输出 e
```

- **Removing leading and trailing whitespace from a String.**

Use the method trim( ). It will create a new String but removes whitespace. Whitespace is defined as all Unicode characters less than or equal to "\u0020".

```
public class TestProg
{
    public static void main(String args[])
    {
        String s = "  \t Hello, world! ";
        System.out.println(s.trim());                  // 输出 Hello, world!
    }
}
```

- **Converting a primitive to a String.**

There are several ways you can do this. In Java programs, you will often see that a primitive is converted by using the concatenation operator on Strings ("" + primitive). A more performant way is to use the method toString on the Object that represents the primitive or to use the static function valueOf in the String class.

Here's an example:

```java
//本例展示基本数据类型和字符串之间的转换
public class TestProg
{
    public static void main(String args[])
    {
        Short     p_short = 4;
        int       p_int = 10;
        long      p_long = 2349945821628L;
        float     p_float = 3.1415F;
        double    p_double = 0.00000000004D;
        boolean   p_boolean = false;
        char      p_char = 'E';
        char      p_array[] = { 'e', 's', 'u', 's' }
        System.out.println(""+p_short);                        // 4
        System.out.println(Short.toString(p_short));           // 4
        System.out.println(String.valueOf(p_short));           // 4 (转换为int!)
        System.out.println(""+p_int);                          // 10
        System.out.println(Integer.toString(p_int));           // 10
        System.out.println(String.valueOf(p_int));             // 10
        System.out.println(""+p_long);                         // 2349945821628
        System.out.println(Long.toString(p_long));             // 2349945821628
        System.out.println(String.valueOf(p_long));            // 2349945821628
        System.out.println(""+p_float);                        // 3.1415
        System.out.println(Float.toString(p_float));           // 3.1415
        System.out.println(String.valueOf(p_float));           // 3.1415
        System.out.println(""+p_double);                       // 4.0E-11
        System.out.println(Double.toString(p_double));         // 4.0E-11
        System.out.println(String.valueOf(p_double));          // 4.0E-11
        System.out.println(""+p_boolean);                      // false
        System.out.println(new Boolean(p_boolean).toString()); // false
        System.out.println(String.valueOf(p_boolean));         // false
        System.out.println(""+p_char);                         // E
        System.out.println(new Character(p_char).toString());  // E
        System.out.println(String.valueOf(p_char));            // E
        System.out.println(""+p_array);                        // esus
        System.out.println(p_array.toString());                // [C@1a
        System.out.println(String.valueOf(p_array));           // esus
    }
}
```

Notice that the Boolean and Character classes do not have a static method toString(). In that case, you have to instantiate an object and call the toString method on it.

> 注意：Boolean 类和字符类都没有 toString() 静态方法。在这种情况下，必须先初始化一个具有 toString 方法的对象实例，并调用改方法。

To test out the difference in performance of using toString( ) or valueOf( ) over ""+p, I ran the following test:

```java
//本例展示 toString 和 valueOf 的性能
public class TestPerformance
{
    public static void main(String args[])
    {
        double p_double = 0.00000000004D;
        long start, end;
        System.out.println("Converting to String using concatenation");
        start = System.currentTimeMillis();
        for (int i=0; i<100000; i++)
            String s = "" + p_double;
        end = System.currentTimeMillis();
        System.out.println("\tElapsed time: " + (end - start));
        System.out.println("Converting to String using valueOf");
        start = System.currentTimeMillis();
        for (int i=0; i<100000; i++)
            String s = String.valueOf(p_double);
        end = System.currentTimeMillis();
        System.out.println("\tElapsed time: " + (end - start));
        System.out.println("Converting to String using toString");
        start = System.currentTimeMillis();
        for (int i=0; i<100000; i++)
            String s = Double.toString(p_double);
        end = System.currentTimeMillis();
        System.out.println("\tElapsed time: " + (end - start));
    }
}
```

The output (times may vary!) shows a performance gain of almost 20%!

Converting to String using concatenation
   Elapsed time: 5270
Converting to String using valueOf
   Elapsed time: 4230
Converting to String using toString
   Elapsed time: 4230

13.2.5 StringBuffer Objects

As suggested at the end of the previous section, String objects are immutable in Java. However, there are also times that you want to change your strings, rather than simply building new strings each time. For such purposes, Java provides the Java.lang.StringBuffer class. (Again, you need not import this class to refer to objects simply as StringBuffer.)

You can create a new StringBuffer in one of two ways, using a zero-parameter constructor, which builds an empty buffer, or using a string as the parameter, which builds a string buffer based on the given string.

> 前一部分已介绍，Java 中的字符串对象是不变的。然而，有时可能想改变字符串，而不是简单地每次创建一个新的字符串。为实现这个目的，Java 提供了 Java.lang.StringBuffer 类。创建一个新的 StringBuffer 可采用下列两种方式之一，一种是使用一个零参数构建函数，它将构建一个空的缓冲区；另一种方式是使用带参数的字符串，以创建一个基于给定字符串的字符缓冲空间。

For example:
```
StringBuffer initiallyEmpty;
StringBuffer sillyWord;
initiallyEmpty = new StringBuffer();
sillyWord = new StringBuffer("abracadabra");
```

The StringBuffer class provides many of the same methods as the String class, including indexOf, substring, and length. However, the StringBuffer class also provides many methods for changing the underlying string.

`append(String str)`: Appends str to the end of the buffer.

`insert(int offset, String str)`: Inserts str starting at position offset. Everything that was at that position is moved to the right.

`delete(int start, int end)`: Delete the characters between start and end-1.

You can also use the toString method to recover a string from a StringBuffer.

```
append(String str)
insert(int offset,String str)
delete(int start,int end)
```

The String class versus the StringBuffer class.

Irritatingly, Java has two classes with similar contents, but different functionality. These are the String and StringBuffer classes. Having these two separate classes probably allows Java to be much more efficient in its use of memory, but it's a minor pain for the developer.

The two classes differ in that instances of the String class cannot be modified in place, while instances of the StringBuffer class can be modified in place. Since instances of the String class cannot be modified in place, calling a method such as toUpper or toLower on a String typically causes another String to be returned. This will force the use of lots of temporary variables. If you're building a string, the best way to do it is to use a StringBuffer and the StringBuffer.append() method.

> String 和 StringBuffer 两个类不同之处在于字符串类的实例不能实地修改，而字符缓冲类的实例则可以实地修改。既然字符串类的实例不能实地修改，则字符串类的一些方法如 toUpper 或 toLower 方法都将导致产生另一新的字符。这将迫使用户使用大量临时变量。如果要构建一个字符串，最好的方式是使用 StringBuffer 类及 StringBuffer.append()方法。

The difference between the StringBuffer's capacity( ) and length( )

The StringBuffer has internally an array of characters in which it keeps the values of the characters. Each time you try to add more characters than this array's length, a new, bigger array must be created and the characters must be copied into it. StringBuffer.length( ) returns the length of string you have stored into the StringBuffer, the value you would normally expect. StringBuffer.capacity( ) returns the size of that internal array, which means how many characters can that StringBuffer keep without having to allocate a new, bigger array.

> StringBuffer 类中 capacity( )与 length( )方法的区别: StringBuffer 类有一个用来存储输入字符数据的内部字符数组。StringBuffer.length( )返回实际字符串的长度, 而 StringBuffer.capacity( )返回内存数组的最大值。

13.2.6 String Analyzing

- **Analyzing a string token-by-token**

Tokenization in Java consists of two separate issues: one is on a character-by-character basis, and the other is done on the basis of a separator character. The former is well-supported in the Java platform, by way of the StringTokenizer class. The latter must be approached algorithmically.

> Java 中的标记有两种独立的方式: 一种基于逐字符的标记分析, 另一种基于单词的标记分析。前者在 Java 平台中通过 StringTokenizer 类支持非常好, 后者必须通过算法实现。

- **Analyzing a string character-by-character**

You will use:
  a String with your input in it
  a char to hold individual chars
  a for-loop
  the String.charAt( ) method
  the String.length( ) method
  the String.indexOf( ) method

The method String.charAt( ) returns the character at an indexed position in the input string.

> String.charAt( )方法返回输入字符串中处于索引位置的字符。
> String.length( )方法返回字符的长度。
> String.indexOf( )方法返回特定字符在字符串中的位置, 如字符串中没有特定字符, 将返回-1。

For example, the following code fragment analyzes an input word character-by-character and prints out a message if the input word contains a coronal consonant:

```java
// 下面两行代码显示了带有字符串常量的字符串构造函数的用法
String input = new String ("mita");
String coronals = new String("sztdSZ");
int index;
char tokenizedInput;
// String 的 length()方法返回一个字符串的长度
// 长度减1是因为索引是从0开始的
for (index = 0; index < input.length() - 1; index++) {
    tokenizedInput = input.charAt(index);
    // 如果没有所包含的字符, indexOf 将返回-1
    // 如果没有返回-1, 那就表示包含字符
    if (coronals.indexOf(tokenizedInput) != -1){
        System.out.print("The word <");
        System.out.print(input);
        System.out.print("contains the coronal consonant <);
        System.out.print(tokenizedInput);
```

```
        System.out.println(">.");
    }
}
```

This produces the output The word <mita> contains the coronal consonant <t>.

- **Analyzing a string word-by-word**

You will use:

　　the StringTokenizer class

　　the StringTokenizer.hasMoreTokens( ) method

　　the StringTokenizer.nextToken( ) method

　　a while-loop

> StringTokenizer 类的主要用途是将字符串以定界符为界，分析为一个个的 token (可理解为单词)，定界符可以自己指定。

```
// 创建一个字符串对象
String input = new String("im an ant?am i an anty");
// 创建一个 tokenizer 对象。注意, 传递给它的字符串是用来切分的字符串
StringTokenizer tokenizer = new StringTokenizer(input);
// StringTokenizer.hasMoreTokens() 返回 true 表示还有一些未提供给你的数据
// there's more data in it that hasn't yet been given to you
while (tokenizer.hasMoreTokens()) {
    // StringTokenizer.nextToken() 返回 StringTokenizer 包含的下一个 token
    // 因此, 第一次调用它的时候, 将返回第一个数据。
    String currentToken = tokenizer.nextToken();
    // ...现在就可以对想要处理的 token 做任意处理!
    checkForCoronalConsonants(currentToken);
}
```

13.3　Sample Examples

Example 13.1

This short example shows how comparisons are made amongst string:

```
//本例展示了字符串之间是如何比较的
class strCmp
{
    public static void main (String args[])
    {
    String str = "Hello";
    String str2 = "Java";
    System.out.println (str.equals(str2));      // false
    System.out.println (str.compareTo(str2));
    // 一个负数, i.e. str 比 str2 小
    System.out.println (str.charAt(0));    // H, i.e. char is position 0
    System.out.println (str.length() + str2.length());  // 5 + 4 = 9
    }
}
```

Example 13.2

```java
//Alter Strings
//本例更改字符串
class altStr
{
    public static void main (String args[])
    {
    String str = "Hello";
    String str2 = "Java";
    str = str.toUpperCase();
    str2 = str2.toLowerCase();
    System.out.println (str);              // HELLO
    System.out.println (str2);             // Java
    str = str.concat(str2);                // str 现在等于 "HELLO Java"
    System.out.println (str);
    str = str.trim();                      // str 现在等于 "HELLOJava"
    System.out.println (str);
    str = str.substring (5,str.length());  // str = "Java"
    System.out.println (str);
    str = str.replace ('a', 'i');          // str = "jivi"
    System.out.println (str);
    str = String.valueOf (3.141);          // str = "3.141"
    System.out.println (str);
    }
}
```

Example 13.3

Use of IndexOf( ) and LastIndexOf( ) methods.

```java
//使用 IndexOf()和 LastIndexOf()方法
class indexTest
{
    public static void main (String args[])
    {
    String str = "This string will be searched";
    String find = "will";
    System.out.println (str.indexOf ('s'));       // 查找字符，没有偏移
    System.out.println (str.lastIndexOf ('s', 4)); // 查找 s, 偏移 9
    System.out.println(str.indexOf(find));        //查找字符串 find, 没有偏移
    System.out.println(str.lastIndexOf(find));
    //在 str 中查找字符 find, 没有偏移
    }
}
```

Example 13.4

```java
//String Buffer capacity
//字符串缓冲区的容量
class altBuf
{
    public static void main (String args[])
    {
        StringBuffer strbuf = new StringBuffer ("Hello");
```

```
                        System.out.println (strbuf);    // Hello
                        strbuf.append ("_world");
                        System.out.println (strbuf);    // Hello_World
                        strbuf.setCharAt (5, ' ');
                        System.out.println (strbuf);    // Hello World
                        strbuf.insert (6, "Java ");
                        System.out.println (strbuf);    // Hello Java World
                }
        }
```

Example 13.5

```
//检查字符
class chChar
{
        public static void main (String args[])
        {
        System.out.println ("'A' is lower case");
        System.out.println (Character.isLowerCase('A'));
        System.out.println ("'A' is upper case");
        System.out.println (Character.isUpperCase('A'));
        System.out.println ("'1' is a digit");
        System.out.println (Character.isDigit('1'));
        System.out.println ("' ' is a space");
        System.out.println (Character.isSpace(' '));
        }
}
```

13.4 Exercise for you

1. Give an example of getByte( ), a toCharArray( ).
2. What is trim( ). Give an example.
3. Give an example to convert data using valueOf( ).

Chapter 14　Networking

Here we discuss the Java.net package, which provides support of networking. This package includes low level API and high level API, low level API manage objects like address, socket and interface. The socket is communication method building among machines. High level API manage objects like URI and connections.

> Java.net 包为实现网络应用程序提供支持。Java.net 包包括低级别 API 和高级别 API，其中低级别 API 处理地址、套接字和接口等对象。套接字是在网络上建立机器之间的通信链接的方法。高级 API 处理 URI 和网络连接。

14.1　Computer Network Basics

An internet work is a collection of individual networks, connected by intermediate networking devices, that functions as a single large network. The networking devices are the vital tools for communication. Whenever you have a set of computers or networking devices to be connected, you make the connections, depending on the physical layout and your requirements Depending on the physical layout or topology of the network, there are three types of networks.

- **LAN**（局域网）

LAN stands for Local Area Network. These networks evolved around the PC revolution. LANs enabled multiple users in a relatively small geographical area to exchange files and messages, as well as access shared resources such as file servers.

- **WAN**（广域网）

WAN stands for Wide Area Network. The interconnection of various LAN's through telephone network, which unites geographically distributed users is achieved through WAN. In short when we log on to the internet, we become a part of a WAN.

- **MAN**（城域网）

MAN stands for Metropolitan Area Network. It is usually the interconnection between various LAN's in a particular geographical area like a metropolitan city like Bombay, Hence the name.

Internetworking evolved as a solution to three key problems: isolated LANs, duplication of resources, and a lack of network management. Isolated LANs made electronic communication between different offices or departments impossible. Duplication of resources meant that the same hardware and software had to be supplied to each office or department, as did a separate support staff. This lack of network management meant that no centralized method of managing and troubleshooting networks existed.

> 互联网存在三个关键问题：孤立的 LAN，重复资源，以及缺乏网络管理。孤立的 LAN 使各不同办公室和部门之间不能进行电子通信。重复资源意味着不得不将同样的硬盘和软件提供给每个办公室和部门。缺乏网络管理意味着各网络出现的问题不能集中解决。

Chapter 14　Networking

● **Open Systems Interconnection Model (OSI)**

The OSI model is the basic model describing the data movement through a network. The Open Systems Interconnection (OSI) reference model describes how information from a software application in one computer moves through a network medium to a software application in another computer. The OSI reference model is a conceptual model composed of seven layers, each specifying particular network functions. The OSI model divides the tasks involved with moving information between networked computers into seven smaller, more manageable task groups. A task or group of tasks is then assigned to each of the seven OSI layers.

> OSI 模型是描述数据在网络中传输的基本模型，它描述了一台计算机的数据从软件应用层通过网络媒介传输到另一台计算机的软件应用层。OSI 划分为 7 层不同的模型，每层分管不同的任务。

The following list details the seven layers of the Open System Interconnection (OSI) reference model:

Layer 7—Application layer (A)
Layer 6—Presentation layer (P)
Layer 5—Session layer (S)
Layer 4—Transport layer (T)
Layer 3—Network layer (N)
Layer 2—Data Link layer (D)
Layer 1—Physical layer (P)

● **Protocol**

The OSI model provides a conceptual framework for communication between computers, but the model itself is not a method of communication. Actual communication is made possible by using communication protocols. In the context of data networking, a protocol is a formal set of rules and conventions that governs how computers exchange information over a network medium. A protocol implements the functions of one or more of the OSI layers. A wide variety of communication protocols exist, but all tend to fall into one of the following groups: LAN protocols, WAN protocols, network protocols, and routing protocols. LAN protocols operate at the network and data link layers of the OSI model and define communication over the various LAN media. WAN protocols operate at the lowest three layers of the OSI model and define communication over the various wide-area media. Routing protocols are network-layer protocols that are responsible for path determination and traffic switching. Finally, network protocols are the various upper-layer protocols that exist in a given protocol suite.

> OSI 提供了计算机之间进行通信时的一种概念框架，但它本身并不是通信的一种方法，仅是一种概念。实际上通信是通过协议来实现的。在网络中，协议是一组规则和约定。它约束了网络中的各计算机怎样通过网络媒介进行通信。协议用来实现 OSI 模型中的一个或多个功能。虽有大量的通信协议存在，但它们大致就是以下几部分：LAN 协议，WAN 协议，网络协议，路由协议。LAN 协议工作在 OSI 的数据链路层，WAN 协议工作在 OSI 的最低三层，路由协议工作在网络层，它主要用于数据传输路径的选择，网络协议工作在 OSI 的最上层。

public class **Socket** extends Java.lang.Object

This class models a client site socket. A socket is a TCP/IP endpoint for network communications conceptually similar to a file handle.

This class does not actually do any work. Instead, it redirects all of its calls to a socket implementation object which implements the SocketImpl interface. The implementation class is instantiated by factory class that implements the SocketImplFactory interface. A default factory is provided, however the factory may be set by a call to the setSocketImplFactory method. Note that this may only be done once per virtual machine. If a subsequent attempt is made to set the factory, a SocketException will be thrown.

> 套接字是两台机器之间的通信端点。Socket 类实际上不做任何事情，它将调用重新映射到实现了 SocketImpl 接口的套接字对象。应用程序通过更改创建套接字实现的套接字工厂可以配置它自身，以创建适合本地防火墙的套接字。

Computers operating over the Internet use protocols at several different levels:

    Application Layer   - http
    Network Layer   - IP
    Transport Layer   - TCP, UDP
    Data Link Layer   - Ethernet

Java Provides a Java.net package that makes it easy to use TCP/IP and UDP/IP. When two applications want to communicate with each other reliably, they should use TCP/IP. Java provides the URL, URLConnection, Socket, ServerSocket classes that allow you to use TCP. Unreliable communication with lower overheads can use UDP/IP. Java provides DatagramPacket, DatagramSocket, MulticastSocket classes for using UDP. Many security firewalls do not allow UDP.

> Java 提供了 Java.net 包，可以很方便地使用 TCP/IP 和 UDP/IP。如两个应用程序需进行可靠通信时，就可以使用 TCP/IP，Java 提供了 URL、URLConnection、Socket、ServerSocket 类。在不需要可靠通信情况下可使用 UDP/IP，Java 提供 DatagramPacket、DatagramSocket、MulticastSocket 类来使用 UDP。但许多安全防火墙不允许使用 UDP。

- **Ports**

Messages transmitted over the Internet carry a 32 bits IP address (which will become 64 bits with IP V6). The IP address identifies the destination computer. The message also carries a 16-bit port address. The port address identifies a specific application within the computer.

When you open a TCP/IP connection, a port is allocated at each end for exclusive use of that connection. Ports 0~1023 are reserved for specific applications, such as email, http and ftp. They are called well-known ports. You should not use them. Ports 1024~65535 are free for all.

> 端口地址标识计算机内的一个指定的应用程序。当打开 TCP/IP 连接时，每个连接会被分配一个端口号。端口 0~1023 是被保留给一些特定的应用程序的，如 email、http 和 ftp。它们不能被用户调用，而端口 1024~65535 是可以为用户使用的。

- **Universal Resource Locators(URL)**

A Universal Resource Locator is an address of a resource on the Internet. The Internet address is called a URL address. The Java object containing a URL address is called a URL object . A URL address contains :

A protocol identifier such as http followed by ://

A resource name that contains

- host name – the name of a machine on the Internet
- file pathname – the pathname of a file on the machine
- port number – to connect to a specific application on that machine
- reference   - to define a specific location within a file

For example: http://www.ist.unomaha.edu/faculty/pdasgupta/index.html
        protocol   machine   file pathname

14.2 URL Objects in Java

An absolute URL can be created from a String:
```
URL games = new URL("http://www.games.com/ ");
```
Once you have a URL, you can create others from it.

Example:
```
URL newgame = new URL(games, "newgame.html ");
```
concatenates the existing URL address and the String argument to yield:

    http://www.games.com/newgame.html

Many html files contain internal labeled reference points called anchors.

Example:
```
URL newgameBottom = new URL(games, "#BOTTOM ");
```
gives you http://www.games.com/newgame.html#BOTTOM.

14.2.1 Creating URL Objects

There is also a URL constructor that takes three strings: a protocol, a machine name, and a path name.

> URL 构造方法包括三个字段：协议，主机名，文件路径。

Example:
```
URL newgame = new URL(" http", "www.games.com", "/newgame.html");
```
yields

    http://www.games.com/newgame.html

A fourth constructor adds a port number.

Example:
```
URL thegame = new URL("http","www.games.com",80,"/newgame.html");
```
yields

    http://www.games.com:80/newgame.html

URL objects are final(write-once) you cannot change them

14.2.2 Query Methods on URL Objects

- getProtocol

- getHost
- getPort (-1 if there is no port)
- getFile
- getRef

You can also get the URL address back again as a String by using the toString method.

Example:

```
String s = myURL.toString();
```

It is likely that you will generate a malformed or incorrect URL address occasionally, and your program must be able to handle it. Trying to create a malformed URL object or trying to connect to an incorrect URL will cause an exception to be raised. The following example containing a try/catch pair shows how to instantiate URL objects:

> 试图创建一个不符合规范的 URL 对象去连接另一个不正确的 URL 会导致异常,需要对其进行异常捕获。

```
try{
    URL myURL = new URL("http://www.games.com");
}
catch(MalformedURLException e){
    …        //这里编写异常处理代码        }
```

The next example extends the previous one to handle the failure of an attempt to connect to a URL, which is quite likely

```
try{
    URL yahoo = new URL("http://www.yahoo.com");
    URLConnection yc = yahoo.openConnection();
}
catch(MalformedURLException e) {
    … //处理非法 URL
}
catch(IOException e){
…//这里处理没有建立连接的情况
}
```

14.2.3　Reading from a URL Connection

Once you have a URL connection, you can read from it and write to it using, for example, the BufferedReader class:

```
//本例演示下载一个 URL 指向的网页
import Java.net.*;
import Java.io.*;
public class URLConnectionReader {
public static void main(String args[]) throws Exception {
    URL yahoo = new URL("http://www.yahoo.com/");
    URLConnection yc = yahoo.openConnection();
    BufferedReader in = new BufferedReader(//提供文本行
    new InputStreamReader(//提供字符
    yc.getInputStream()));
```

```
    String inputLine;
    while((inputLine = in.readLine()) != null)
        System.out.println(inputLine);
        yc.close();        }         }
```

14.2.4 URL Operations

Converting Between a URL and a URI

```
        URI uri = null;
        URL url = null;
        // 创建一个 URI
        try{
            uri = new URI("file://D:/almanac1.4/Ex1.Java");
        }
        catch (URISyntaxException e) {
        }
        // 将一个绝对 URI 转变为一个 URL
        try {
            url = uri.toURL();
        }
        catch (IllegalArgumentException e) {
            // URI 不是绝对的
        }
        catch (MalformedURLException e) {
        }
        // 将一个 URL 转变为一个 URI
        try {
            uri = new URI(url.toString());
        }
        catch (URISyntaxException e) {
        }
```

Parsing a URL

```
        try {
            URL url = new URL("http://hostname:80/index.html#_top_");
            String protocol = url.getProtocol();  // http
            String host = url.getHost();          // hostname
            int port = url.getPort();             // 80
            String file = url.getFile();          // index.html
            String ref = url.getRef();            // _top_
        }
        catch (MalformedURLException e) {
        }
```

Sending a POST Request Using a URL

```
        try {
            // 构造数据
            String data = URLEncoder.encode("key1", "UTF-8") + "=" + URLEncoder.
                    encode("value1", "UTF-8");
            data += "&" + URLEncoder.encode("key2", "UTF-8") + "=" + URLEncoder.
                    encode("value2", "UTF-8");
            // 发送数据
```

```java
            URL url = new URL("http://hostname:80/cgi");
            URLConnection conn = url.openConnection();
            conn.setDoOutput(true);
            OutputStreamWriter wr = new OutputStreamWriter(conn.getOutputStream());
            wr.write(data);
            wr.flush();
            //获取响应
            BufferedReader rd = new BufferedReader(new InputStreamReader(conn.
                            getInputStream()));
            String line;
            while ((line = rd.readLine()) != null) {
            //处理读取的行
            }
            wr.close();
            rd.close();
        }
        catch (Exception e) {
        }
```

Getting Text from a URL

```java
        try {
            //为一个要访问的网页创建一个 URL
            URL url = new URL("http://hostname:80/index.html");
            //读取服务器返回的所有文本
            BufferedReader in = new BufferedReader(new InputStreamReader(url.
                            openStream()));
            String str;
            while ((str = in.readLine()) != null) {
            // str 是文本的一行，readLine 表示读一个新的字符串行
            }
            in.close();
        }
        catch (MalformedURLException e) {
        }
        catch (IOException e) {
        }
```

Getting an Image from a URL

```java
        try {
            // 为本地的图片位置创建一个 URL
            URL url = new URL("http://hostname:80/image.gif");
            // 获取图片
            Java.awt.Image image = Java.awt.Toolkit.getDefaultToolkit().getDefault
                            Toolkit().createImage(url);
        }
        catch (MalformedURLException e) {
        }
        catch (IOException e) {
        }
```

Getting a Jar File Using a URL

```java
        try {
            // 创建一个指向一个网络上的 JAR 文件的 URL
```

```java
        URL url = new URL("jar:http://hostname/my.jar!/");
        // 创建一个指向本地文件系统的 JAR 文件的 URL
        url = new URL("jar:file:/c:/almanac/my.jar!/");
        // 获取 JAR 文件
        JarURLConnection conn = (JarURLConnection)url.openConnection();
        JarFile jarfile = conn.getJarFile();
        // 当没有在 URL 中指明实体时，实体名是空的
        String entryName = conn.getEntryName();   // null
        // 创建一个指向一个 JAR 文件中的某个实体的 URL
        url = new URL("jar:file:/c:/almanac/my.jar!/com/mycompany/MyClass.class");
        // 获取 JAR 文件
        conn = (JarURLConnection)url.openConnection();
        jarfile = conn.getJarFile();
        // 获取实体的名称，这必须和 URL 中指定的一致
        entryName = conn.getEntryName();
        // 获取 JAR 实体
        JarEntry jarEntry = conn.getJarEntry();
    }
    catch (MalformedURLException e) {
    }
    catch (IOException e) {
    }
```

Accessing a Password-Protected URL

```java
    // 安装用户的认证书
    Authenticator.setDefault(new MyAuthenticator());
    // 访问网页
    try {
        // 为指定的网页创建一个 URL
        URL url = new URL("http://hostname:80/index.html");
        // 读取服务器返回的所有文本
        BufferedReader in = new BufferedReader(new InputStreamReader(url.
                    openStream()));
        String str;
        while ((str = in.readLine()) != null) {
        // str 仅是文本的一行，readLine 表示读取新的字符串串行
        }
        in.close();
    }
    catch (MalformedURLException e) {
    }
    catch (IOException e) {
    }
    public class MyAuthenticator extends Authenticator {
        // 当访问一个受密码保护的 URL 的时候，本方法将被调用
        protected PasswordAuthentication getPasswordAuthentication() {
            // 获取请求的相关信息
            String promptString = getRequestingPrompt();
            String hostname = getRequestingHost();
            InetAddress ipaddr = getRequestingSite();
            int port = getRequestingPort();
            // 从 user 中获取用户名
            String username = "myusername";
```

```
            // 从 user 中获取密码
            String password = "mypassword";
            // 返回信息
            return new PasswordAuthentication(username, password.toCharArray());
        }
    }
```

Getting the Response Headers from an HTTP Connection

```
    try {
        // 为一个 URL 创建一个 URLConnection 对象
        URL url = new URL("http://hostname:80");
        URLConnection conn = url.openConnection();
        // 列出来自服务器的所有响应头
        //注意：对于 getHeaderFileKey()的首次调用将隐式地给服务器发送 HTTP 请求
        for (int i=0; ; i++) {
            String headerName = conn.getHeaderFieldKey(i);
            String headerValue = conn.getHeaderField(i);
            if (headerName == null && headerValue == null) {
                // 没有更多的头
                break;
            }
            if (headerName == null) {
                // 头部的值包含服务器 HTTP 的版本
            }
        }
    }
    catch (Exception e) {
    }
```

Here's a sample of headers from a website:

Key=Value

null=HTTP/1.1 200 OK

Server=Netscape-Enterprise/4.1

Date=Mon, 20 March 2006 09:23:26 GMT

Cache-control=public

Content-type=text/html

Etag="9fa67d2a-58-71-3bbdad3283"

Last-modified=Fri, 05 Oct 2001 12:53:06 GMT

Content-length=115

Accept-ranges=bytes

Connection=close

Getting the Cookies from an HTTP Connection

When the server wants to set a cookie in the client, it includes a response header of the form:

   Set-Cookie: cookie-value; expires=date; path=path; domain=domain-name; secure

cookie-value is some arbitrary string data that should be returned to the server in future URL requests. The life time of the cookie is specified by expires. If expires is not specified, the cookie expires at the end of the session. When a URL request is made, the cookie should be sent along only if domain-name matches the end of the fully-qualified host name of the URL request and path matches the beginning of the path of the URL request. If secure is specified, the cookie should be sent to the server only through HTTPS.

> 服务器要想在客户端设置 cookie，必须包含如下形式的 Set-Cookie 响应：cookie-value; expiers=date;path=path;comain=domain-name;security。cookie-value 是一些任意的字符串数据，在后面的 URL 请求中它将被返回给服务器。cookie 的生命期通过 expires 指定。如没有指定 expieres，则 cookie 的生命期将至整个会话结束为止。当 URL 请求产生时，只有域名和 URL 请求的主机名及其路径全部相同时，cookie 才会被设置。如安全级别被指定，则 cookie 只能通过 HTTPS 发送给服务器端。

```java
try {
    //为一个 URL 创建一个 URLConnection
    URL url = new URL("http://hostname:80");
    URLConnection conn = url.openConnection();
    //获取从服务器来的所有 cookies 值
    //注意：首次调用 getHeaderFieldKey，将隐式地向服务器发送 HTTP 请求
    for (int i=0; ; i++) {
        String headerName = conn.getHeaderFieldKey(i);
        String headerValue = conn.getHeaderField(i);
        if (headerName == null && headerValue == null) {
            //没有更多的头部
            break;
        }
        if ("Set-Cookie".equalsIgnoreCase(headerName)) {
            //解析 COOKIE
            String[] fields = headerValue.split(";\\s*");
            String cookieValue = fields[0];
            String expires = null;
            String path = null;
            String domain = null;
            boolean secure = false;
            //解析每个字段
            for (int j=1; j<fields.length; j++) {
                if ("secure".equalsIgnoreCase(fields[j])) {
                    secure = true;
                }
                else if (fields[j].indexOf('=') > 0) {
                    String[] f = fields[j].split("=");
                    if ("expires".equalsIgnoreCase(f[0])) {
                        expires = f[1];
                    }
                    else if ("domain".equalsIgnoreCase(f[0])) {
                        domain = f[1];
                    }
                    else if ("path".equalsIgnoreCase(f[0])) {
                        path = f[1];
                    }
                }
            }
            //保存 COOKIE
        }
```

```
        }
    }
    catch (MalformedURLException e) {
    }
    catch (IOException e) {
    }
```

Here's a sample of cookies from two websites:

    B=a43ka6gu6f4n4&b=2; expires=Thu, 20 Apr 2010 20:00:00 GMT;
        path=/; domain=.yahoo.com

    PREF=ID=e51:TM=686:LM=86:S=BL-w0; domain=.google.com; path=/;
        expires=Sun, 17-Jan-2038 19:14:07 GMT

Getting the Response Headers from an HTTP Connection

```
try {
  //为一个 URL 创建一个 URLConnection 对象
  URL url = new URL("http://hostname:80");
  URLConnection conn = url.openConnection();
  //列出所有来自服务器的响应头
  //注意：首次调用 getHeaderFieldKey 方法将隐式地向服务器发送 HTTP 请求
  for (int i=0; ; i++) {
    String headerName = conn.getHeaderFieldKey(i);
    String headerValue = conn.getHeaderField(i);
    if (headerName == null && headerValue == null) {
      //没有更多的协议头
      break;
    }
    if (headerName == null) {
      //头部值中包含了服务器 HTTP 协议版本信息
    }
  }
}
catch (Exception e) {
}
```

Here's a sample of headers from a website:

    Key=Value

    null=HTTP/1.1 200 OK

    Server=Netscape-Enterprise/4.1

    Date=Mon, 11 Feb 2002 09:23:26 GMT

    Cache-control=public

    Content-type=text/html

    Etag="9fa67d2a-58-71-3bbdad3283"

    Last-modified=Fri, 05 Oct 2001 12:53:06 GMT

    Content-length=115

    Accept-ranges=bytes

    Connection=close

Sending a Cookie to an HTTP Server

```
try {
    //为一个 URL 创建一个 URLConnection 对象
    URL url = new URL("http://hostname:80");
    URLConnection conn = url.openConnection();
    //设置要发送的 COOKIE 值
    conn.setRequestProperty("Cookie", "name1=value1; name2=value2");
    //发送请求给服务器
    conn.connect();
}
catch (MalformedURLException e) {
}
catch (IOException e) {
}
```

Preventing Automatic Redirects in a HTTP Connection

By default, when you make an HTTP connection using URLConnection, the system automatically follows redirects until it reaches the final destination. This example demonstrates how to prevent automatic redirection.

> 在默认情况下,当使用 URLConnection 建立 HTTP 连接后,系统会自动重定向直到找到目标地址。

```
//对所有 HTTP 请求取消自动重定向
HttpURLConnection.setFollowRedirects(false);
//对某个特定的连接取消自动重定向
try {
    //为一个 URL 创建一个 URLConnection 对象
    URL url = new URL("http://hostname:80");
    URLConnection conn = url.openConnection();
    //仅对本连接取消自动重定向
    HttpURLConnection httpConn = (HttpURLConnection)conn;
    httpConn.setInstanceFollowRedirects(false);
    //发送请求给服务器
    conn.connect();
}
catch (MalformedURLException e) {
}
catch (IOException e) {
}
```

Getting the IP Address of a Hostname

```
try {
    InetAddress addr = InetAddress.getByName("Javaalmanac.com");
    byte[] ipAddr = addr.getAddress();
    // 转为点形式
    String ipAddrStr = "";
    for (int i=0; i<ipAddr.length; i++) {
        if (i > 0) {
            ipAddrStr += ".";
        }
        ipAddrStr += ipAddr[i]&0xFF;
```

```
            }
        }
        catch (UnknownHostException e) {
        }
```

Getting the Hostname of an IP Address

This example attempts to retrieve the hostname for an IP address. Note that getHostName() may not succeed, in which case it simply returns the IP address.

```
        try {
            // 通过文本形式表示的 IP 地址获取主机名
            InetAddress addr = InetAddress.getByName("127.0.0.1");
            // 通过一个包含 IP 地址的字节数组获取主机名
            byte[] ipAddr = new byte[]{127, 0, 0, 1};
            addr = InetAddress.getByAddress(ipAddr);
            // 获取主机名
            String hostname = addr.getHostName();
            // 获取标准主机名
            String hostnameCanonical = addr.getCanonicalHostName();
        }
        catch (UnknownHostException e) {
        }
```

Getting the IP Address and Hostname of the Local Machine

```
        try {
            InetAddress addr = InetAddress.getLocalHost();
            // 获取 IP 地址
            byte[] ipAddr = addr.getAddress();
            // 获取主机名
            String hostname = addr.getHostName();
        }
        catch (UnknownHostException e) {
        }
```

How to display a particular web page from an applet

An applet can instruct a web browser to load a particular page, using the showDocument method of the Java.applet.AppletContext class. If you want to display a web page, you first have to obtain a reference to the current applet context.

The following code snippet shows you how this can be done. The show page method is capable of displaying any URL passed to it.

> 下面程序段演示了显示页面的方法。

```
    import Java.net.*;
    import Java.awt.*;
    import Java.applet.*;
    public class MyApplet extends Applet
    {
        //在此编写 Applet 代码
        //显示一个网页
        public void showPage ( String mypage )
        {
            URL myurl = null;
```

```
            //创建一个 URL 对象
            try
            {
                myurl = new URL ( mypage );
            }
            catch (MalformedURLException e)
            {
                //非法 URL
            }
            //显示 URL
            if (myurl != null)
            {
                getAppletContext().showDocument (myurl);
            }
        }
    }
```

How to display more than one page from an applet

The showDocument method of the AppletContext interface is overloaded - meaning that it can accept more than one parameter. It can accept a second parameter, which represents the name of the browser window that should display a page.

> AppletContext 接口的 showDocument 方法是可重载的,这意味着它能接收多个参数,即可以接收另一个需要显示页面的浏览器窗口名的参数。

For example,

```
    myAppletContext.showDocument (myurl, "frame1")
```

will display the document in frame1. If there exists no window named frame1, then a brand new window will be created.

How to fetch files using HTTP

The easiest way to fetch files using HTTP is to use the Java.net.URL class. The openStream( ) method will return an InputStream instance, from which the file contents can be read. For added control, you can use the openConnection( ) method, which will return a URLConnection object.

> 用 HTTP 请求页面可用 Java.net.URL 类,该类的 opentStream( )方法可返回一个输入流实例,该实例的文件可读。如要控制下载页面,可用 openConnection( )方法,它能返回 URLConnection 对象。

Here's a brief example that demonstrates the use of the Java.net.URL.openStream( ) method to return the contents of a URL specified as a command line parameter.

```
import Java.net.*;
import Java.io.*;
public class URLDemo
{
    public static void main(String args[]) throws Exception
    {
        try
        {
            //检查确保是否输入了一个命令参数
            if (args.length != 1)
```

```java
        {
            //输出消息，暂停然后退出
            System.err.println ("Invalid command parameters");
            System.in.read();
            System.exit(0);
        }
        //创建一个 URL 对象
        URL url = new URL(args[0]);
        //为读获取一个输入流
        InputStream in = url.openStream();
        //创建一个带缓冲的输入流提高效率
        BufferedInputStream bufIn = new BufferedInputStream(in);
        //重复直到文件末尾
        for (;;)
        {
            int data = bufIn.read();
            // 检查 EOF
            if (data == -1)
                break;
            else
                System.out.print ( (char) data);
        }
    }
    catch (MalformedURLException mue)
    {
        System.err.println ("Invalid URL");
    }
    catch (IOException ioe)
    {
        System.err.println ("I/O Error - " + ioe);
    }
}
```

How to use a proxy server for HTTP requests

When a Java applet under the control of a browser (such as Netscape or Internet Explorer) fetches content via a URLConnection, it will automatically and transparently use the proxy settings of the browser.

> 当 Java applet 在浏览器(如 Netscape 或 IE)控制下通过 URLConnection 提取文件时，它将自动使用浏览器的代理设置。

If you're writing an application, however, you'll have to manually specify the proxy server settings. You can do this when running a Java application or you can write code that will specify proxy settings automatically for the user (providing you allow the users to customize the settings to suit their proxy servers).

To specify proxy settings when running an application, use the -D parameter:

   jre -DproxySet=true -DproxyHost=myhost -DproxyPort=myport MyApp

Alternately, your application can maintain a configuration file, and specify proxy settings before using a URLConnection:

  //修改系统属性

```
Properties sysProperties = System.getProperties();
//指定代理设置
sysProperties.put("proxyHost", "myhost");
sysProperties.put("proxyPort", "myport");
sysProperties.put("proxySet",  "true");
```

14.3 Sockets in Java

A URL connection is okay for reading a Web page but sockets are more useful when two programs must talk to each other. A socket is one endpoint of a two-way communication link between two programs running on the network. A socket is bound to a specific port number. Incoming messages with that port number are delivered to that Socket.

The Java.net package in the Java library provides a class, Socket, that implements the client side of a two way connection between your Java program and other program on the network. The ServerSocket class implements a socket that servers can use to listen for, and accept connections to clients.

> URL 连接对访问网页是可以的，但当两个程序进行通信时最好用 socket。套接字是两台机器之间的通信端点。套接字一定要指定端口号。接收到带有端口号的消息必须传递给相应的套接字。Java 类库中提供了 Socket 类来实现客户端和服务器端的程序进行通信。ServerSocket 类实现服务器监听和建立到客户端的连接。

14.3.1 Establishing a Connection

The client is the processor that initiates the connection, The server is the processor that accepts the connection. There may be no difference between them subsequently. The server has a port on which it listens for connection requests. The client creates a port and sends a request to the server's port. The server creates a new port and allocates it to this connection. It then establishes the connection between this new port and the client's port. This allows the server to continue to listen for more. Figure 14.1 shows connection request from client.

> 客户端发起连接请求，服务器端接收连接请求。服务器有一个端口号用来接听连接请求，客户端创建一个端口号并用来发送给服务器的端口号。服务器创建一个端口号分配给该连接，然后用其创建的端口号和客户端的端口号就可以建立一个连接。

Figure 14.1 Connection Request

14.3.2 The Client Side of a Socket Connection

All client socket connections contain the same six steps:
- Open a socket to the server
- Open an input stream to the socket
- Open an output stream to the socket

- Read from and write to the stream according to the protocol agreed with the server
- Close the streams
- Close the socket

A Simple Example Program:

Our example program opens a socket connection to another computer using port number 4444, connected to an echo program. The program opens an input stream and an output stream to the port. The program reads text from the user's keyboard and sends it out to the server using the output stream. The echo program on the other computer sends the same text back after adding the string "Echo::" to the string that was sent. The program reads the text from the input stream and displays it. Finally, the program closes its streams and its socket.

> 下例使用端口号 4444 建立 socket 连接和另一台计算机进行通信的响应程序。该例程为端口打开了输入输出流。该例程使用输出流从键盘输入字符文本传递给服务器端。另一台计算机的响应程序接收到文本后在其字符流后添加"Echo::"后再发回给客户端。程序读取输入流的内容并将其显示。最终程序关闭输入输出流和 socket。

Example: The Client Side

```java
import Java.io.*;
import Java.net.*;
public class EchoClient{
public static void main(String args[]) throws IOException{
    Socket echoSocket = null;
    PrintWriter out = null;
    BufferedReader in = null;
    // 首先定义并创建一个套接字和输入输出流
    try{
        echoSocket = new Socket("lloth.ist.unomaha.edu", 4444);
        out = new PrintWriter(echoSocket.getOutputStream(), true);
        // "true" 表示 PrintWriter 发送一行数据，而非等缓冲满
        in = new BufferedReader(new InputStreamReader(echoSocket.getInputStream()));
        // 通过中间媒介 InputStreamReader，连接到 echoSocket
    }
    catch(UnknownHostException e){
        System.err.println("Don't know about host: lloth.ist.unomaha.edu.");
        System.exit(1);
    }
    catch(UnknownHostException e){
        System.err.println("Couldn't get connection to: lloth.ist.unomaha.edu");
        System.exit(1);
    }
    //使用 echoSocketBufferedReader 发送并接收
    BufferedReader stdIn = new
    BufferedReader(new InputStreamReader(System.in));
    String userInput;
    while((userInput = stdIn.readLine()) != null){
        //从键盘读入多行文本通过 System.in
        out.println(userInput);
```

```
            System.out.println("echo: " + in.readLine());
            //读入文本,并输出
        }
        stdIn.close();
        out.close();
        in.close();
        echoSocket.close();
        //关闭流和套接字
    }
}
```

Well-behaved programs clean up after themselves. However, note that you do not need to close System.in or System.out.

Example: The Server Side

```
import Java.io.*;
import Java.net.*;
public class KnockKnockServer{
    public static void main(String args[]) throws IOException {
        ServerSocket serverSocket = null;
        try{
            serverSocket = new ServerSocket(4444);
            //服务器首先在端口 4444 打开一个套接字
        }
        catch(IOException e){
            System.err.println("Could not listen on port: 4444.");
            System.exit(1);
        }
        Socket clientSocket = null;
        try{
            clientSocket = serverSocket.accept();
            //服务器随后调用 serverSocket 来接收等待一个客户连接。当有一个客户连接,
            //serverSocket.accept 将返回一个新套接字,该套接字可使服务器能够用其来与客户通信
        }
        catch(IOException e){
            System.err.printlln("Accept failed. ");
            System.exit(1);
        }
        //准备好使用套接字
        PrintWriter  out = new PrintWriter(clientSocket.getOutputStream(),true);
        BufferedReader in = new BufferedReader( new
                        InputStreamReader(clientSocket.getInputStream()));
        //首先为新客户套接字 clientSocket 创建一个 PrintWriter 和 BufferedReader
        String inputLine, outputLine;
        KnockKnockProtocol kkp = new KnockKnockProtocol();
        //同时创建一个新实例 kkp 来与客户通信
        outputLine = kkp.processInput(null);
        out.println(outputLine);
        //从协议获取首条消息,并将其发送给客户
        while((inputLine = in.readLine()) != null){
```

```java
            //直到输入是一个NULL或者输出是Bye
            //通过套接字获取客户的一条消息
            //in.readLine 连接客户到clientSocket
            outputLine = kkp.processInput(inputLine);
            //将消息发给协议,并获取一个响应
            out.println(outputLine);
            //通过套接字将响应发送给客户
            //断开和客户的连接
            if (outputLine.equals("Bye"))    break;
        }
        //关闭连接
        out.close();
        in.close();
        clientSocket.close();
        serverSocket.close();
    }
}
```

The KnockKnockProtocol class

```java
public class KnockKnockProtocol{
    final static String echoString = "Echo:: ";
    // method to return a String after prepending
    // echoString to it
    public String processInputLine(String s){
        return echoString+s;
    }
}
```

14.3.3　Socket Operations

Creating a Client Socket

```java
    //创建一个套接字,没有设置超时
    try {
        InetAddress addr = InetAddress.getByName("Java.sun.com");
        int port = 80;
        //直到连接成功,否则该构造函数将阻塞
        Socket socket = new Socket(addr, port);
    }
    catch (UnknownHostException e) {
    }
    catch (IOException e) {
    }
    //创建一个套接字,带有超时设置
    try {
        InetAddress addr = InetAddress.getByName("Java.sun.com");
        int port = 80;
        SocketAddress sockaddr = new InetSocketAddress(addr, port);
        //创建一个未绑定的套接字
        Socket sock = new Socket();
        //当超过timeoutMs,本方法将阻塞
```

```
        //如果超时发生，将抛出SocketTimeoutException
        int timeoutMs = 2000;    // 2 seconds
        sock.connect(sockaddr, timeoutMs);
}
catch (UnknownHostException e) {
}
catch (SocketTimeoutException e) {
}
catch (IOException e) {
}
```

Creating a Server Socket

```
try {
    int port = 2000;
    ServerSocket srv = new ServerSocket(port);
    //等待客户端的连接
    Socket socket = srv.accept();
}
catch (IOException e) {
}
```

Reading Text from a Socket

```
try {
    BufferedReader rd = new BufferedReader(new InputStreamReader(socket.
                    getInputStream()));
    String str;
    while ((str = rd.readLine()) != null) {
        process(str);
    }
    rd.close();
}
catch (IOException e) {
}
```

Writing Text to a Socket

```
try {
    BufferedWriter wr = new BufferedWriter(new OutputStreamWriter(socket.
                    getOutputStream()));
    wr.write("aString");
    wr.flush();
}
catch (IOException e) {
}
```

Sending a POST Request Using a Socket

```
try {
    // 构造数据
    String data = URLEncoder.encode("key1", "UTF-8") + "=" + URLEncoder.
                encode("value1", "UTF-8");
    data += "&" + URLEncoder.encode("key2", "UTF-8") + "=" + URLEncoder.
                encode("value2", "UTF-8");
    // 为主机创建一个套接字
    String hostname = "hostname.com";
```

```java
            int port = 80;
            InetAddress addr = InetAddress.getByName(hostname);
            Socket socket = new Socket(addr, port);
            // 发送协议头
            String path = "/servlet/SomeServlet";
            BufferedWriter wr = new BufferedWriter(new OutputStreamWriter(socket.
                        getOutputStream(), "UTF8"));
            wr.write("POST "+path+" HTTP/1.0\r\n");
            wr.write("Content-Length: "+data.length()+"\r\n");
            wr.write("Content-Type: application/x-www-form-urlencoded\r\n");
            wr.write("\r\n");
            // 发送数据
            wr.write(data);
            wr.flush();
            // 获取响应
            BufferedReader rd = new BufferedReader(new InputStreamReader(socket.
                        getInputStream()));
            String line;
            while ((line = rd.readLine()) != null) {
                // 处理数据行
            }
            wr.close();
            rd.close();
        }
        catch (Exception e) {
        }
```

Sending a Datagram

```java
    public static void send(InetAddress dst, int port, byte[] outbuf, int len) {
        try {
            DatagramPacket request = new DatagramPacket(outbuf, len, dst, port);
            DatagramSocket socket = new DatagramSocket();
            socket.send(request);
        }
        catch (SocketException e) {
        }
        catch (IOException e) {
        }
    }
```

Receiving a Datagram

```java
        try {
            byte[] inbuf = new byte[256];    // 默认尺寸
            DatagramSocket socket = new DatagramSocket();
            // 等待包
            DatagramPacket packet = new DatagramPacket(inbuf, inbuf.length);
            socket.receive(packet);
            // 数据现在在 inbuf 中
            int numBytesReceived = packet.getLength();
        }
        catch (SocketException e) {
        }
```

```
        catch (IOException e) {
        }
```

Joining a Multicast Group

```
        public void join(String groupName, int port) {
            try {
                MulticastSocket msocket = new MulticastSocket(port);
                group = InetAddress.getByName(groupName);
                msocket.joinGroup(group);
            }
            catch (IOException e) {
            }
        }
```

Receiving from a Multicast Group

Once you've created a multicast socket and joined the group, all datagrams sent to its corresponding multicast address will be available to be read from the socket. You can read from the socket just like you would from a unicast socket.

> 一旦创建多点传送socket并加入其组后，所有发送到其多点传送通信地址的数据报都能从socket接收到。

```
        public void read(MulticastSocket msocket, byte[] inbuf) {
            try {
                DatagramPacket packet = new DatagramPacket(inbuf, inbuf.length);
                //等待数据包
                msocket.receive(packet);
                //数据现在在 inbuf 中
                int numBytesReceived = packet.getLength();
            }
            catch (IOException e) {
            }
        }
```

Sending to a Multicast Group

You can send to a multicast socket using either a DatagramSocket or a MulticastSocket. What makes it multicast is the address that is in the datagram. If the address is a multicast address, the datagram will reach the multicast members in the group. You only need to use MulticastSocket if you want to control the time-to-live of the datagram.

> 通过 DatagramSocket 或 MulticastSocket 类可以发送数据给多点传送 socket。多点传送的地址放在数据报中，如地址是多点传送的地址，则数据报将会送达到多点传送的组群中。如果需控制数据报的发送接收时间周期，那么只需使用 MulticastSocket 类即可。

```
        byte[] outbuf = new byte[1024];
        int port = 1234;
        try {
            DatagramSocket socket = new DatagramSocket();
            InetAddress groupAddr = InetAddress.getByName("228.1.2.3");
            DatagramPacket packet = new DatagramPacket(outbuf, outbuf.length,
                            groupAddr, port);
            socket.send(packet);
```

```
        }
        catch (SocketException e) {
        }
        catch (IOException e) {
        }
```

DatagramSocket allows a server to accept UDP packets, whereas ServerSocket allows an application to accept TCP connections. It depends on the protocol you're trying to implement. If you're creating a new protocol, here's a few tips.

- DatagramSockets communicate using UDP packets. These packets don't guarantee delivery - you'll need to handle missing packets in your client/server.
- ServerSockets communicate using TCP connections. TCP guarantees delivery, so all you need to do is have your applications read and write using a socket's InputStream and OutputStream.

> DatagramSocket 允许服务器接收 UDP 包，而 ServerSocket 允许应用程序接受 TCP 连接。这依据试图执行的协议。如创建一新的协议，下面有一些建议：
> DatagramSockets 通信使用 UDP 包，但该协议包不保证传送的正确性，这样需在客户端/服务器解决丢包的问题。
> ServerSockets 通信使用 TCP 连接，而 TCP 连接保证传送的正确性，故只需注意 socket 的输入输出流的正确性即可。

Getting the IP address of a machine from its hostname

The InetAddress class is able to resolve IP addresses for you. Obtain an instance of InetAddress for the machine, and call the getHostAddress( ) method, which returns a string in the xxx.xxx.xxx.xxx address form.

> InetAddress 类可解决 IP 地址问题，通过为机器创建一个 InetAddress 的实例，就可调用 getHostAddress( )方法，该方法可返回形如 xxx.xxx.xxx.xxx 的地址。

```
        InetAddress inet = InetAddress.getByName("www.davidreilly.com");
        System.out.println ("IP : " + inet.getHostAddress());
```

Hostname lookup for an IP address

The InetAddress class contains a method that can return the domain name of an IP address. You need to obtain an InetAddress class, and then call its getHostName( ) method. This will return the hostname for that IP address. Depending on the platform, a partial or a fully qualified hostname may be returned.

> InetAddress 类包含一个能返回 IP 地址主机名的方法 getHostName( )。

```
        InetAddress inet = InetAddress.getByName("209.204.220.121");
        System.out.println ("Host: " + inet.getHostName());
```

If you're using a DatagramSocket, every packet that you receive will contain the address and port from which it was sent.

```
        DatagramPacket packet = null;
        // 接收下一个套接字
        myDatagramSocket.receive ( packet );
        // 打印地址和端口
```

```
System.out.println ("Packet from : " + packet.getAddress().getHostAddress()
+ ':' + packet.getPort());
```

If you're using a ServerSocket, then every socket connection you accept will contain similar information. The Socket class has a getInetAddress( ) and getPort( ) method which will allow you to find the same information.

```
Socket mySock = myServerSocket.accept();
// 打印地址和端口
System.out.println ("Connection from : " +
mySock.getInetAddress().getHostAddress() + ':' + mySock.getPort());
```

Finding out the current IP address of your machine

The InetAddress has a static method called getLocalHost( ) which will return the current address of the local machine. You can then use the getHostAddress( ) method to get the IP address.

```
InetAddress local = InetAddress.getLocalHost();
// 打印地址
System.out.println ("Local IP : " + local.getHostAddress());
```

Why can't applet connect via sockets, or bind to a local port

Applets are subject to heavy security constraints when executing under the control of a browser. Applets are unable to access the local file-system, to bind to local ports, or to connect to a computer via sockets other than the computer from which the applet is loaded. While it may seem to be an annoyance for developers, there are many good reasons why such tight constraints are placed on applets. Applets could bind to well known ports, and service network clients without authorization or consent. Applets executing within firewalls could obtain privileged information, and then send it across the network. Applets could even be infected by viruses, such as the Java StrangeBrew strain. Applets might become infected without an applet author's knowledge and then send information back that might leave hosts vulnerable to attack.

Applets 在浏览器下运行时对安全性的要求极高，且 Applets 不能访问本地文件系统，对本地端口的访问也有限制，它不能通过 socket 进行网络连接。这对开发者来说似乎是一个障碍，但有许多方面可以说明这些限制对 Applets 的好处。Applets 对公认的端口有限制，而服务器和客户端不需要获得授权。Applets 在防火墙下也能获得许可去对网络进行访问，且 Applets 甚至极易被病毒感染。由于 Applets 易感染病毒从而使运行它的主机极易受到网络攻击。

Signed applets may be allowed greater freedom by browsers than unsigned applets, which could be an option. In cases where an applet must be capable of network communication, HTTP can be used as a communication mechanism. An applet could communicate via Java.net.URLConnection with a CGI script, or a Java servlet. This has an added advantage - applets that use the URLConnection will be able to communicate through a firewall.

签名的 applets 比未签名的 applets 具有更大的自由度，它具有可选择性。当 applet 必须进行网络通信时，可以将 HTTP 作为其通信机制。Applet 可通过 Java.net.URLConnection 在 CGI 脚本或 Java 脚本支持下进行通信。这是 applet 新增的功能，即可通过 URLConnection 透过防火墙进行通信。

What are socket options

Socket options give developers greater control over how sockets behave. Most socket behavior is controlled by the operating system, not Java itself, but as of JDK1.1, you can control several socket options, including SO_TIMEOUT, SO_LINGER.

> TCP_NODELAY、SO_RCVBUF 和 SO_SNDBUF 套接字选项使程序员可以更大限度地控制 socket 的行为，大多 socket 行为都会受操作系统的限制，而不是 Java 本身，但从 JDK1.1 起，Java 提供了选项套接字，包括 SO_TIMEOUT、SO_LINGER、TCP_NODELAY、SO_RCVBUF 和 SO_SNDBUF。

These are advanced options, and many programmers may want to ignore them. That's OK, but be aware of their existence for the future. You might like to specify a timeout for read operations, to control the amount of time a connection will linger for before a reset is sent, whether Nagle's algorithm is enabled/disabled, or the send and receive buffers for datagram sockets.

> 目前，对高级选项，程序员可以忽略它们，但在将来，要做好高级选项存在的准备。当连接耗费大量时间，在重新连接要求发送之前，程序不知连接能否激活，或不知数据报 socket 发送和接收缓冲是否成功的情况下，可以为 socket 读操作设置连接超时的选项。

14.4　Sample Examples

Example 14.1

```java
//获取本地主机信息
import Java.net.*;
import Java.io.*;
public class JavaNet1
{
    public static void main(String args[]) throws IOException
    {
        InetAddress ia = InetAddress.getLocalHost(); //注意这里！
        System.out.println("Host and address: " + ia);
        System.out.println("Host name: " + ia.getHostName());
        String s = ia.toString();
        System.out.println("ia address: " + s.substring(s.indexOf("/") + 1));
    }
}
```

Example 14.2

```java
//从键盘输入主机名，然后获取主机 IP 地址
import Java.net.*;
import Java.io.*;
class JavaNet2
{
    public static void main(String args[]) throws IOException
    {
        System.out.print("Enter the host name:");
        BufferedReader b = new BufferedReader(new InputStreamReader(System.in));
```

```
                String s = b.readLine();
                try
                {
                    InetAddress i = InetAddress.getByName(s);
                    System.out.println("Ip address: " + i);
                }
                catch(Exception e)
                {
                    System.out.println("can't find the ip address,invalid name");
                }
        }
    }
```

Example 14.3

```
//通过域名获取主机 IP 地址
import Java.net.*;
import Java.io.*;
class JavaNet3
{
    public static void main(String args[]) throws IOException
    {
        try
        {
            InetAddress[]address=InetAddress.getAllByName("Java.sun.com");
            for(int i=0; i<address.length; ++i)
            {
                System.out.println(address[i]);
            }
        }
        catch(Exception e)
        {
        System.out.println("The Internet  connection is un available");
        }
    }
}
```

Example 14.4

```
//通过 URL 进行数据获取下载
import Java.net.*;
import Java.io.*;
import Java.util.Date;
import Java.text.DateFormat;
import Java.text.ParseException;
public class DateURLConnection extends URLConnection {
    DatagramSocket sock = null;
    public DateURLConnection(URL u) {
        super(u);
    }
    public void connect() throws IOException {
        if (connected) {
            return;
        }
```

```java
        InetAddress dst = InetAddress.getByName(getURL().getHost());
        byte[] outbuf = new byte[1];
        outbuf[0] = '\n';
        int port;
        if ((port = getURL().getPort()) == -1)
            port = 13;        // daytime
        DatagramPacket request =
            new DatagramPacket(outbuf, outbuf.length, dst, port);
        try {
            sock = new DatagramSocket();
            sock.send(request);
            connected = true;
        }
        catch (SocketException e) {
            sock = null;
            throw e;
        }
    }
    // Override instead letting content handler be selected
    // based on content-type
    public Object getContent() throws IOException {
        if (!connected)
            connect();
        byte[] inbuf = new byte[256];   // default size
        DatagramPacket reply = new DatagramPacket(inbuf, inbuf.length);
        sock.receive(reply);
        sock.close();
        sock = null;
        connected = false;
        String dateStr = new String(reply.getData());
        if (dateStr != null) {
            try {
                DateFormat df = DateFormat.getDateInstance(DateFormat.FULL);
                return (df.parse(dateStr));
            }
            catch (ParseException e) {
                System.err.println("Date string: " + dateStr);
                e.printStackTrace();
            }
        }
        throw new ProtocolException("Not conforming to date protocol");
    }
}
```

Example 14.5
Communication of Server and Client

```java
    //实现主机和客户机的通信
import Java.net.*;
import Java.io.*;
public class SimpleClient
{
    public static void main(String args[]) throws IOException
```

Chapter 14 Networking

```java
        {
            int c;
            Socket s1;
            InputStream s1_In;
            DataInputStream din_s1;
            OutputStream s1_out;
            DataOutputStream dout_s1;
            BufferedReader b = new BufferedReader(new InputStreamReader(System.in));
            s1 = new Socket(InetAddress.getLocalHost(),5432);
            String s1_recv = new String("");
            String s1_send = new String("");
            s1_out = s1.getOutputStream();
            dout_s1 = new DataOutputStream(s1_out);
            s1_In = s1.getInputStream();
            din_s1 = new DataInputStream(s1_In);
            while(!s1_send.equals("Bye") && !s1_recv.equals("Bye"))
            {
                try
                {
                    s1_out = s1.getOutputStream();
                    dout_s1 = new DataOutputStream(s1_out);
                    s1_send = b.readLine();
                    dout_s1.writeUTF(s1_send);
                    s1_In = s1.getInputStream();
                    din_s1 = new DataInputStream(s1_In);
                    s1_recv = din_s1.readUTF();
                    System.out.println("Server: " + s1_recv);
                }
                catch(IOException e)
                {
                    System.out.println("exception caught");
                }
            }
            din_s1.close();
            dout_s1.close();
            s1.close();
            s1_In.close();
            s1_out.close();
        }
    }
```

Server

```java
    import Java.net.*;
    import Java.io.*;
    public class SimpleServer
    {
        public static void main(String args[]) throws IOException
        {
            ServerSocket S = null;
            Socket s1;
            String S_send = new String("");
            String S_recv = new String("");
            OutputStream S_out;
            InputStream S_in;
            DataOutputStream dout_S;
```

```java
            DataInputStream din_S;
            try
            {
                S = new ServerSocket(5432);
            }
            catch(IOException e)
            {
                System.out.println("exception caught");
            }

            s1 = S.accept();
            BufferedReader b = new BufferedReader(new InputStreamReader(System.in));
            S_in = s1.getInputStream();
            din_S = new DataInputStream(S_in);
            S_out = s1.getOutputStream();
            dout_S = new DataOutputStream(S_out);
            while(!S_send.equals("Bye") && !S_recv.equals("Bye"))
            {
                try
                {
                    S_in = s1.getInputStream();
                    din_S = new DataInputStream(S_in);
                    S_recv = din_S.readUTF();
                    System.out.println("Client: " + S_recv);

                    S_out = s1.getOutputStream();
                    dout_S = new DataOutputStream(S_out);
                    S_send = b.readLine();
                    dout_S.writeUTF(S_send);
                }
                catch(IOException e)
                {
                    System.out.println("exception caught 2");
                }
            }
            dout_S.close();
            din_S.close();
            S_in.close();
            S_out.close();
            s1.close();
        }
    }
```

14.5　Exercise for you

1. Make one server and 4 clients. They should communicate to each other at the same time.
2. Make one server and one client on another machine. The server should detect the IP address of that client machine and start chatting.
3. Make a program so that one server can send some .avi files to another server/ client.

Chapter 15   Applets

Applet is the power behind the Java's popularity on the Internet and web programming. We get the applet support from java.applet.Applet class. Java has two flavors, one is the application and another is the applet. Here we discuss the applet. The applet classes are contained in the java.applet package. Applet contains several methods that give detailed control over the execution of the applet. In addition, java.applet also defines three interfaces. They are applet context, audio clip and appletstub. Lets start to create compile and run applets here.

> Applet 是 Java 程序风靡互联网的强大动力。java.applet.Applet 类为 Applet 应用小程序提供了支持。Java 程序分两类，一类是应用程序，另一类是 applet(即小程序)。Applet 类包含于 Java.applet 包中，它提供了 applet 实现的各方法。此外，Java.applet 还定义了三个接口，即 applet context，audio clip 及 appletstub。

15.1 Applet Overview

- Java programs consists of applications and applets.
- Applications are executed from MSDOS prompt by using the interpreter
- Applets are executed with the help of the appletviewer utility or the Net browser. Applets
- Compilation stages are same in both the cases.

What are Applets
- Applets are dynamic (animations and graphics).
- Applets are interactive programs (via GUI components).

What you need to execute an Applet
- Browser (IE 4.0 or Netscape Navigator 4.0 or higher)
- Appletviewer utility in the JDK.

Features of an Applet
- Provides facility for frames
- Event handling facility and userInteraction
- GUI user interface
- Graphics and multimedia

Your First Applet

Type in the following code using any text editor or in DOS editor.

```
import java.awt.* ;
import java.applet.*;
public classhello extends Applet {
    public void paint(Graphics g) {
        g.drawString("Welcome to Java Applets",20,20); } }
```

Save the file as hello.java and compile it by using Javac. Now, type in the following HTML code in your editor and save the file as hello.html.

```
<applet code =
"hello.class" width = 200 height =
150></applet>
```

Execute the HTML file by giving appletviewer hello.html. Another way to run the applet is to give the above HTML coding as a comment after the two import statements. Execute the applet as appletviewer hello.java .

Concepts and Explanation
- All applets are the subclasses of Applet class. All applets must import the Java.applet package
- The Paint( ) method is defined by AWT Component class. It takes one parameter of type Graphics.
- The drawString is the member of the Graphics class of the AWT package.
- This method is responsible for printing the String "welcome to Java applets".

Facts
- Applets do not begin execution at main( ) method. However, it is possible to execute the applets by using the Java interpreter (by using the main( ) method) if you extend your class with Frame.
- I/O Streams do not provide much scope for applets.
- It is not possible for an applet to access the files on the users hard disk.
- It is not possible to access the source code of the applets. It can be accessed only from the original server.

> 所有applets都是Applet类的子类，编写Java小应用程序必须要导入java.applet包。Applets的执行入口方法并不是main。如果小应用程序类继承了Frame，则可以通过解释器运行。I/O流对applets的支持较为有限。applet小应用程序不能访问用户本地硬盘上的文件，Applet源代码一般不能被访问，只有运行该applet的服务器才能访问。

15.2 Life Cycle, Graphics, Fonts, Colors

15.2.1 Life Cycle of an Applet

There are some important methods during life cycle of an applet as shown in Table 15.1.

Table 15.1 Methods for Applet

Method	Class	Description
init( )	Applet	First method to be called, initialize variables in this method
start( )	Applet	Called when restarted after being stopped.occurs after init( )
stop( )	Applet	Called when the applet leaves the webpage
destroy( )	Applet	Called when the applet wants to be removed out of memory
paint( )	Component	Called when the applet needs to be drawn

The repaint( ) method - If the applet wants to be repainted again, then this method is called. Useful for animation purposes.

- **The Life and Death of an Applet**

On a typical Java-enhanced Web page, Java applets start and stop frequently as the user switches among pages. As pages are loaded into the browser, viewed and then left, applets are started and terminated in response to messages passed to the Applet class. The four most important methods in the life of an applet are init, start, stop, and destroy. They are described in table 15.2.

> 程序是下载到用户浏览器后运行的,因此它的生命周期主要是由 4 个方法来控制的: init,start,stop,destroy。

Table15.2   Important Methods Over the Life of an Applet

Method	Purpose
init	Called when the applet is initially loaded
start	Called each time the user loads the host Web page
stop	Called each time the user leaves the host Web page
destroy	Called to release resources allocated by the applet

● **The init Method**

When an applet is first loaded by either a browser or a stand-alone interpreter such as is provided with Visual J++, the init method is called. This allows the applet to execute any startup code that is necessary. Usually an applet will resize itself at this point. The applet may also acquire and load resources such as images or sounds at this point. The init method will be called only once for each time when the applet is loaded.

> init 方法在 applet 应用程序第一次被浏览器或独立的 applet 解释器如 Visual J++加载时被调用。在该方法中一般用来执行必需的初始化工作,包括确定 applet 的大小、为 applet 加载图片声音资源等。每次加载 applet 方法都只执行一次 init 方法。

● **The destroy Method**

The destroy method is called when the browser is closing an applet. Ideally this method should be used to release any resources that have been allocated during the life of the applet; however, because Java performs garbage collection, it is not necessary to release resources.

> destroy 方法一般是在浏览器关闭 applet 时调用,主要用来释放 applet 应用程序在运行期间分配到的资源。然而由于 Java 平台具有自动垃圾回收功能,可以不用该方法来释放资源。

● **The start and stop Methods**

The start method is similar to init except that start is called each time the Web page in which the applet is embedded is loaded. The most frequent use of the start method is to create a new thread. Similarly, the stop method is called whenever the browser moves off the Web page that hosts the applet.

> start 方法在包含 applet 的网页每次被加载时都要被调用,这是 start 方法和 init 方法唯一的区别。一般 start 方法的运行都会产生一个新进程。同样,stop 方法每当浏览器关闭运行 applet 的网页时被调用。

The distinction between a method that gets called when an applet is first loaded and when its host page is first loaded might seem like a fine distinction, but understanding this distinction enables you to write more efficient applets. In Netscape Navigator 3 and Microsoft Internet Explorer 3, when a page containing a Java applet is first loaded, the init and start methods are called. Following a link from the page calls the applet's stop method. Backing up to the page calls the applet's start method, but not its init method.

In Netscape Navigator, reloading a page that contains an applet will stop and then start the applet. Performing the equivalent function in Microsoft Internet Explorer, however, causes the stop, destroy, init, and start methods all to be invoked in that order.

15.2.2 Parameter Passing

- Use a special parameter tag <param> in the HTML file. This tag takes two attributes, namely Name and Value.
- Use a method getParameter( ) inside the init( ) method, which takes one argument (i.e., the string representing the name of the parameter being looked for). Give this name a value in the HTML coding.

> 给 applet 小应用程序传递参数可以通过在 HTML 文件中使用一个特殊的参数标签 <param>。该标签有两个属性，即 Name 和 Value 属性。在 init( )方法内使用带有一个参数的 getParameter( )方法。

15.2.3 Graphics Class

Graphics class in the Java.awt package contains methods for drawing strings, lines and rectangles, ovals, polygon, fill rect, fill oval, etc.

Methods

- To draw a string use the drawString(String str,int X,int Y) method, where str is the name of the string and X and Y are the coordinates for where the string is to be printed.
- To draw a line use drawLine(int x1,int y1,int x2,int y2), where x1 and y1 are the starting point coordinates and x2 and y2 are the ending point coordinates.
- To draw a rectangle use drawRect(int x1,int y1,int width,int height), where x1 and y1 are the starting point coordinates and width and height are the measurements for rectangle
- To draw a RoundRect use drawRoundRect(int x1,int y1,int width,int height,width1,height1), where x1 and y1 are the starting point coordinates and width and height are the measurements for rectangle and width1 and height1 are the angles of the corners.
- To draw an oval use drawOval(int x1,int y1,int width,int height), where x1 and y1 are the coordinates of the top corners and width and height are the respective measurements of the oval.
- To draw an arc use drawArc(int x1,int y1,int width,int height,angle1, angle2), where x1 and y1 are the coordinates of the top corners and width and height are the respective measurements of the arc and angle1 and angle2 are the starting arc and ending arcs (in degrees).

15.2.4 Font Class

The Font class in the Java.awt package contains methods for displaying fonts.
(1) Declare the Font Name, style and the size using the Font Constructor.

(2) Finally, pass the font's object by using the setFont method in the Java.awt package.
Sample
```
Font f = new Font("Courier",Font.BOLD+Font.Italic,16);
g.setFont(f);
```

15.2.5 Color Class

The Color class in the Java.awt package contains methods for dealing with colors.
Methods
- setColor(Color.gray) sets the string color to gray .
- setBackground(Color.red) sets the background color to red.

15.3 User Interface Components

How to organize so many classes of user interface? The Class Hierarchy of UI components is shown in figure 15.1:

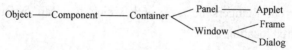

Figure 15.1 Class Hierarchy of UI Components

Usage of Label
(1) Label( ) - Creates an empty label
(2) Label(String) - Creates a label with the given string
(3) Label(String, align) - Creates a string label with the specified alignment (RIGHT,LEFT,CENTER)

Usage of Button
(1) Button( ) - Creates a button without a label
(2) Button(String str) - Creates a button with the string

Usage of Checkbox( )
(1) Checkbox( ) - Creates a checkbox without a label
(2) Checkbox(String str) - Creates a checkbox with a string label in it

Usage of CheckboxGroup
Create a Checkbox Group and add the group's object to the individual checkboxes to get a radio-button style of interface. Only one box can be selected at a time.

Usage of TextField
(1) TextField( ) - Creates a empty Text Field
(2) TextField(String str) - Creates a Text Field with the specified String
(3) TextField(String str,align) - Creates a Text Field with the specified String with the alignment
(4) The setText( ) method is used in connection with the Text Fields. For instance, to set a text in Choice to the text field, the method setText( ) method is used

Usage of Text Area
(1) TextArea( ) - Creates empty Text Area
(2) TextArea(rows,charcters) - Creates empty Text Area, with the specified rows and charcters.

(3) TextArea(String str,rows,charcters) - Creates a default String Text Area, with the specified rows and charcters

Usage of Choice

(1) Create a Choice object (Choice c = new Choice( ))

(2) Add the individual items to the Choice by add method and connecting the Choice object

(3) Finally, add the choice object to the container.

Choice ch = new Choice( ); ch.add("Java"), ch.add("XML"); add(ch)

Usage of Lists

(1) Create a List Object (List l = new List( ))

(2) Add the individual items to the List by add method and connecting the List object

(3) Finally, add the List object to the Container.

- You can select multiple items from the list box. But only one from a Choice.
- The setEditable(Boolean) method is used to edit the text inside a choice component.

All these above components together form a GUI Interface. You can create any type of user-friendly applications you want by making use of the above components. In the next section, we will take a look at Layout Mangers in Java, with which you can dynamically place the above components at your desired location.

15.4 Applet Fundamentals

You learn about the four methods that control the life span of an applet. You also learn how to write code to respond to events generated by an applet's graphical user interface. In this chapter you also see how to use applet parameters to control the behavior of a program. Finally, you learn about Uniform Resource Locators (URLs) and how to display simple graphics.

15.4.1 The Applet Class

To create your own applets, you derive a class from Java's Applet class and then add the desired functionality. The Applet class contains methods for actions such as interacting with the browser environment in which it is running, for retrieving images, for playing sound files, and for resizing its host window. In this section you learn about the most frequently used methods in this important class.

> 必须通过继承 Applet 类才能创建自己的 applet，然后就可以添加需要的功能。Applet 类中包含处理与 applet 运行所在浏览器进行交互，检索图片，播放声音文件，调整主窗口大小等功能的方法。

```
import java.awt.*;
import java.applet.*;
public class SimpleApplet extends Applet {
    public void paint (Graphics g) {
        g.drawString("First Applet", 50, 50 );
    }
}
```

AWT is Abstract Window Toolkit.

AWT contains support for a window-based, graphical interface.

Paint method defined by AWT. Called whenever the applet must redraw its output.

Applet Tag

```
<applet code="SimpleApplet.class" width=200 height=200>
</applet>
```

else

C:\>appletviewer MyHTML.html

15.4.2 Applet Architecture

An applet waits until an event occurs. The AWT notifies the applet about an event by calling event handler that has been provided by the applet. The applet takes appropriate action and then quickly return control to AWT.

```
import java.awt.*;
import java.applet.*;
/*
<applet code="AppletSkel.class" width=300 height=200>
</applet>
*/
public class AppletSkel extends Applet {
   //Called First
   public void init() {
      // initialization
   }
 /*Called second, after init(). Also called whenever the applet is restarted.*/
   public void start() {
      // start or resume execution
   }
   // Called when the applet is stopped.
   public void stop() {
      // Suspends execution
   }
   /*Called when applet is terminated. This is last method executed.*/
   public void destroy() {
      //perform shutdown activities
   }
   //Called when an applet's window must be restored.
   public void paint(Graphics g) {
      //redisplay contents of window
   }
}
```

Order(执行顺序)

When applet begins:

```
init()
start()
paint()
```

When an applet is terminated:

```
stop()
destroy()
```

Simple Applet Display Methods:

```
void drawstring(String message, int x, int y)
void setBackground(Color newColor)
void setForeground(Color newColor)
```

Sample Program

```java
import java.awt.*;
import java.applet.*;
/*
<applet code="Sample.class" width=300 height=200>
</applet>
*/
public class Sample extends Applet {
String msg;
public void init() {
    setBackground(Color.cyan);
    setForeground(Color.red);
    msg = "Inside init() -";
}
public void start() {
    msg += "Inside start() -";
}
public void paint(Graphics g) {
    msg += "Inside paint().";
    g.drawString(msg, 20, 30);
}
}
```

15.4.3 Requesting Repainting

Applet must quickly return control to the AWT run-time system.

So to change a particular information itself, we can not make a loop in the paint method that repeatedly updates it.

So, whenever your applet needs to update the information displayed in its window, it simply calls **repaint( )**.

> Applet 必须迅速将控制返回给 AWT 运行时系统。如要更改特定的信息，不能在 paint 方法中利用循环来重复更新。无论 applet 应用程序何时需更新窗口显示信息，它都会调用 repaint( )方法。

```
void repaint() //entire window
```
or
```
void repaint(int left, int top, int width, int height)
//specifies region to be repainted
```

It is possible for a method other than paint( ) or update( ) to output to an applet's window. To do so, it must obtain a graphics context by calling getGraphics( ) (defined by Component) and then use this to output to the window.

A Simple Banner Applet

```java
import java.awt.*;
import java.applet.*;
```

```
/*
<applet code="SampleBanner.class" width=300 height=200>
</applet>
*/
public class SampleBanner extends Applet implements Runnable{
    String msg = "A Simple Moving Banner.";
    Thread t = null;
    int state;
    boolean stopFlag;
    public void init() {
        setBackground(Color.cyan);
        setForeground(Color.red);
    }
    public void start() {
        t = new Thread(this);
        stopFlag = false;
        t.start();
    }
    public void run () {
        char ch;
        for( ; ; ) {
            try {
                repaint();
                Thread.sleep(250);
                ch = msg.charAt(0);
                msg = msg.subString(1, msg.length());
                msg += ch;
                if(stopFlag)
                break;
            } catch(InterruptedException e) {}
        }
    }
    public void stop() {
        stopFlag = true;
        t = null;
    }
    public void paint(Graphics g) {
        g.drawString(msg, 50, 30);
        showStatus("This is shown in the status window.");
    }
}
```

Passing Parameters to Applets

```
<applet code="ParamDemo.class" width=300 height=80>
<param name=fontName value=Courier>
<param name=fontSize value=14>
<param name=counter value=2>
</applet>
```

How to use these parameters

```
String fontName;
String param;
```

```
    int fontSize;
fontName = getParameter("fontName");
if(fontName == null)
fontName = "Not Found";
param = getParameter("fontSize");
try {
    if (param != null)
        fontSize = Integer.parseInt(param);
    else
        fontSize = 0;
}catch(NumberFormatException e) {
    fontSize = -1;
}
```

The best way to understand how the init, start, stop, and destroy methods interact is to build an applet to demonstrate their use. Table 15.3 shows the method calls under Navigator and Internet Explorer. The class test1 includes a private string member named history that is empty initially. As the various methods are called, text is added to this string to indicate the sequence in which the methods were called. Parentheses are placed around the start and stop messages to improve readability.

Table 15.3 Applet method calls under Navigator and Internet Explorer

What	Navigator	Internet Explorer
Page first loaded	init + start	init + start
Forward then backward	stop + start	stop + start
Reload/Refresh	stop + start	stop + destroy + init + start

Example 15.1
```
//Applet 生命周期演示
import java.applet.*;
import java.awt.*;
public class test1 extends Applet
{
    private String history = "";
    public test1()
    {
    }
    public void init()
    {
        history = history + "init ";
        resize(320, 240);
    }
    public void start()
    {
        history = history + "(start ";
    }
    public void stop()
    {
        history = history + "stop) ";
    }
    public void destroy()
    {
```

```
        history = history + "destroy ";
    }
    public void paint(Graphics g)
    {
        g.drawString(history, 10, 10);
    }
}
```

In this case, Netscape Navigator was used to run this applet. After the applet was loaded, the Reload button was pressed repeatedly. This had the effect of stopping and starting the applet each time, Closing Navigator will stop the applet and then destroy it.

15.5 Working with URLs and Graphics

One of the most common things for a Web page to do is display a graphic. Because of this, the Applet class includes a method for retrieving an image from a URL (a Uniform Resource Locator, the way in which files are addressed on the Internet). The signature for getImage is as follows:

```
public Image getImage(URL url)
```

Accessing a URL（访问 URL）

Because getImage loads an image from a URL, you must first create an instance of a URL. This can be done by using the following URL constructor: getImage

> 要从 URL 下载图像，必须通过 URL 类的构造方法生成 URL 的实例对象，构造方法如下：

```
public URL(String spec) throws MalformedURLException
```

This constructor is simply passed to the URL as a string, as in the following example:

```
myURL = new URL("http://mtngoat/Java/test.jpg");
```

However, because the constructor can throw a MalformedURLException exception, the constructor must be enclosed in a try...catch block that catches the exception. This could be done as follows:

```
try {
    myURL = new URL("http://mtngoat/Java/test.jpg");
}
catch (MalformedURLException e) {
    // handle error
}
```

An Example of Displaying an Image

The getImage method retrieves an image from a URL but does not display it. To display the retrieved image, use the Graphics.drawImage method in the paint method of the applet.

> getImage 方法根据指定的 URL 查找图片，但不显示该图片，需要调用 Graphics.drawImage 方法实现。

Example 15.2

```
//Applet 显示一幅图片
import java.applet.*;
import java.awt.*;
```

```
import java.net.*;
public class test2 extends Applet
{
    private Image m_image;
    private URL m_URL;
    private String error = "";
    public void init()
    {
        resize(600, 440);
        try {
          m_URL = new
            URL("http://spider.innercite.com/~lcohn/savannah.jpg");
          m_image = getImage(m_URL);
        }
        catch (MalformedURLException e) {
          error = "Image is unavailable.";
        }
    }
    public void paint(Graphics g)
    {
        g.drawString("Here's a cute baby:", 10,10);
        if (error.length() == 0)
            g.drawImage(m_image, 10, 30, this);
        else
            g.drawString(error, 20, 30);
    }
}
```

15.6   Using the instanceof Keyword in Java

When you are typecasting and using instanceof keyword in Java, there are several things you should keep in mind. Consider the following example program:

```
class Point { int x, y; }
class Element { int atomicNumber; }
class Test {
    public static void main(String[] args) {
        Point p = new Point();
        Element e = new Element();
        if (e instanceof Point) { // compile-time error
            System.out.println("I get your point!");
            p = (Point)e;    // compile-time error
        }
    }
}
```

This example results in two compilation errors. The cast (Point)e is incorrect because no instance of Element, or any of its possible subclasses (none are shown here), could possibly be an instance of any subclass of Point. The instanceof expression is incorrect for exactly the same reason. If, on the other hand, the class Point were a subclass of Element, such as:

```
class Point extends Element { int x, y; }
```

the cast would work, though it would require a run-time check. The cast (Point)e would not throw an exception because it would not be executed if the value of e could not correctly be cast to type Point. The use instanceof expression would then be sensible and valid.

15.7 Sample Examples

Example 15.3

```java
//在页面上输出"Hello, world!"字符串
import java.applet.*;
import java.awt.*;          //Don't forget the ; !!!
public class appletDemo extends Applet
{
    public void paint(Graphics g)
    {
        g.drawString("Hello,world!\n",11,11);
    }
}
/*
<applet code = "appletDemo.class" width = 600 height = 160>
</applet>
*/
```

Example 15.4

```java
//Applet 生命周期演示
import java.awt.*;
import java.applet.*;
public class appletDemo2 extends Applet
{
    String out;
    public void init()
    {
        setBackground(Color.blue);
        setForeground(Color.yellow);
        System.out.println("init");
        out = "init->";
    }
    public void start()
    {
        out += "start->";
    }
    public void stop()
    {
        System.out.println("stop");
        out += "stop->";
    }
    public void display()
    {
        System.out.println("display");
        out += "display";
```

```java
    }
    public void paint(Graphics g)
    {
        System.out.println("paint");
        out += "paint->";
        g.drawString(out,111,111);
    }
}
/*
<applet code = "appletDemo2.class" width = 600 height = 160>
</applet>
*/
```

Example 15.5

```java
//A applet life circle demostration program,pay more attention!
//Applet 生命周期演示
import java.applet.Applet;
import java.awt.Graphics;
public class appletLifeCircle extends Applet
{
    String s = "The ";
    public void init()
    {
        s += "New ";
        System.out.println(s);
    }
    public void start()
    {
        s += "Applet "; //when the applet's inited or the applet window is
                       //activated from the minimized state!
        System.out.println(s);
    }
    public void stop()
    {
        s += "Stops ";  //when minimize the applet window!
        System.out.println("Stops");
    }
    public void destroy()
    {
        s += "Here ";   //when close the applet window!
        System.out.println("Here");
    }
    public void paint(Graphics g)
    {
        g.drawRect(25,25,100,100);
        s += "Draw ";   //when the rectangle has been drawn
        g.drawString(s,25,40);
    }
}
/*
<applet code = "appletLifeCircle.class" width = 160 height = 160>
```

```
</applet>
*/
```

Example 15.6

```java
//A simple audio program
//一个简单的声音程序
import java.awt.*;
import java.awt.event.*;
import java.net.*;
import java.applet.*;
public class audioDemo extends Applet implements ActionListener
{
    AudioClip A;
    public void init()
    {
        String S = "";
        URL CodeBase = getCodeBase();
        A = getAudioClip(CodeBase,"ycg.mp3");
        setLayout(new BorderLayout());
        Panel menu = new Panel();
        Button play = new Button("Play");
        Button loop = new Button("Loop");
        Button stop = new Button("Stop");
        menu.add(play);
        menu.add(loop);
        menu.add(stop);
        play.addActionListener(this);
        loop.addActionListener(this);
        stop.addActionListener(this);
        add(menu);
    }
    public void actionPerformed(ActionEvent e)
    {
        String check = e.getActionCommand();
        if(check.equals("play"));
            A.play();
        if(check.equals("loop"));
            A.loop();
        if(check.equals("stop"));
            A.stop();
    }
}
/*
<applet code = "audioDemo.class" width = 160 height = 100>
</applet>
*/
```

Example 15.7

```java
//A simple thread program with applet
//一个简单的Applet线程序
import java.applet.*;
import java.awt.*;
```

```java
public class threadAppletDemo1 extends Applet
{
    char c = 'a';
    Font f;
    public void init()
    {
        f = new Font("TimesNewRoman",Font.BOLD,16);
        setFont(f);
    }
    public void paint(Graphics g)
    {
        while(true)
        {
            if(c >= 'z')
                c = 'a';

            ++c;
            g.drawString(" " + c,11,24);
            try
            {
                Thread.sleep(1000);
            }
            catch(InterruptedException e)
            {
                System.out.println(e.getMessage());
            }
        }
    }
}
/*
<applet code = "threadAppletDemo1.class" width = 600 height =  160>
</applet>
*/
```

15.8 Exercise for you

1. Create an audio applet and run one .avi file when the applet runs.
2. Create an html form and embed an applet in to that where it will show you the page properties of the html (like what the font is used, what's the size,etc.).
3. Make a banner applet which can move.

Chapter 16　Swing GUI Introduction

This chapter presents the basic classes in the Swing package and teaches the basic techniques for using these classes to define GUIs. GUIs are windowing interfaces that handle user input and output. GUI is pronounced "gooey" and stands for graphical user interface. Entire books have been written on Swing, so we will not have room to give you a complete description of Swing in this chapter. However, we will teach you enough to allow you to write a variety of windowing interfaces.

> 本章介绍 Swing 包中的基础类，并介绍如何使用这些类设计图形用户界面——GUI。GUI 是处理用户输入和输出的窗口界面。GUI 代表图形用户界面。关于 Swing 编程的书籍有很多，本章没有足够的篇幅对 Swing 编程进行完整详细描述。然而，本章会教你足够的知识让你能够编写出各种窗口界面。

The AWT(Abstract Window Toolkit) package is an older package designed for doing windowing interfaces. Swing can be viewed as an improved version of the AWT. However, Swing did not completely replace the AWT package. Some AWT classes are replaced by Swing classes, but other AWT classes are needed when using Swing. We will use classes from both Swing and the AWT.

Swing GUIs are designed using a particular form of object-oriented programming that is known as event-driven programming. Our first section begins with a brief overview of event-driven programming.

> Swing 图形用户界面设计采用了特殊的面向对象编程形式，这种编程形式称为事件驱动编程。16.1 节将给出事件驱动编程的简要概述。

16.1　Event-Driven Programming

Event-driven programming is a programming style that uses a signal-and-response approach to programming. Signals to objects are things called events, a concept we explain in this section.

> 事件驱动编程是一种编程风格，它使用"信号—响应方式"进行编程。传给对象的信号称为"事件"。

Swing programs use events and event handlers. An event is an object that acts as a signal to another object known as a listener. The sending of the event is called firing the event. The object that fires the event is often a GUI component, such as a button. The button fires the event in response to being clicked. The listener object performs some action in response to the event. For example, the listener might place a message on the screen in response to a particular button being clicked. A given component may have any number of listeners, from zero to several listeners. Each listener might respond to a different kind of event, or multiple listeners might respond to the same events.

In Swing GUIs an event often represents some action such as clicking a mouse, dragging the mouse, pressing a key on the keyboard, clicking the close-window button on a window, or any other action that is expected to elicit a response. A listener object has methods that specify what will happen when events of various kinds are received by the listener. These methods that handle events are called event handlers. You the programmer will define (or redefine) these event-handler methods. The relationship between an event-firing object, such as a button, and its event-handling listener is shown diagrammatically in Figure 16.1.

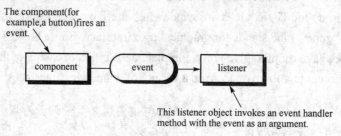

Figure 16.1 Event Firing and an Event Listener

Event-driven programming is very different from most programming you've seen before now. All our previous programs consisted of a list of statements executed in order. There were loops that repeat statements and branches that choose one of a list of statements to execute next. However, at some level, each run of a program consists of a list of statements performed by one agent (the computer) that executes the statements one after the other in order.

> 事件驱动编程与大多数编程有很大的不同。以前的程序包括一个顺序执行的语句列表。有循环语句和分支选择语句，它们选择下一条要执行的语句。然而，在一定程度上，每个程序的运行都是由一个代理(计算机)从可执行的语句列表中选择语句顺序执行。

Event-driven programming is a very different game. In event-driven programming, you create objects that can fire events, and you create listener objects to react to the events. For the most part, your program does not determine the order in which things happen. The events determine that order. When an event-driven program is running, the next thing that happens depends on the next event. It's as though the listeners were robots that interact with other objects (possibly other robots) in response to events (signals) from these other objects. You program the robots, but the environment and other robots determine what any particular robot will actually end up doing.

If you have never done event-driven programming before, one aspect of it may seem strange to you: You will be writing definitions for methods that you will never invoke in any program. This will likely feel a bit strange at first, because a method is of no value unless it is invoked. So, somebody or something other than you, the programmer, must be invoking these methods. That is exactly what does happen. The Swing system automatically invokes certain methods when an event signals that the method needs to be called.

> 如果从未做过事件驱动的编程，它的一个方面可能让你觉得很奇怪：你所写的方法定义，在你写的任何程序中，都没有调用它。一开始，你可能会觉得这有点怪，因为不被其他程序调用的方法是没有价值的。因此，一定有其他程序员写的程序，必须调用这些方法。确实是这样的。当调用这些方法的信号由事件发出时，Swing 系统会自动调用这些方法。

Event-driven programming with the Swing library makes extensive use of inheritance. The classes you define will be derived classes of some basic Swing library classes. These derived classes will inherit methods from their base class. For many of these inherited methods, library software will determine when these methods are invoked, but you will override the definition of the inherited method to determine what will happen when the method is invoked.

16.2 Event Handling

The Java Virtual Machine works by sending messages to Java applications whenever specific events occur. For example, if the user presses a key, a message named KEY_PRESS is passed to the application. If a Java application wants to respond in a particular way to a message, a *handler* for that message is written. The complete list of Java messages as shown in table 16.1.

> 只要有指定事件发生，Java 虚拟机就会发送消息给 application 程序。如用户按键，则会有一个 KEY_PRESS 的消息传送给 application。如 Java application 需对消息做特别响应，则需要对 *handler* 重写代码。

Table 16.1 Java messages

WINDOW_DESTROY	MOUSE_DRAG
WINDOW_EXPOSE	SCROLL_LINE_UP
WINDOW_ICONIFY	SCROLL_LINE_DOWN
WINDOW_DEICONIFY	SCROLL_PAGE_UP
WINDOW_MOVED	SCROLL_PAGE_UP
KEY_PRESS	SCROLL_ABSOLUTE
KEY_RELEASE	LIST_SELECT
KEY_ACTION	LIST_DESELECT
KEY_ACTION_RELEASE	ACTION_EVENT
MOUSE_DOWN	LOAD_FILE
MOUSE_UP	SAVE_FILE
MOUSE_MOVE	GOT_FOCUS
MOUSE_ENTER	LOST_FOCUS
MOUSE_EXIT	

16.2.1 The Component Class

In Java, events are generated in response to a user's interaction with the graphical user interface (GUI) of a program. As events are generated, they are sent to the component in or over which they occurred. The Component class is an abstract class, so instances of it cannot be created directly. However, it serves as the superclass for almost all parts of a program's user interface. This allows events to be sent to the specific part of the user interface that will be best able to interpret the message.

> Java 中，在用户和图形用户界面交互的过程中会产生各种事件以响应用户的各种操作，当事件产生时它将被发送给相应的组件。Component 类是一个抽象类，不能直接生成其实例化对象。它是图形用户接口几乎其他所有组件类的父类，使得可以将各种事件更好地传递给用户界面中特定的组件。

For example, if the user presses the mouse button down while the mouse pointer is over a button, the MOUSE_DOWN event is sent to the button object. This works because Button is a subclass of Component. If the button object chooses not to handle the event, it can be passed up to the object on which the button is located, usually a Window or Dialog object. That object will then have the option of handling the event, ignoring it, or passing it up to a higher level. The example above shows a class hierarchy of the subclasses of Component, which is a very active superclass.

> 例如，由于 Button 类是 Component 类的子类，当用户在按钮上单击鼠标会触发 MOUSE_DOWN 事件，该事件就传递给按钮对象。如按钮对象不去响应该事件，则该事件通常会被 Window 对象或对话框对象响应。每个组件对象都可以选择处理或忽略处理某事件，还可以将该事件传递给上一级类对象处理。

16.2.2 The Event Class

When an event occurs, it is passed to the event handling methods of a component as an instance of the Event class. This class is defined in the Java.awt package. At the time the event occurs, an instance of Event is created and member variables are used to store information about the event and the state of the system at the time. For example, if a KEY_PRESS event occurs, it may be useful to know whether the Shift key was being held down at the time. Similarly, when the mouse is clicked, it is usually important to know the x and y coordinates of the mouse pointer when the click occurred. All relevant information about an event is stored in the member variables of the Event class. Table 16.2 shows the member variables of the Event class:

> 当事件发生时，该事件会被封装成 Event 类的对象，并被传递给组件的事件处理方法处理。Event 类位于 Java.awt 包中。事件处理时，事件类对象用来保存事件的信息和系统的当前状态。例如，当 KEY_PRESS 事件触发，系统会检查是否同时有 shift 键被按下。同样，当单击鼠标时系统需要检测到鼠标所在的位置，即 X 和 Y 坐标。所有事件的相关信息都会保存在事件类的成员变量中。

Table 16.2 The member variables of Event

Member	Purpose
arg	Contains arbitrary, event-specific data
ClickCount	The number of consecutive mouse clicks. (For example, 2 indicates a double-click)
evt	The next event
Id	A number representing the type of event
Key	The key that was pressed if this is a keyboard event
Modifiers	Indicates the state of the modifier keys (Shift, Ctrl, and so on)
Target	The target component for the event
When	The time at which the event occurred expressed in seconds since midnight January 1, 1970 (GMT)
X	The x coordinate where the event occurred
Y	The y coordinate where the event occurred

In addition to the member variables shown in Table, the Event class includes some public methods you will find useful in handling events. These are shown in table 16.3. Many of these public

Chapter 16 Swing GUI Introduction

variables and methods are described in detail and used in examples in the following sections of this chapter where mouse and keyboard event handling are covered.

Table 16.3   The public, nonconstructor member methods of Event

Member	Purpose
boolean controlDown	Returns true if the Ctrl key was down
boolean metaDown	Returns true if the meta key was down
boolean shiftDown	Returns true if the Shift key was down
String toString	Returns a formatted string showing information about the event, which can be useful for display while debugging
void translate	Translates the x and y coordinates of an event relative to a component

16.2.3 Event Handling for the Mouse

While handleEvent provides a very flexible and powerful way of catching and acting on Java events, it is sometimes tedious to have to look inside the supplied Event object for the necessary values. Because mouse and keyboard events are so common, Java includes a simple set of methods customized for use with these events as shown in table 16.4.

> handleEvent 是 Java 事件处理中灵活、强有力的事件处理方法，在该方法中有时需要获取 Event 对象内部的一些变量信息，同时 Java 也为鼠标和键盘事件处理提供了一套可以用户自定义的事件处理方法。

Table 16.4   Methods for handling Mouse Events Directly

Member	Purpose
mouseDown	Called when the mouse button is pressed
mouseDrag	Called when the mouse pointer is dragged while the button is held down
mouseEnter	Called for a component when the mouse pointer is moved over the component
mouseExit	Called for a component when the mouse pointer moves off the component
mouseMove	Called when the mouse pointer is moved while the button is not held down
mouseUp	Called when the mouse button is released

The signature for each of these methods is the same. For example, the signature for mouseDrag is as follows:

```
public boolean mouseDrag(Event evt, int x, int y)
```

In handling a mouse event, the first thing a method usually looks at is the location of the mouse pointer (as given by its x and y coordinates) when the event occurred. Because these values are so frequently used, they are passed directly to the mouse event-handling methods. There is an additional advantage to using the event-specific methods instead of the generic handleEvent. By using these functions, you can reduce the clutter that is caused by involving many case statements inside handleEvent.

> 在处理鼠标事件中，方法首先要知道鼠标所在的位置(即 x、y 坐标)，主要是因为鼠标事件中经常需要用到这两个值。使用特定的事件处理方法比使用统一的 handleEvent 更好，它能减少 handleEvent 方法中使用太多 case 语句造成的结构冗余。

Because of the simplifications to the application, it requires only the use of the mouseDrag method. Similar to the MOUSE_DRAG case of handleEvent, mouseDrag stores the location of the mouse pointer and then repaints the application. Of course, because x and y have been provided as parameters they are used directly instead of evt.x and evt.y.

> 可以将该 application 简化，而只使用 mouseDrag 方法，与 handleEvent 方法中的 MOUSE_DRAG 情况类似，mouseDrag 方法存储鼠标的位置并实现重画操作。当然，由于 x、y 坐标直接作为参数传递给 mouseDrag 而不使用 evt.x 和 evt.y。

In this case, mouseDown, mouseUp, and mouseDrag are all used. The application will respond by drawing a line from the location where the mouse button was pushed down, to its current location while the mouse is dragged, and then remove the line once the mouse button is released.

16.2.4 Keyboard Event Handling

Directly handling keyboard events is similar to handling mouse events. Two methods, keyDown and keyUp, are provided. These methods are summarized in table 16.5. The signature for each method is the same and is as follows:

```
public boolean keyDown(Event evt, int key)
```

Table 16.5 Methods for Directly Handling Keyboard Events

Member	Purpose
keyDown	Called when a key is pressed down
keyUp	Called when a key is released

In addition to the keys that generate displayable characters, Java supports the usual set of navigational and function keys. The Event class defines values for the following special keys as shown in table 16.6.

Table 16.6 Values for Java keys

HOME	RIGHT	F7
END	F1	F8
PGUP	F2	F9
PGDN	F3	F10
UP	F4	F11
DOWN	F5	F12
LEFT	F6	

Each of these values can be accessed through the event class. For example, Event.LEFT represents the left arrow key on the keyboard.

The Shift, Control, and Meta Modifiers

Sometimes a key is augmented by the use of shift, control, or other modifiers. In Java you can distinguish between an unmodified key and one combined with the Shift or Ctrl key.

Earlier in this chapter you learned that the Event class contains a member named modifiers. When an event is generated, modifiers hold the state of the Shift, Ctrl, and meta keys. To discern the state of these keys, you can either look directly at modifiers or you can use the provided utility methods controlDown, shiftDown, and metaDown.

> 前面已知 Event 类包含 modifiers 成员，当事件触发时，modifiers 保存了 shift、ctrl、meta 键的状态，通过直接查看 modifiers 的值或 controlDown、shiftDown、metaDown 常量值来查看各个键的状态。

16.3 Buttons, Events, and Other Swing Basics

In this section we present enough about Swing to allow you to do some simple GUI programs.

This example contains a Swing program that produces a simple window. The window contains nothing but a button on which is written "Click to end program." If the user follows the instructions and clicks the button with his or her mouse, the program ends.

> 这个例子包含一个产生一个简单窗口的 Swing 程序。窗口只包含一个"单击结束程序"按钮。如果用户按照指示，用鼠标单击按钮，程序将结束。

The import statements give the names of the classes used and which package they are in. What we and others call the Swing library is the package named javax.swing. The AWT library is the package java.awt. Note that one package name contains an "x" and one does not.

This program is a simple class definition with only a main method. The first line in the main method creates an object of the class JFrame. That line is reproduced below:

```
JFrame firstWindow = new JFrame();
```

This is an ordinary declaration of a variable named firstWindow and an invocation of the no-argument constructor for the class JFrame. A JFrame object is a basic window. A JFrame object includes a border and the usual three buttons for minimizing the window down to an icon, changing the size of the window, and closing the window. These buttons are shown in the upper-right corner of the window, which is typical, but if your operating system normally places these buttons someplace else, that is where they will likely be located in a JFrame on your computer.

The initial size of the JFrame window is set using the JFrame method setSize, as follows:

```
firstWindow.setSize(WIDTH, HEIGHT);
```

In this case WIDTH and HEIGHT are defined int constants. The units of measure are pixels, so the window produced is 300 pixels by 200 pixels. (The term pixel is defined in the box entitled "Pixel.") As with other windows, you can change the size of a JFrame by using your mouse to drag a corner of the JFrame window.

> 在这种情况下，宽度和高度定义为 int 型常量。度量单位是像素，因此产生的窗口是 300×200 像素。与其他窗口一样，可以改变 JFrame 的大小，使用鼠标拖动 JFrame 窗口中的任意一个角落。

The buttons for minimizing the window down to an icon and for changing the size of the window behave as they do in any of the other windows you have used. The minimization button shrinks the window down to an icon. (To restore the window, you click the icon.) The second button changes the size of the window back and forth from full screen to a smaller size. The lose-window button can behave in different ways depending on how it is set by your program.

The behavior of the close-window button is set with the JFrame method setDefaultCloseOperation. The line of the program that sets the behavior of the close-window button is reproduced below:

```
firstWindow.setDefaultCloseOperation(JFrame.DO_NOTHING_ON_CLOSE);
```

In this case the argument JFrame.DO_NOTHING_ON_CLOSE is a defined constant named DO_NOTHING_ON_CLOSE, which is defined in the JFrame class. This sets the closewindow button so that when it is clicked nothing happens (unless we programmed something to happen, which we have not done).

> 在这种情况下，参数 JFrame.DO_NOTHING_ON_CLOSE 是一个在 JFrame 类中定义为 DO_NOTHING_ON_CLOSE 的常量。当单击关闭窗口按钮时，什么也不会发生。

The method setDefaultCloseOperation takes a single int argument, and each of the constants is an int constant. However, do not think of them as int values. Think of them as policies for what happens when the user clicks the close-window button. It was convenient to name these policies by int values. However, they could just as well have been named by char values or String values or something else. The fact that they are int values is an incidental detail of no real importance.

Descriptions of some of the most important methods in the class JFrame are given below. Some of these methods will not be explained until later in this chapter.

The class JFrame is in the javax.swing package.

```
public JFrame()
```
Constructor used to create an object of the class JFrame.

```
public JFrame(String title)
public void setDefaultCloseOperation(int operation)
public void setSize(int width, int height)
public void setTitle(String title)
public void add(Component componentAdded)
public void setLayout(LayoutManager manager)
public void setJMenuBar(JMenuBar menubar)
public void dispose()
```

A JFrame can have components added, such as buttons, menus, and text labels. For example, the following line adds the JButton object named end-Button to the JFrame named firstWindow:

```
firstWindow.add(endButton);
```

The description of how the JButton named endButton is created and programmed will be given in the two subsections entitled "Buttons" and "Action Listeners and Action Events" a little later in this section.

> 将在后面介绍名为 endButton 的 JButton 是如何创建和编程的。

We end this subsection by jumping ahead to the last line of the program, which is

```
firstWindow.setVisible(true);
```

This makes the JFrame window visible on the screen. At first glance this may seem strange. Why not have windows automatically become visible? Why would you create a window if you did not want it to be visible? The answer is that you may not want it to be visible at all times. You have certainly experienced windows that disappear and reappear. To hide the window, which is not desirable in this example, you would replace the argument true with false.

> 这使得 JFrame 窗口显示在屏幕上。这似乎有些奇怪。为什么不能让 Windows 自动成为可见呢？如果不希望它是可见的，为什么要创建一个窗口呢？答案是，你可能不希望它在任何时候都可见。你一定看过自动消失和重新出现的窗口。要隐藏窗口，可以将参数 true 换成 false。

Example 16.1

```java
//A First Swing Demonstration Program 第一个 Swing 演示程序
import javax.swing.JFrame;
import javax.swing.JButton;
public class FirstSwingDemo
{
    public static final int WIDTH = 300;
    public static final int HEIGHT = 200;
    public static void main(String[] args)
    {
        JFrame firstWindow = new JFrame();
        firstWindow.setSize(WIDTH, HEIGHT);
        firstWindow.setDefaultCloseOperation(
        JFrame.DO_NOTHING_ON_CLOSE);
        JButton endButton = new JButton("Click to end program.");
        EndingListener buttonEar = new EndingListener();
        endButton.addActionListener(buttonEar);
        firstWindow.add(endButton);
        firstWindow.setVisible(true);
    }
}
import java.awt.event.ActionListener;
import java.awt.event.ActionEvent;
public class EndingListener implements ActionListener
{
    public void actionPerformed(ActionEvent e)
    {
        System.exit(0);
    }
}
```

The program above produces a simple window, as shown in figure 16.2.

The setVisible Method

Many classes of Swing objects have a setVisible method. The setVisible method takes one argument of type boolean. If w is an object, such as a JFrame window, that can be displayed on the screen, then the call

```java
w.setVisible(true);
```

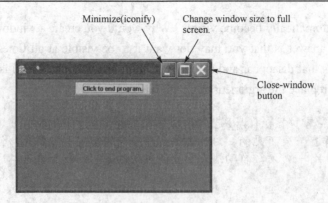

Figure 16.2   A First Swing Demonstration Program

will make w visible. The call

```
w.setVisible(false);
```

will hide w.

Syntax

```
Object_For_Screen.setVisible(Boolean_Expression);
```

Example

```
public static void main(String[] args)
{
    JFrame firstWindow = new JFrame();
    .firstWindow.setVisible(true);
}
```

16.3.1 Buttons

A button object is created in the same way that any other object is created, but you use the class JButton. For example, the following example creates a button:

```
JButton endButton = new JButton("Click to end program.");
```

The argument to the construct, in this case "Click to end program.", is a string that will be written on the button when the button is displayed. If you look at the picture of the GUI created, you will see that the button is labeled "Click to end program."

We have already discussed adding components, such as buttons, to a JFrame. The button is added to the JFrame by the following line:

```
firstWindow.add(endButton);
```

In the next subsection we explain the lines involving the methodaddActionListener.

16.3.2 Action Listeners and Action Events

Clicking a button with your mouse (or activating certain other items in a GUI) createsan object known as an event and sends the event object to another object (or objects) known as the listener(s). This is called firing the event. The listener then performs some action. When we say that the event is "sent" to the listener object, what we really mean is that some method in the listener object is invoked with the event object as the argument. This invocation happens automatically. Your Swing GUI class definition will not normally contain an invocation of this method. However, your Swing GUI class definition does need to do two things:

First, for each button, it needs to specify what objects are listeners that will respond to events fired by that button; this is called registering the listener.

Second, it must define the methods that will be invoked when the event is sent to the listener. Note that these methods will be defined by you, but in normal circumstances, you will never write an invocation of these methods. The invocations will take place automatically.

> 用鼠标单击一个按钮(或激活一个 GUI 上的某些其他控件)将创建一个事件对象,这个事件对象将被发送给称为侦听器的一个对象。这就是所谓的触发事件。侦听器负责执行一些动作。当说该事件"发送"到侦听器对象,真正的意思是,一些侦听器对象中的方法是使用事件对象作为参数的。此调用自动发生。Swing GUI 类的定义通常不会包含调用此方法。但是,Swing GUI 类的定义需要做两件事情:
> 首先,为每个按钮指定一个侦听器。这个侦听器响应该按钮触发的事件,这就是所谓的注册侦听器。
> 其次,它必须定义事件发送到侦听器时被调用的方法。注意,这些方法将会由你定义,但在正常情况下,你将永远不会写这些方法的调用。它们的调用将自动进行。

The following lines create an EndingListener object named buttonEar and register buttonEar as a listener to receive events from the button named endButton:

```
EndingListener buttonEar = new EndingListener();
endButton.addActionListener(buttonEar);
```

The second line says that buttonEar is registered as a listener to endButton, which means buttonEar will receive all events fired by endButton.

Different kinds of components require different kinds of listener classes to handle the events they fire. A button fires events known as action events, which are handled by listeners known as action listeners.

An action listener is an object whose class implements the ActionListener interface. For example, the class EndingListener implements the ActionListener interface. The ActionListener interface has only one method heading that must be implemented, namely the following:

```
public void actionPerformed(ActionEvent e)
```

In the class EndingListener above, the actionPerformed method is defined as follows:

```
public void actionPerformed(ActionEvent e)
{
    System.exit(0);
}
```

If the user clicks the button endButton, that sends an action event to the action listener for that button. But buttonEar is the action listener for the button endButton, so the action event goes to buttonEar. When an action listener receives an action event, the event is automatically passed as an argument to the method actionPerformed and the method actionPerformed is invoked. If the event is called e, then the following invocation takes place in response to endButton firing e:

```
buttonEar.actionPerformed(e);
```

In this case the parameter e is ignored by the method actionPerformed. The method actionPerformed simply invokes System.exit and thereby ends the program. So, if the user clicks endButton (the one labeled "Click to end program."), the net effect is to end the program and so the window goes away.

> 在这种情况下，参数 e 将被 actionPerformed 方法忽略。方法 actionPerformed 简单地调用 System.exit，从而结束程序。所以，如果用户单击 endButton（一个标有"单击结束程序"的按钮），程序结束，窗口消失。

Note that you never write any code that says buttonEar.actionPerformed(e);This action does happen, but the code for this is embedded in some class definition inside the Swing and/or AWT libraries. Somewhere the code says something like

```
bla.actionPerformed(e);
```

and somehow buttonEar gets plugged in for the parameter bla and this invocation of actionPerformed is executed. But, all this is done for you. All you do is define the method actionPerformed and register buttonEar as a listener for endButton.

Note that the method actionPerformed must have a parameter of type ActionEvent, even if your definition of actionPerformed does not use this parameter. This is because the invocations of actionPerformed were already programmed for you and so must allow the possibility of using the ActionEvent parameter e. As you will see, in other Swing GUIs the method actionPerformed does often use the event e to determine which button was clicked. This first example is a special, simple case because there is only one button. Later in this chapter we will say more about defining the actionPerformed method in more complicated situations. 3 4 5

> 方法 actionPerformed 必须有一个 ActionEvent 类型的参数，即使你定义的 actionPerformed 不使用这个参数。这是因为调用 actionPerformed 的程序已经为你写好了，所以必须允许使用 ActionEvent 参数 e 的可能性存在。正如你将看到的，在其他 Swing 图形用户界面的 actionPerformed 方法中经常使用，以确定哪个按钮被单击的事件 e。这个例子是一个特殊的、简单的例子，因为只有一个按钮。本章后面会讲解有关定义 action-Performed 方法的更复杂的情况。

A GUI program is normally based on a kind of infinite loop. There may not be a Java loop statement in a GUI program, but nonetheless the GUI program need not ever end. The windowing system normally stays on the screen until the user indicates that it should go away (for example, by clicking the "Click to end program." button in Example before). If the user never asks the windowing system to go away, it will never go away. When you write a Swing GUI program, you need to use System.exit to end the program when the user (or something else) says it is time to end the program. Unlike the kinds of programs we saw before this chapter, a Swing program will not end after it has executed all the code in the program. A Swing program does not end until it executes a System.exit. (In some cases, the System.exit may be in some library code and need not be explicitly given in your code.)

16.3.3 Labels

We have seen how to add a button to a JFrame. If you want to add some text to your JFrame, use a label instead of a button. A label is an object of the class JLabel. A label is little more than a line of text. The text for the label is given as an argument to the JLabel constructor as follows:

```
JLabel greeting = new JLabel("Hello");
```

The label greeting can then be added to a JFrame just as a button is added. For example, the following might appear in a constructor for a derived class of JFrame:

```
JLabel greeting = new JLabel("Hello");
add(greeting);
```

The next Programming Example includes a label in a JFrame GUI.

16.3.4 Color

You can set the color of a JFrame (or other GUI object). To set the background color of a JFrame, you use getContentPane( ).setBackground(Color); For example, the following will set the color of the JFrame named someFrame to blue:

```
someFrame.getContentPane().setBackground(Color.BLUE);
```

Alternatively, if you set the color in the constructor for the JFrame, the invocation takes the form

```
getContentPane().setBackground(Color.BLUE);
```

which is equivalent to

```
this.getContentPane().setBackground(Color.BLUE);
```

The next Programming Example shows a JFrame object (in fact, two of them) with color. The method invocation getContentPane( ) returns something called the content pane of the JFrame. So,

```
getContentPane().setBackground(Color.BLUE);
```

actually sets the color of the content pane to blue. The content pane is the "inside" of the JFrame, so coloring the content pane has the effect of coloring the inside of the JFrame. However, you can think of

```
getContentPane().setBackground(Color);
```

as a peculiarly spelled method invocation that sets the color of the JFrame. (In this book, we will not be referring to the content pane of a JFrame except when we want to color the JFrame, so we will explain the content pane no further.)

You use getContentPane only when you give color to a JFrame. As you will see, to set the color of some component in a JFrame, such as a button, you simply use the method setBackground with the button or other component as the calling object. You will see examples of adding color to components in Section 16.3.

> 当给一个 JFrame 指定颜色时，可以使用 getContentPane 方法。正如所看到的，在给一个 JFrame 中的一些组件如按钮设置颜色时，只需用按钮或其他组件来调用方法 setBackground。

What kind of thing is a color when used in a Java Swing class? Like everything else in Java, a color is an object—in this case, an object that is an instance of the class Color. The class Color is in the java.awt package. (Note that the package name is java.awt, not javax.awt.)

> 当用在 Java Swing 类中时，一种颜色是什么东西？就像 Java 中其他东西一样，颜色也是一个对象。在这种情况下，一种颜色就是 Color 类的一个实例。Color 类定义在 java.awt 包中（注意包的名称是 java.awt 中，不是 javax.awt）。

In a later chapter, you will see how you can define your own colors, but for now we will use the colors that are already defined for you, such as Color.BLUE, which is a constant named BLUE that is defined in the class Color. The constant, of course, represents the color blue. If you set the background of a JFrame to Color.BLUE, then the JFrame will have a blue background. The type of the constant Color.BLUE and other such constants is Color. The list of color constants defined for you are given in below. The next Programming Example has an example of a constructor with one parameter of type Color.

The Color Constants
- Color.BLACK
- Color.BLUE
- Color.CYAN
- Color.DARK_GRAY
- Color.GRAY
- Color.GREEN
- Color.LIGHT_GRAY
- Color.MAGENTA
- Color.ORANGE
- Color.PINK
- Color.RED
- Color.WHITE
- Color.YELLOW

The class Color is in the java.awt package.

This example shows a class for GUIs with a label and a background color. We have already discussed the use of color for this window. The label is used to display the text string "Close-window button works." The label is created as follows:

```
JLabel aLabel = new JLabel("Close-window button works.");
```

The label is added to the JFrame with the method add as shown in the following line:

```
add(aLabel);
```

The GUI class ColoredWindow programs the close-window button as follows:

```
setDefaultCloseOperation(JFrame.EXIT_ON_CLOSE);
```

This way, when the user clicks the close-window button, the program ends. Note that if the program has more than one window, and the user clicks the close-window button in any one window of the class ColoredWindow, then the entire program ends and all windows go away.

Note that we set the title of the JFrame by making it an argument to super rather than an argument to setTitle. This is another common way to set the title of a JFrame.

If you run the program DemoColoredWindow, then the two windows will be placed one on top of the other. To see both windows, you need to use your mouse to move the top window.

Example 16.2

```
//A JFrame with Color 带演示的窗体
import javax.swing.JFrame;
import javax.swing.JLabel;
```

```java
import java.awt.Color;
public class ColoredWindow extends JFrame
{
    private static final long serialVersionUID = -5579202859750338270L;
    public static final int WIDTH = 300;
    public static final int HEIGHT = 200;
    public ColoredWindow(Color theColor)
    {
        super("No Charge for Color");
        setSize(WIDTH, HEIGHT);
        setDefaultCloseOperation(JFrame.EXIT_ON_CLOSE);
        getContentPane().setBackground(theColor);
        JLabel aLabel = new JLabel("Close-window button works.");
        add(aLabel);
    }
    public ColoredWindow()
    {
        this(Color.GREEN);
    }
}
import java.awt.Color;
public class DemoColoredWindow
{
    public static void main(String[] args)
    {
        ColoredWindow w1 = new ColoredWindow();
        w1.setVisible(true);
        ColoredWindow w2 = new ColoredWindow(Color.GRAY);
        w2.setVisible(true);
    }
}
```

The program above produces two simple window, as shown in figure 16.3.

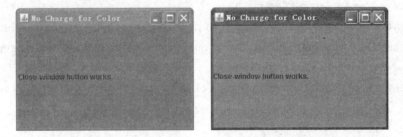

Figure 16.3   A JFrame with Color Resulting GUI

16.4   Containers and Layout Managers

There are two main ways to create new classes from old classes. One way is to use inheritance; this is known as the Is-A relationship. For example, an object of the class ColoredWindow is a JFrame because ColoredWindow is a derived class of the class JFrame. The second way to create a new class from an existing class (or classes) is to have instance variables of an already existing class type; this is known as composition or the Has-A relationship. The Swing library has already set things up so you can easily use composition. The actual code for declaring instance variables is in the Swing library classes, such as the class JFrame. Rather than declaring instance variables, you add components to a JFrame using the add method. This does ultimately set some instance variables, but that is done automatically when you use the add method. In this section, we discuss adding and arranging components in a GUI or subpart of a GUI.

Thus far, we have only added one component, either a button or a label, to a JFrame. You can add more than one component to a JFrame. To do so, you use the add method multiple times, but the add method simply tells which components are added to the JFrame; it does not say how they are arranged, such as side by side or one above the other. To describe how the components are arranged, you need to use a layout manager.

> 到目前为止，我们只添加了组件、按钮或标签到JFrame。可以添加多个组件到JFrame。要做到这点，可以使用add方法多次，但是add方法只是简单地告诉哪些组件添加到JFrame，它不说是如何布局的，如并排或上下排列。为了描述组件是如何布局的，需要使用布局管理器。

In this section, we will see that there are other classes of objects besides JFrames that can have components added with the add method and arranged by a layout manager. All these classes are known as container classes.

> 除了JFrames外还有其他类的对象，可以用add方法添加组件和使用布局管理器排列组件。所有这些类被称为容器类。

16.4.1 Border Layout Managers

Figure 16.4 contains an example of a GUI that uses a layout manager to arrange three labels in a JFrame. The labels are arranged one below the other on three lines.

A layout manager is added to the JFrame class in Figure 16.4 with the following line:

```
setLayout(new BorderLayout());
```

BorderLayout is a layout manager class, so new BorderLayout( ) produces a new anonymous object of the class BorderLayout. This BorderLayout object is given the task of arranging components (in this case, labels) that are added to the JFrame.

It may help to note that the above invocation of setLayout is equivalent to the following:

```
BorderLayout manager = new BorderLayout();
setLayout(manager);
```

Example 16.3

```java
//The BorderLayout Manager 边界布局管理器使用示例
import javax.swing.JFrame;
import javax.swing.JLabel;
import java.awt.BorderLayout;
public class BorderLayoutJFrame extends JFrame
{
    private static final long serialVersionUID = -1244791694895810528L;
    public static final int WIDTH = 500;
    public static final int HEIGHT = 400;
```

```
        public BorderLayoutJFrame()
        {
            super("BorderLayout Demonstration");
            setSize(WIDTH, HEIGHT);
            setDefaultCloseOperation(JFrame.EXIT_ON_CLOSE);
            setLayout(new BorderLayout());
            JLabel label1 = new JLabel("First label");
            add(label1, BorderLayout.NORTH);
            JLabel label2 = new JLabel("Second label");
            add(label2, BorderLayout.SOUTH);
            JLabel label3 = new JLabel("Third label");
            add(label3, BorderLayout.CENTER);
        }
    }
    public class BorderLayoutDemo
    {
        public static void main(String[] args)
        {
            BorderLayoutJFrame gui = new BorderLayoutJFrame();
            gui.setVisible(true);
        }
    }
```

The program above produces a simple window, as shown in figure 16.4.

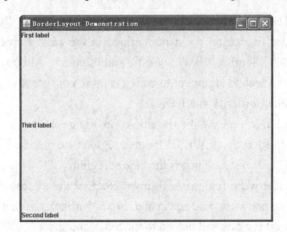

Figure 16.4 The BorderLayout Manager Resulting GUI

A BorderLayout manager places labels (or other components) into the five regions: BorderLayout.NORTH, BorderLayout.SOUTH, BorderLayout.EAST, BorderLayout.WEST, and BorderLayout.CENTER. These five regions are arranged as shown in Figure 16.5. The outside box represents the JFrame (or other container to which you will add things). None of the lines in the diagram will be visible unless you do something to make them visible. We drew them in to show you where each region is located.

In figure 16.5, we added labels as follows:

```
        JLabel label1 = new JLabel("First label");
        add(label1, BorderLayout.NORTH);
        JLabel label2 = new JLabel("Second label");
        add(label2, BorderLayout.SOUTH);
```

```
JLabel label3 = new JLabel("Third label");
add(label3, BorderLayout.CENTER);
```

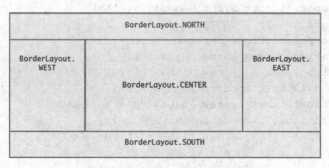

Figure 16.5　BorderLayout Regions

When you use a BorderLayout manager, you give the location of the added component as a second argument to the method: add, as in the following:

```
add(label1, BorderLayout.NORTH);
```

The labels (or other components to be added) need not be added in any particular order, because the second argument completely specifies where the label is placed.

> 标签(或其他要被添加的组件)无需按任何指定顺序添加,因为第二个参数指定了标签放在哪里。

BorderLayout.NORTH, BorderLayout.SOUTH, BorderLayout.EAST, BorderLayout.WEST, and BorderLayout.CENTER are five string constants defined in the class BorderLayout. The values of these constants are "North", "South", "East", "West", and "Center". Although you can use a quoted string such as "North" as the second argument to add, it is more consistent with our general style rules to use a defined constant like BorderLayout.NORTH.

You need not use all five regions. For example, in Figure 16.5 we did not use the regions BorderLayout.EAST and BorderLayout.WEST. If some regions are not used, any extra space is given to the BorderLayout.CENTER region, which is the largest region.

(The space is divided between regions as follows: Regions are allocated space in the order first north and south, second east and west, and last center. So, in particular, if there is nothing in the north region, then the east and west regions will extend to the top of the space.)

From this discussion, it sounds as though you can place only one item in each region, but later in this chapter, when we discuss panels, you will see that there is a way to group items so that more than one item can (in effect) be placed in each region.

> 从这个讨论,听起来好像在每个区域只能放置一个项目。但在本章的后面,你会看到有一个项目分组的方式,可以使多个项目同时放置在任意一个区域。

There are some standard layout managers defined for you in the java.awt package, and you can also define your own layout managers. However, for most purposes, the layout managers defined in the standard libraries are all that you need, and we will not discuss how you can create your own layout manager classes.

> 在java.awt包中定义了一些标准的布局管理器。你也可以定义自己的布局管理器。但是，对于大多数用途，标准库中定义的布局管理器能够满足你所有的需求，本章将不讨论如何创建自己的布局管理器类。

16.4.2 Flow Layout Managers

The FlowLayout manager is the simplest layout manager. It arranges components one after the other, going from left to right, in the order in which you add them to the JFrame (or other container class) using the method add. For example, if the class in Display 17.7 had used the FlowLayout manager instead of the BorderLayout manager, it would have used the following code:

```
setLayout(new FlowLayout());
JLabel label1 = new JLabel("First label");
add(label1);
JLabel label2 = new JLabel("Second label");
add(label2);
JLabel label3 = new JLabel("Third label");
add(label3);
```

Note that if we had used the FlowLayout manager, as in the preceding code, then the add method would have only one argument. With a FlowLayout manager, the items are displayed in the order they are added, so that the labels above would be displayed all on one line as follows:

First label Second label Third label

You will see a number of examples of GUIs that use the FlowLayout manager class later in this chapter.

16.4.3 Grid Layout Managers

A GridLayout manager arranges components in a two-dimensional grid with some number of rows and columns. With a GridLayout manager, each entry is the same size. For example, the following says to use a GridLayout manager with aContainer, which can be a JFrame or other container:

```
aContainer.setLayout(new GridLayout(2, 3));
```

The two numbers given as arguments to the constructor GridLayout specify the number of rows and columns. This would produce the following sort of layout as figure 16.6 shown.

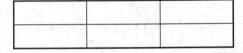

Figure 16.6 GridLayout

The lines will not be visible unless you do something special to make them visible. They are just included here to show you the region boundaries.

> 图16.6中的行是不可见的，除非添加代码，使它们可见。它们只是在这里展示区域的边界。

When using a GridLayout manager, each component is stretched so that it completely fills its grid position.

Although you specify a number of rows and columns, the rules for the number of rows and columns is more complicated than what we have said so far. If the values for the number of rows and the number of columns are both nonzero, then the number of columns will be ignored. For example, if the specification is new GridLayout(2, 3), then some sample sizes are as follows: If you add six items, the grid will be as shown. If you add seven or eight items, a fourth column is automatically added, and so forth. If you add fewer than six components, there will be two rows and a reduced number of columns.

There is another way to specify that the number of columns is to be ignored. You can specify that the number of columns is to be ignored by setting the number of columns to zero, which will allow any number of columns. So a specification of (2, 0) is equivalent to (2, 3), and in fact is equivalent to (2, n) for any nonnegative value of n. Similarly, you can specify that the number of rows is to be ignored by setting the number of rows to zero, which will allow any number of rows.

When using the GridLayout class, the method add has only one argument. The items are placed in the grid from left to right, first filling the top row, then the second row, and so forth. You are not allowed to skip any grid position (although you will later see that you can add something that does not show and so gives the illusion of skipping a grid position).

> 当使用 GridLayout 类，add 方法只有一个参数。该项目是由左到右的网格，放置在最上面一行首先填充，然后第二行。不得跳过任何的 grid（虽然可以添加一个不被显示的东西，从而有跳过一个 grid 的错觉）。

Example 16.4

```java
//The GridLayout Manager 网格布局管理器使用示例
import javax.swing.JFrame;
import javax.swing.JLabel;
import java.awt.GridLayout;
public class GridLayoutJFrame extends JFrame
{
    private static final long serialVersionUID = -6247169824901937988L;
    public static final int WIDTH = 500;
    public static final int HEIGHT = 400;
    public static void main(String[] args)
    {
        GridLayoutJFrame gui = new GridLayoutJFrame(2, 3);
        gui.setVisible(true);
    }
    public GridLayoutJFrame(int rows, int columns )
    {
        super();
        setSize(WIDTH, HEIGHT);
        setTitle("GridLayout Demonstration");
        setDefaultCloseOperation(JFrame.EXIT_ON_CLOSE);
        setLayout(new GridLayout(rows, columns ));
        JLabel label1 = new JLabel("First label");
        add(label1);
        JLabel label2 = new JLabel("Second label");
        add(label2);
        JLabel label3 = new JLabel("Third label");
        add(label3);
```

```
            JLabel label4 = new JLabel("Fourth label");
            add(label4);
            JLabel label5 = new JLabel("Fifth label");
            add(label5);
        }
    }
```

Note that we have placed a demonstration main method in the class definition in Example 16.4. This is handy, but is not typical. Normally, a Swing GUI is created and displayed in a main method (or other method) in some class other than the class that defines the GUI. However, it is perfectly legal and sometimes convenient to place a main method in the GUI class definition so that it is easy to display a sample of the GUI. Note that the main method that is given in the class itself is written in the same way as a main method that is in some other class. In particular, you need to construct an object of the class, as in the following line from the main method in Example 16.4:

```
GridLayoutJFrame gui = new GridLayoutJFrame(2, 3);
```

The program of Example 16.4 produces a simple window, as shown in figure 16.7.

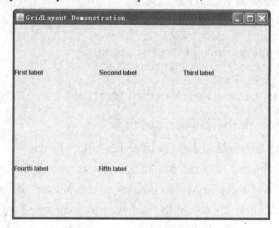

Figure 16.7   The GridLayout Manager RESULTING GUI

16.4.4 Panels

A GUI is often organized in a hierarchical fashion, with window-like containers, known as panels, inside of other window-like containers. A panel is an object of the class JPanel, which is a very simple container class that does little more than group objects. It is one of the simplest container classes, but an extremely useful one. A JPanel object is analogous to the braces used to combine a number of simpler Java statements into a single larger Java statement. It groups smaller objects, such as buttons and labels, into a larger component (the JPanel). You can then put the JPanel object in a JFrame. Thus, one of the main functions of JPanel objects is to subdivide a JFrame (or other container) into different areas.

For example, when you use a BorderLayout manager, you can place components in each of the five locations BorderLayout.NORTH, BorderLayout.SOUTH, BorderLayout.EAST, BorderLayout.WEST, and BorderLayout.CENTER. But what if you want to put two components at the bottom of the screen in the BorderLayout.SOUTH position? To do this, you would put the two components in a panel and then place the panel in the BorderLayout.SOUTH position.

例如，当使用 BorderLayout 管理器时，可以将组件放置在 5 个地点：BorderLayout.NORTH、BorderLayout.SOUTH、BorderLayout.EAST、BorderLayout.WEST 和 BorderLayout.CENTER。但是，如果要把两个组件放置在位于屏幕底部的 BorderLayout.SOUTH，该怎么做？要做到这一点，可以把这两个组件放置在一个面板上，然后把该面板放置在 BorderLayout.SOUTH。

You can give different layout managers to a JFrame and to each panel in the JFrame. Because you can add panels to other panels and each panel can have its own layout manager, this enables you to produce almost any kind of overall layout of the items in your GUI.

可以给一个 JFrame、JFrame 中的每个面板指定不同的布局管理器。因为可以添加面板到其他面板，每个面板都可以有自己的布局管理器，这样能够在 GUI 项目中生成任意的总体布局。

For example, if you want to place two buttons at the bottom of your JFrame GUI, you might add the following to the constructor of your JFrame GUI:

```
setLayout(new BorderLayout());
JPanel buttonPanel = new JPanel();
buttonPanel.setLayout(new FlowLayout());
JButton firstButton = new JButton("One");
buttonPanel.add(firstButton);
JButton secondButton = new JButton("Two");
buttonPanel.add(secondButton);
add(buttonPanel, BorderLayout.SOUTH);
```

The next Programming Example makes use of panels within panels.

When run, the GUI defined in Example 16.5 looks as shown in the first view. The entire background is light gray, and there are three buttons at the bottom of the GUI labeled "Green", "White", and "Gray". If you click any one of the buttons, a vertical stripe with the color written on the button appears. You can click the buttons in any order. In the last three views in figure 16.8, figure 16.9 and figure 16.10, we show what happens if you click the buttons in left-to-right order.

首次运行时，第一眼看起来，Example 16.5 定义的 GUI 看起来如下面的图所示。整个背景是浅灰色，并在 GUI 的底部有三个按钮，按钮标有"绿色"、"白"和"灰色"。如果单击任何一个按钮，会出现一个按钮颜色的竖条纹。可以按任意顺序单击按钮。在图 16.8 到图 16.10 中，展示从左到右单击按钮，程序运行界面发生的变化。

The green, white, and gray stripes are the JPanels named greenPanel, whitePanel, and grayPanel. At first the panels are not visible because they are all light gray, so no borders are visible. When you click a button, the corresponding panel changes color and so is clearly visible.

名为 greenPanel、whitePanel、grayPanel 的绿色、白色、灰色条纹，都是面板（JPanel）。起初面板是不可见的，因为它们都是浅灰色的，所以没有边界是可见的。当按一下按钮，相应的面板颜色会发生变化，所以清晰可见了。

Notice how the action listeners are set up. Each button registers the this parameter as a listener, as in the following line:

```
        redButton.addActionListener(this);
```
Because this line appears inside of the constructor for the class PanelDemo, the this parameter refers to PanelDemo, which is the entire GUI. Thus, the entire JFrame (the entire GUI) is the listener, not the JPanel. So when you click one of the buttons, it is the actionPerformed method in PanelDemo that is executed.

When a button is clicked, the actionPerformed method is invoked with the action event fired as the argument to actionPerformed. The method actionPerformed recovers the string written on the button with the following line:
```
        String buttonString = e.getActionCommand();
```
The method actionPerformed then uses a multiway if-else statement to determine if buttonString is "Green", "White", or "Gray" and changes the color of the corresponding panel accordingly. It is common for an actionPerformed method to be based on such a multiway if-else statement, although we will see another approach in the subsection entitled "Listeners as Inner Classes" later in this chapter.

Example 16.5 also introduces one other small but new technique. We gave each button a color. We did this with the method setBackground, using basically the same technique that we used in previous examples. You can give a button or almost any other item a color using setBackground. Note that you do not use getContentPane when adding color to any component other than a JFrame.

Example 16.5
```java
//Using Panels 面板使用示例
import javax.swing.JFrame;
import javax.swing.JPanel;
import java.awt.BorderLayout;
import java.awt.GridLayout;
import java.awt.FlowLayout;
import java.awt.Color;
import javax.swing.JButton;
import java.awt.event.ActionListener;
import java.awt.event.ActionEvent;
public class PanelDemo extends JFrame implements ActionListener
{
private static final long serialVersionUID=7019347414299879833L;
    public static final int WIDTH = 300;
    public static final int HEIGHT = 200;
    private JPanel greenPanel;
    private JPanel whitePanel;
    private JPanel grayPanel;
    public static void main(String[] args)
    {
        PanelDemo gui = new PanelDemo();
        gui.setVisible(true);
    }
    public PanelDemo()
    {
        super("Panel Demonstration");
        setSize(WIDTH, HEIGHT);
        setDefaultCloseOperation(JFrame.EXIT_ON_CLOSE);
        setLayout(new BorderLayout());
        JPanel biggerPanel = new JPanel();
        biggerPanel.setLayout(new GridLayout(1, 3));
```

```java
            greenPanel = new JPanel();
            greenPanel.setBackground(Color.LIGHT_GRAY);
            biggerPanel.add(greenPanel);
            whitePanel = new JPanel();
            whitePanel.setBackground(Color.LIGHT_GRAY);
            biggerPanel.add(whitePanel);
            grayPanel = new JPanel();
            grayPanel.setBackground(Color.LIGHT_GRAY);
            biggerPanel.add(grayPanel);
            add(biggerPanel, BorderLayout.CENTER);
            JPanel buttonPanel = new JPanel();
            buttonPanel.setBackground(Color.LIGHT_GRAY);
            buttonPanel.setLayout(new FlowLayout());
            JButton greenButton = new JButton("Green");
            greenButton.setBackground(Color.GREEN);
            greenButton.addActionListener(this);
            buttonPanel.add(greenButton);
            JButton whiteButton = new JButton("White");
            whiteButton.setBackground(Color.WHITE);
            whiteButton.addActionListener(this);
            buttonPanel.add(whiteButton);
            JButton grayButton = new JButton("Gray");
            grayButton.setBackground(Color.GRAY);
            grayButton.addActionListener(this);
            buttonPanel.add(grayButton);
            add(buttonPanel, BorderLayout.SOUTH);
        }
        public void actionPerformed(ActionEvent e)
        {
            String buttonString = e.getActionCommand();
            if (buttonString.equals("Green"))
                greenPanel.setBackground(Color.GREEN);
            else if (buttonString.equals("White"))
                whitePanel.setBackground(Color.WHITE);
            else if (buttonString.equals("Gray"))
                grayPanel.setBackground(Color.GRAY);
            else
                System.out.println("Unexpected error.");
        }
    }
```

The program above produces four simple window, as shown in figure 16.8, figure 16.9, figure 16.10 and figure 16.11.

Figure 16.8 Using Panels RESULTING GUI (When first run)

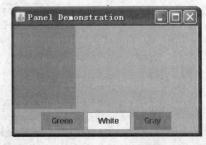

Figure 16.9 Using Panels RESULTING GUI (After clicking Green button)

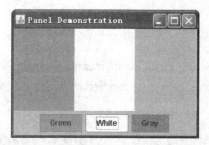

Figure 16.10   Using Panels RESULTING GUI
(After clicking White button)

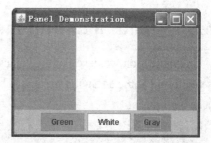

Figure 16.11   Using Panels Resulting GUI
(After clicking Gray button)

16.4.5   The Container Class

The class called Container is in the java.awt package. Any descendent class of the class Container can have components added to it (or, more precisely, can have components added to objects of the class). The class JFrame is a descendent class of the class Container, so any descendent class of the class JFrame can serve as a container to hold labels, buttons, panels, or other components.

> 容器类在 java.awt 包中。容器类的任何派生类，都可以添加组件(更确切地说，可以添加组件到容器类的对象)。类 JFrame 是一个派生的容器类，因此任何 JFrame 的派生类都可以作为一个容器来保存标签、按钮、面板或其他组件。

Similarly, the class JPanel is a descendent of the class Container, and any object of the class JPanel can serve as a container to hold labels, buttons, other panels, or other components. Note that the Container class is in the AWT library and not in the Swing library. This is not a major issue, but it does mean that the import statement for the Container class is

import java.awt.Container;

A container class is any descendent class of the class Container. The class JComponent serves a similar roll for components. Any descendent class of the class JComponent is called a JComponent or sometimes simply a component. You can add any JComponent object to any container class object.

The class JComponent is derived from the class Container, so you can add a JComponent to another JComponent. Often, this will turn out to be a viable option; occasionally it is something to avoid.

The classes Component, Frame, and Window are AWT classes that some readers may have heard of. We include them for reference value, but we will have no need for these classes.

> 组件、框架和窗口，是一些读者可能已经听说过的 AWT 类。我们将不使用这些类，仅参考这些类的使用。

When you are dealing with a Swing container class, you have three kinds of objects to deal with:
(1) The container itself, probably some sort of panel or window-like object
(2) The components you add to the container, like labels, buttons, and panels
(3) A layout manager, which positions the components inside the container

You have seen examples of these three kinds of objects in almost every JFrame class we have defined. Almost every complete GUI you build, and many subparts of the GUIs you build, will be made up of these three kinds of objects.

16.5 Menus and Buttons

In this section we describe the basics of Swing menus. Swing menu items (menu choices) behave essentially the same as Swing buttons. They generate action events that are handled by action listeners, just as buttons do.

> 本节将描述 Swing 菜单的基础知识。Swing 菜单项(菜单选择)在行为本质上和 Swing 按钮是相同的。它们产生动作，动作侦听器处理事件，就像按钮一样。

16.5.1 Menu Bars, Menus, and Menu Items

When adding menus, you use the three Swing classes JMenuItem, JMenu, and JMenuBar. Entries on a menu are objects of the class JMenuItem. These JMenuItems are placed in JMenus, and then the JMenus are typically placed in a JMenuBar. Let's look at the details.

> 当添加菜单时，可以使用三个 Swing 类：JMenuItem，JMenu，JMenuBar。菜单项类 JMenuItem 的对象是菜单的入口。这些菜单项放置在菜单上，菜单放置在菜单栏上。

A menu is an object of the class JMenu. A choice on a menu is called a menu item and is an object of the class JMenuItem. A menu item is identified by the string that labels it, such as "Green", "White", or "Gray" in the menu in Example 16.6. You can add as many JMenuItems as you wish to a menu. The menu lists the items in the order

Example 16.6

```java
//A GUI with a Menu 带菜单的图形用户接口程序示例
import javax.swing.JFrame;
import javax.swing.JPanel;
import java.awt.GridLayout;
import java.awt.Color;
import javax.swing.JMenu;
import javax.swing.JMenuItem;
import javax.swing.JMenuBar;
import java.awt.event.ActionListener;
import java.awt.event.ActionEvent;
public class MenuDemo extends JFrame implements ActionListener
{
    private static final long serialVersionUID=6411455966934962764L;
    public static final int WIDTH = 300;
    public static final int HEIGHT = 200;
    private JPanel greenPanel;
    private JPanel whitePanel;
    private JPanel grayPanel;
    public static void main(String[] args)
    {
        MenuDemo gui = new MenuDemo();
        gui.setVisible(true);
    }
    public MenuDemo()
    {
```

Chapter 16 Swing GUI Introduction

```java
        super("Menu Demonstration");
        setSize(WIDTH, HEIGHT);
        setDefaultCloseOperation(JFrame.EXIT_ON_CLOSE);
        setLayout(new GridLayout(1, 3));
        greenPanel = new JPanel();
        greenPanel.setBackground(Color.LIGHT_GRAY);
        add(greenPanel);
        whitePanel = new JPanel();
        whitePanel.setBackground(Color.LIGHT_GRAY);
        add(whitePanel);
        grayPanel = new JPanel();
        grayPanel.setBackground(Color.LIGHT_GRAY);
        add(grayPanel);
        JMenu colorMenu = new JMenu("Add Colors");
        JMenuItem greenChoice = new JMenuItem("Green");
        greenChoice.addActionListener(this);
        colorMenu.add(greenChoice);
        JMenuItem whiteChoice = new JMenuItem("White");
        whiteChoice.addActionListener(this);
        colorMenu.add(whiteChoice);
        JMenuItem grayChoice = new JMenuItem("Gray");
        grayChoice.addActionListener(this);
        colorMenu.add(grayChoice);
        JMenuBar bar = new JMenuBar();
        bar.add(colorMenu);
        setJMenuBar(bar);
    }
    public void actionPerformed(ActionEvent e)
    {
        String buttonString = e.getActionCommand();
        if (buttonString.equals("Green"))
            greenPanel.setBackground(Color.GREEN);
        else if (buttonString.equals("White"))
            whitePanel.setBackground(Color.WHITE);
        else if (buttonString.equals("Gray"))
            grayPanel.setBackground(Color.GRAY);
        else
            System.out.println("Unexpected error.");
    }
}
```

The program above produces four simple window, as shown in figure 16.12, figure 16.13, figure 16.14 and figure 16.15.

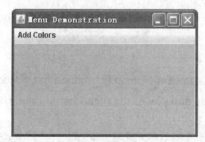

Figure 16.12 Using Panels RESULTING GUI

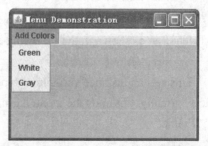

Figure 16.13 Using Panels RESULTING GUI (after clicking Add Colors in the menu bar)

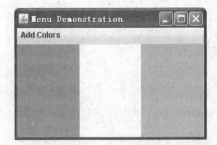
Figure 16.14　Using Panels RESULTING GUI (after choosing Green and White on the menu)

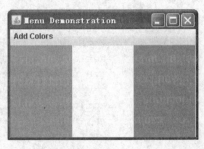
Figure 16.15　Using Panels RESULTING GUI (after choosing all the colors on the menu)

In which they are added. The following code, taken from the constructor in Example 16.6, creates a new JMenu object named colorMenu and then adds a JMenuItem labeled "Red". Other menu items are added in a similar way.

```
JMenu colorMenu = new JMenu("Add Colors");
JMenuItem greenChoice = new JMenuItem("Green");
greenChoice.addActionListener(this);
colorMenu.add(greenChoice);
```

Note that, just as we did for buttons before in this chapter we have registered the this parameter as an action listener for each menu item. Defining action listeners and registering listeners for menu items are done in the exact same way as for buttons. In fact, the syntax is even the same.

You add a JMenuItem to an object of the class JMenu using the method add in exactly the same way that you add a component, such as a button, to a container object. Moreover, if you look at the preceding code, you will see that you specify a string for a JMenuItem in the same way that you specify a string to appear on a button.

> 添加一个菜单项对象到菜单类 JMenu 的一个对象中，使用的方法和添加一个按钮组件到一个容器对象所使用的方法完全相同。此外，为菜单项指定一个字符串，与为一个按钮指定一个字符串的方式是一样的。

A menu bar is a container for menus, typically placed near the top of a windowing interface. You add a menu to a menu bar using the method add in the same way that you add menu items to a menu. The following code from the constructor creates a new menu bar named bar and then adds the menu named colorMenu to this menu bar:

```
JMenuBar bar = new JMenuBar();
bar.add(colorMenu);
```

There are two different ways to add a menu bar to a JFrame. You can use the method setJMenuBar, as shown in the following code:

```
setJMenuBar(bar);
```

This sets an instance variable of type JMenuBar so that it names the menu bar named bar. Saying it less formally, this adds the menu bar named bar to the JFrame and places the menu bar at the top of the JFrame.

> 这里设置了一个 JMenuBar 类型的实例变量。因此，它将菜单命名为 bar。添加了名为 bar 的菜单到 JFrame，并将其放置在 JFrame 的顶部菜单栏。

Alternatively, you can use the add method to add a menu bar to a JFrame (or to any other container). You do so in the same way that you add any other component, such as a label or a button.

> 可以使用 add 方法添加菜单栏到 JFrame(或任何其他容器)。可以以同样的方式,添加任意组件,如标签或按钮。

16.5.2 The AbstractButton Class

The classes JButton and JMenuItem are derived classes of the abstract class named AbstractButton. All of the basic properties and methods of the classes JButton and JMenuItem are inherited from the class AbstractButton. That is why objects of the class JButton and objects of the class JMenuItem are so similar. Some of the methods for the class AbstractButton are listed below. All these methods are inherited by both the class JButton and the class JMenuItem. (Some of these methods were inherited by the class AbstractButton from the class JComponent, so you may sometimes see some of the methods listed as "inherited from JComponent.")

Some Methods in the Class AbstractButton

The abstract class AbstractButton is in the javax.swing package. All of these methods are inherited by both of the classes JButton and JMenuItem.

```
public void setBackground(Color theColor)
```
sets the background color of this component.
```
public void addActionListener(ActionListener listener)
```
adds an ActionListener.
```
public void removeActionListener(ActionListener listener)
```
removes an ActionListener.
```
public void setActionCommand(String actionCommand)
```
sets the action command.
```
public String getActionCommand()
```
returns the action command for this component.
```
public void setText(String text)
```
makes text the only text on this component.
```
public String getText()
```
returns the text written on the component, such as the text on a button or the string for a menu item.
```
public void setPreferredSize(Dimension preferredSize)
```
sets the preferred size of the button or label. Note that this is only a suggestion to the layout manager. The layout manager is not required to use the preferred size. The following special case will work for most simple situations. The int values give the width and height in pixels.
```
public void setPreferredSize(new Dimension(int width, int height))
public void setMaximumSize(Dimension maximumSize)
```
sets the maximum size of the button or label. Note that this is only a suggestion to the layout manager. The layout manager is not required to respect this maximum size. The following special cases will work for most simple situations. The int values give the width and height in pixels.
```
public void setMaximumSize(new Dimension(int width, int height))
```

```
public void setMinimumSize(Dimension minimumSize)
```
sets the minimum size of the button or label. Note that this is only a suggestion to the layout manager.

The layout manager is not required to respect this minimum size.

Although we do not discuss the Dimension class, the following special case is intuitively clear and will work for most simple situations. The int values give the width and height in pixels.

```
public void setMinimumSize(
      new Dimension(int width, int height))
```

16.5.3 The setActionCommand Method

When the user clicks a button or menu item, which fires an action event that normally goes to one or more action listeners where it becomes an argument to an actionPerformed method. This action event includes a String instance variable that is known as the action command for the button or menu item and that is retrieved with the accessor method getActionCommand. The action command in the event is copied from an instance variable in the button or menu item object. If you do nothing to change it, the action command is the string written on the button or the menu item. The method setActionCommand for the class AbstractButton can be used with any JButton or JMenuItem to change the action command for that component. Among other things, this will allow you to have different action commands for two buttons, two menu items, or a button and menu item even though they have the same string written on them.

The method setActionCommand takes a String argument that becomes the new action command for the calling button or menu item. For example, consider the following code:

```
JButton nextButton = new JButton("Next");
nextButton.setActionCommand("Next Button");
JMenuItem chooseNext = new JMenuItem("Next");
chooseNext.setActionCommand("Next Menu Item");
```

If we had not used setActionCommand in the preceding code, then the button nextButton and the menu item chooseNext would both have the action command "Next" and so we would have no way to tell which of the two components nextButton and chooseNext an action event "Next" came from. However, using the method setActionCommand, we can give them the different action commands "Next Button" and "Next Menu Item".

> 如果在前面的代码中没有使用 setActionCommand，那么按钮 nextButton 和菜单项 chooseNext 都有"下一步"动作命令，所以没有任何办法判断动作"下一步"来自组件 nextButton 和 chooseNext 中的哪一个。然而，使用方法 setActionCommand，可以为它们指定不同的动作命令"下一步按钮"和"下一个菜单项"。

The action command for a JButton or JMenuItem is kept as the value of a private instance variable for the JButton or JMenuItem. The method setActionCommand is simply an ordinary mutator method that changes the value of this instance variable.

> JButton 或 JMenuItem 对象的动作命令是作为 JButton 或 JMenuItem 对象的私有实例变量的值保存的。方法 setActionCommand 仅仅是一个普通的 mutator 方法，用来改变这个实例变量的值。

An alternate approach to defining action listeners is given in the next subsection. That technique is, among other things, another way to deal with multiple buttons or menu items that have the same thing written on them.

16.5.4 Listeners as Inner Classes

In all of our previous examples, our GUIs had only one action listener object to deal with all action events from all buttons and menus in the GUI. The opposite extreme also has much to recommend it. You can have a separate ActionListener class for each button or menu item, so that each button or menu item has its own unique action listener. There is then no need for a multiway if-else statement. The listener knows which button or menu item was clicked because it listens to only one button or menu item.

The approach outlined in the previous paragraph does have one down side: You typically need to give a lot of definitions of ActionListener classes. Rather than putting each of these classes in a separate file, it is much cleaner to make them private inner classes. This has the added advantage of allowing the ActionListener classes to have access to private instance variables and methods of the outer class.

In Example 16.7 we have redone the GUI in Example 16.6 using the techniques of this subsection.

Example 16.7

```java
//Listeners as Inner Classes 内部类作为监听器
import java.awt.Color;
import java.awt.GridLayout;
import java.awt.event.ActionEvent;
import java.awt.event.ActionListener;
import javax.swing.JFrame;
import javax.swing.JMenu;
import javax.swing.JMenuBar;
import javax.swing.JMenuItem;
import javax.swing.JPanel;
public class InnerListenersDemo extends JFrame
{
    private static final long serialVersionUID = -5317693210496226906L;
    public static final int WIDTH = 300;
    public static final int HEIGHT = 200;
    private JPanel greenPanel;
    private JPanel whitePanel;
    private JPanel grayPanel;
    private class greenListener implements ActionListener
    {
        public void actionPerformed(ActionEvent e)
        {
            greenPanel.setBackground(Color.GREEN);
        }
    } //End of greenListener inner class
    private class WhiteListener implements ActionListener
    {
        public void actionPerformed(ActionEvent e)
```

```java
            {
                whitePanel.setBackground(Color.WHITE);
            }
        } //End of WhiteListener inner class
        private class grayListener implements ActionListener
        {
            public void actionPerformed(ActionEvent e)
            {
                grayPanel.setBackground(Color.GRAY);
            }
        } //End of grayListener inner class
        public static void main(String[] args)
        {
            InnerListenersDemo gui = new InnerListenersDemo();
            gui.setVisible(true);
        }
        public InnerListenersDemo()
        {
            super("Menu Demonstration");
            setSize(WIDTH, HEIGHT);
            setDefaultCloseOperation(JFrame.EXIT_ON_CLOSE);
            setLayout(new GridLayout(1, 3));
            greenPanel = new JPanel();
            greenPanel.setBackground(Color.LIGHT_GRAY);
            add(greenPanel);
            whitePanel = new JPanel();
            whitePanel.setBackground(Color.LIGHT_GRAY);
            add(whitePanel);
            grayPanel = new JPanel();
            grayPanel.setBackground(Color.LIGHT_GRAY);
            add(grayPanel);
            JMenu colorMenu = new JMenu("Add Colors");
            JMenuItem greenChoice = new JMenuItem("Green");
            greenChoice.addActionListener(new greenListener());
            colorMenu.add(greenChoice);
            JMenuItem whiteChoice = new JMenuItem("White");
            whiteChoice.addActionListener(new WhiteListener());
            colorMenu.add(whiteChoice);
            JMenuItem grayChoice = new JMenuItem("Gray");
            grayChoice.addActionListener(new grayListener());
            colorMenu.add(grayChoice);
            JMenuBar bar = new JMenuBar();
            bar.add(colorMenu);
            setJMenuBar(bar);
        }
    }
```

16.6 Text Fields and Text Areas

You have undoubtedly interacted with windowing systems that provide spaces for you to enter text information such as your name, address, and credit card number. In this section, we show you how to add these fields for text input and text output to your Swing GUIs.

> 毫无疑问，在互动的窗口系统，提供空间为你输入文字信息，如姓名、地址、信用卡号码。本节将展示如何增加这些领域的文字输入和文本输出到 Swing 的图形用户界面。

16.6.1 Text Areas and Text Fields

A text field is an object of the class JTextField and is displayed as a field that allows the user to enter a single line of text. The following creates a text field named name in which the user will be asked to enter his or her name:

```
private JTextField name;
...
name = new JTextField(NUMBER_OF_CHAR);
```

The variable name is a private instance variable. The creation of the JTextField in the last of the previous lines takes place inside the class constructor. The number NUMBER_OF_CHAR that is given as an argument to the JTextField constructor specifies that the text field will have room for at least NUMBER_OF_CHAR characters to be visible. The defined constant NUMBER_OF_CHAR is 30, so the text field is guaranteed to have room for at least 30 characters. You can type any number of characters into a text field but only a limited number will be visible; in this case, you know that at least 30 characters will be visible.

A Swing GUI can read the text in a text field and so receive text input, and if that is desired, can produce output by causing text to appear in the text field. The method getText returns the text written in the text field. For example, the following will set a variable named inputString to whatever string is in the text field name at the time that the getText method is invoked:

```
String inputString = name.getText();
```

Both JTextField and JTextArea are derived classes of the abstract class JTextComponent. Most of the methods for JTextField and JTextArea are inherited from JTextComponent and so JTextField and JTextArea have mostly the same methods with the same meanings except for minor redefinitions to account for having just one line or multiple lines.Below code describes some methods in the class JTextComponent. All of these methods are inherited and have the described meaning in both JTextField and JTextArea.

> JTextField 和 JTextArea 都是抽象类 JTextComponent 的派生类。JTextField 和 JTextArea 中的方法大多是从 JTextComponent 继承，因此在 JTextField 和 JTextArea 类中，大多数方法具有相同含义，除了少数用来计算单行字符串还是多行字符串的重载方法外。下面代码清单给出了类 JTextComponent 的一些方法。在 JTextField 和 JTextArea 中所有这些方法都是继承自父类，并具有父类中给定的含义。

You can set the line-wrapping policy for a JTextArea using the method setLineWrap. The method takes one argument of type boolean. If the argument is true, then at the end of a line, any additional characters for that line will appear on the following line of the text area. If the argument is false, the extra characters will be on the same line and will not be visible. For example, the following sets the line wrap policy for the JTextArea object named theText so that at the end of a line, any additional characters for that line will appear on the following line:

```
theText.setLineWrap(true);
```

You can specify that a JTextField or JTextArea cannot be written in by the user. To do so, use the method setEditable, which is a method in both the JTextField and JTextArea classes. If theText names an object in either of the classes JTextField or JTextArea, then the following

```
theText.setEditable(false);
```

will set theText so that only your GUI program can change the text in the text component theText; the user cannot change the text. After this invocation of setEditable, if the user clicks the mouse in the text component named theText and then types at the keyboard, the text in the text component will not change.

To reverse things and make theText so that the user can change the text in the text component, use true in place of false, as follows:

```
theText.setEditable(true);
```

If no invocation of setEditable is made, then the default state allows the user to change the text in the text component.

16.6.2 Window Listeners

In Paragraph before we used the method setDefaultCloseOperation to program the closewindow button in a JFrame. This allows for only a limited number of possibilities for what happens when the close-window button is clicked. When the user clicks the close-window button (or either of the two accompanying buttons), the JFrame fires an event known as a window event. A JFrame can use the method setWindowListener to register a window listener to respond to such window events. A window listener can be programmed to respond to a window event in any way you wish. Window events are objects of the class WindowEvent. A window listener is any class that satisfies the WindowListener interface.

The method headings in the WindowListener interface are given below. If a class implements the WindowListener interface, it must have definitions for all seven of these method headings. If you do not need all of these methods, then you can define the ones you do not need to have empty bodies, like this:

```
public void windowDeiconified(WindowEvent e)
```

Methods in the WindowListener Interface

The WindowListener interface and the WindowEvent class are in the package java.awt.event.

```
public void windowOpened(WindowEvent e)
```
invoked when a window has been opened.

```
public void windowClosing(WindowEvent e)
```
invoked when a window is in the process of being closed. Clicking the close-window button causes an invocation of this method.

```
public void windowClosed(WindowEvent e)
```
invoked when a window has been closed.

```
public void windowIconified(WindowEvent e)
```
invoked when a window is iconified. When you click the minimize button in a JFrame , it is iconified.

```
public void windowDeiconified(WindowEvent e)
```

invoked when a window is deiconified. When you activate a minimized window, it is deiconified.

```
public void windowActivated(WindowEvent e)
```
invoked when a window is activated. When you click in a window, it becomes the activated window. Other actions can also activate a window.

```
public void windowDeactivated(WindowEvent e)
```
invoked when a window is deactivated. When a window is activated, all other windows are deactivated.

Other actions can also deactivate a window.

16.7 Sample Examples

Example 16.8

```java
//创建用户登录窗口界面
import java.awt.*;
public class LoginUI
{
    public static void main(String arg[]){
        Frame f = new Frame("User Login");
        f.setSize(280,150);
        f.setBackground(Color.lightGray);
        f.setLocation(300,240);
        f.setLayout(new FlowLayout());
        Label t1 = new Label("Userid");
        TextField tf1 = new TextField("user1",20);
        Label t2 = new Label("password");
        TextField tf2 = new TextField(20);
        Button b1 = new Button("OK");
        Button b2 = new Button("Cancel");
        f.add(t1);
        f.add(tf1);
        f.add(t2);
        f.add(tf2);
        f.add(b1);
        f.add(b2);
        f.setVisible(true);
    }
}
```

Example 16.9

```java
//文本输入控件JTextField和文本域输入控件JTextArea应用演示
import javax.swing.JFrame;
import javax.swing.JTextField;
import javax.swing.JPanel;
import javax.swing.JLabel;
import javax.swing.JButton;
import java.awt.GridLayout;
import java.awt.BorderLayout;
import java.awt.FlowLayout;
import java.awt.Color;
```

```java
import java.awt.event.ActionListener;
import java.awt.event.ActionEvent;
public class TextFieldDemo extends JFrame implements ActionListener
{
    private static final long serialVersionUID=1639612630726455577L;
    public static final int WIDTH = 400;
    public static final int HEIGHT = 200;
    public static final int NUMBER_OF_CHAR = 30;
    private JTextField name;
    public static void main(String[] args)
    {
        TextFieldDemo gui = new TextFieldDemo();
        gui.setVisible(true);
    }
    public TextFieldDemo()
    {
        super("Text Field Demo");
        setSize(WIDTH, HEIGHT);
        setDefaultCloseOperation(JFrame.EXIT_ON_CLOSE);
        setLayout(new GridLayout(2, 1));
        JPanel namePanel = new JPanel();
        namePanel.setLayout(new BorderLayout());
        namePanel.setBackground(Color.WHITE);
        name = new JTextField(NUMBER_OF_CHAR);
        namePanel.add(name, BorderLayout.SOUTH);
        JLabel nameLabel = new JLabel("Enter your name here:");
        namePanel.add(nameLabel, BorderLayout.CENTER);
        add(namePanel);
        JPanel buttonPanel = new JPanel();
        buttonPanel.setLayout(new FlowLayout());
        buttonPanel.setBackground(Color.GREEN);
        JButton actionButton = new JButton("Click me");
        actionButton.addActionListener(this);
        buttonPanel.add(actionButton);
        JButton clearButton = new JButton("Clear");
        clearButton.addActionListener(this);
        buttonPanel.add(clearButton);
        add(buttonPanel);
    }
    public void actionPerformed(ActionEvent e)
    {
        String actionCommand = e.getActionCommand();
        if (actionCommand.equals("Click me"))
            name.setText("Hello " + name.getText());
        else if (actionCommand.equals("Clear"))
            name.setText("");
        else
            name.setText("Unexpected error.");
    }
}
```

16.8 Exercise for you

1. Create a frame window and use GridLayout manager to arranges four components.
2. Create a shopping cart for ice creams of 5 different flavors and calculate the total price of them.
3. How you can create one window on which you can draw lines and images on by the help of your mouse?

Chapter 17 Programming with JDBC

Database programming has traditionally been a difficult job. You are faced with dozens of available database products, and each one talks to your applications in its own private language. If your application needs to talk to a new database engine, you have to teach it (and yourself) a new language. As Java programmers, however, you should not worry about such translation issues. Java is supposed to bring you the ability to "write once, compile once, and run anywhere," so it should bring it to you with database programming, as well.

> 数据库编程历来是一件困难的工作。面对数十个可用的数据库产品，而且每一个数据库产品都用自己的语言和应用程序交流。如果应用程序需要连接一个新的数据库引擎，需要教它（和你自己）一种新的数据库语言。然而，作为 Java 程序员，不需要担心这样的问题。Java 的 "编写一次，一次编译，随处运行" 的能力，同样也适用数据库编程。

SQL was a key first step in simplifying database access. Java's JDBC API builds on that foundation and provides you with a shared language through which your applications can talk to database engines. Following in the tradition of its other multi-platform APIs, such as the AWT, JDBC provides you with a set of interfaces that create a common point at which database applications and database engines can meet. This chapter will discuss the basic interfaces that JDBC provides.

17.1 JDBC Introduction

Working with leaders in the database field, Sun developed a single API for database access—JDBC. As part of this process, they kept three main goals in mind:
- JDBC should be a SQL-level API.
- JDBC should capitalize on the experience of existing database APIs.
- JDBC should be simple.

> - JDBC 应该是一个 SQL 级别的 API。
> - JDBC 应该利用现有数据库 API 的经验。
> - JDBC 应该简单。

A SQL-level API means that JDBC allows you to construct SQL statements and embed them inside Java API calls. In short, you are basically using SQL. But JDBC lets you smoothly translate between the world of the database and the world of the Java application. Your results from the database, for instance, are returned as Java objects, and access problems get thrown as exceptions. Later in the book, you will go a step further and talk about how you can completely hide the existence of the database from a Java application using a database class library.

> 一个SQL级别的API意味着JDBC允许你构建SQL语句,并且将它们嵌入到Java API调用中。简单说就是使用SQL。JDBC让你顺畅地在数据库和Java应用程序之间传递数据。比如,你从数据库获取的结果,是作为Java对象返回的,同时访问时候出现的问题,是作为异常抛出。

Because of the confusion caused by the proliferation of proprietary database access APIs, the idea of a universal database access API to solve this problem is not new. In fact, Sun drew upon the successful aspects of one such API, Open DataBase Connectivity (ODBC). ODBC was developed to create a single standard for database access in the Windows environment. Although the industry has accepted ODBC as the primary means of talking to databases in Windows, it does not translate well into the Java world. First of all, ODBC is a C API that requires intermediate APIs for other languages. But even for C developers, ODBC has suffered from an overly complex design that has made its transition outside of the controlled Windows environment a failure. ODBC's complexity arises from the fact that complex, uncommon tasks are wrapped up in the API with its simpler and more common functionality. In other words, in order for you to understand a little of ODBC, you have to understand a lot.

> ODBC是被开发出来创建一个单一的访问Windows下数据库的标准。虽然业界已经接受ODBC作为Windows下访问数据库的主要手段,但是在Java下工作的并不理想。首先,ODBC是一个C类型的接口,其他语言使用的时候要求有中间的衔接API。因此ODBC在Java开发中可以使用,但是并不是很理想的,最理想的是通过JDBC访问数据库。

In addition to ODBC, JDBC is heavily influenced by existing database programming APIs, such as X/OPEN SQL Call Level Interface (CLI). Sun wanted to reuse the key abstractions from these APIs, which would ease acceptance by database vendors and capitalize on the existing knowledge capital of ODBC and SQL CLI developers. In addition, Sun realized that deriving an API from existing ones can provide quick development of solutions for database engines that support the old protocols. Specifically, Sun worked in parallel with Intersolv to create an ODBC bridge that maps JDBC calls to ODBC calls, thus giving Java applications access to any database management system (DBMS) that supports ODBC.

JDBC attempts to remain as simple as possible while providing developers with maximum flexibility. A key criterion employed by Sun is simply asking whether database access applications read well. The simple and common tasks use simple interfaces, while more uncommon or bizarre tasks are enabled through specialized interfaces.

> JDBC试图保持尽可能简单,同时提供最大的灵活性。Sun应用的一个关键标准是数据库访问应用程序的可读性是否好。简单和通用的任务使用简单的接口,而不常见或离奇的任务通过专门的接口对数据库进行访问。

17.2 Connecting to the Database

Now I am going to dive into the details about JDBC calls and how to use them. The examples in this book should run on your system regardless of the database or driver you use. The one phase when it is hard to achieve portability is the first step of connecting, because you have to specify a driver. I'll discuss that first to get it out of the way.

When you write a Java database applet or application, the only driver-specific information JDBC requires from you is the database URL. You can even have your application derive the URL at runtime—based on user input or applet parameters.

> 当编写一个 Java 数据库 applet 或应用程序时，JDBC 需要的唯一的驱动程序特定信息是数据库 URL。甚至可以让应用程序在运行时基于用户输入或从 applet 参数中获得数据库 URL。

Using the database URL and whatever properties your JDBC driver requires (generally a user ID and password), your application will first request a java.sql.Connection implementation from the DriverManager. The DriverManager in turn will search through all of the known java.sql.Driver implementations for the one that connects with the URL you provided. If it exhausts all the implementations without finding a match, it throws an exception back to your application.

Once a Driver recognizes your URL, it creates a database connection using the properties you specified. It then provides the DriverManager with a java.sql.Connection implementation representing that database connection. The DriverManager then passes that Connection object back to the application. In your code, the entire database connection process is handled by this oneliner:

```
Connection con = DriverManager.getConnection(url, uid, password);
```

Of course, you are probably wondering how the JDBC DriverManager learns about a new driver implementation. The DriverManager actually keeps a list of classes that implement the java.sql.Driver interface. Somehow, somewhere, something needs to register the Driver implementations for any potential database drivers it might require with the DriverManager. JDBC requires a Driver class to register itself with the DriverManager when it is instantiated. The act of instantiating a Driver class thus enters it in the DriverManager's list. Instantiating the driver, however, is only one of several ways to register a driver:

17.2.1 A Simple Database Connection

Example 17. 1 puts the process of connecting to the database into a more concrete format.

Example 17. 1

```java
//简单的访问数据库的示例
import java.sql.Connection;
import java.sql.DriverManager;
import java.sql.SQLException;
public class SimpleConnection {
  static public void main(String args[]) {
      Connection connection = null;
      // Process the command line
      if( args.length != 4 ) {
          System.out.println("Syntax: java SimpleConnection " +
                  "DRIVER URL UID PASSWORD");
          return;
      }
      try { // 加载数据库驱动程序
```

```
        Class.forName(args[0]).newInstance( );
    }
    catch( Exception e ) {
        e.printStackTrace( );
        return;
    }
    try {
        connection = DriverManager.getConnection(args[1], args[2], args[3]);
        System.out.println("Connection successful!");
        // 可以在此处添加查询或更新数据的代码
    }
    catch( SQLException e ) {
        e.printStackTrace( );
    }
    finally {
     if( connection != null ) {
       try { connection.close( ); }
       catch( SQLException e ) {
         e.printStackTrace( );
       }
     }
    }
   }
  }
```

In connecting to the database, this example catches a SQLException. This is a sort of catch-all exception for database errors. Just about any time something goes wrong between JDBC and the database, JDBC throws a SQLException. In addition to the information you commonly find in Java exceptions, SQLException provides database-specific error information, such as the SQLState value and vendor error code. In the event of multiple errors, the JDBC driver "chains" the exceptions together. In other words, you can ask any SQLException if another exception preceded it by calling getNextException( ).

> 在连接数据库的时候，本例子捕捉了一个 SQLException。该类型的 catch-all 异常捕获机制用来捕获数据库错误。在任何时候任何错误发生在 JDBC 和数据库之间的时候，JDBC 将抛出一个 SQLException 错误。除了提供通常在 Java 中碰到的普通异常信息，SQLException 还提供了数据库特定的错误信息，比如 SQLState 值和数据库生产商定义的错误代码。

17.2.2 The JDBC Classes for Creating a Connection

As Example 17.1 illustrates, JDBC uses one class (java.sql.DriverManager) and two interfaces (java.sql.Driver and java.sql.Connection) for connecting to a database:

> 就像 Example 17.1 说明的那样，JDBC 使用一个类(java.sql.DriverManager)和两个接口(java.sql.Driver and java.sql.Connection)来连接数据库：

java.sql.Driver

Unless you are writing your own custom JDBC implementation, you should never have to deal with this class from your application. It simply gives JDBC a launching point for database connectivity by responding to DriverManager connection requests and providing information about the implementation in question.

java.sql.DriverManager

Unlike most other parts of JDBC, DriverManager is a class instead of an interface. Its main responsibility is to maintain a list of Driver implementations and present an application with one that matches a requested URL. The DriverManager provides registerDriver() and deregisterDriver( ) methods, which allow a Driver implementation to register itself with the DriverManager or remove itself from that list. You can get an enumeration of registered drivers through the getDrivers() method.

java.sql.Connection

The Connection class represents a single logical database connection. In other words, you use the Connection class for sending a series of SQL statements to the database and managing the committing or aborting of those statements.

17.3 Basic Database Access

Now that you are connected to the database, you can begin making updates and queries. The most basic kind of database access involves writing JDBC code when you know ahead of time whether the statements you are sending are updates (INSERT, UPDATE, or DELETE) or queries (SELECT). In the next paragraph, you will discuss more complex database access that allows you to execute statements of unknown types.

> 既然已经连接到数据库，就可以开始进行数据更新和查询操作了。当提前知道你所发送的语句是更新(INSERT，UPDATE 或 DELETE)还是查询(SELECT)，就能够编写最基本的数据库访问 JDBC 代码。下节将讨论更复杂的数据库访问，以便可以执行未知类型的数据库语句。

Basic database access starts with the Connection object you created in the previous section. When this object first gets created, it is simply a direct link to the database. You use a Connection object to generate implementations of java.sql.Statement tied to the same database transaction. After you have used one or more Statement objects generated by your Connection, you can use it to commit or rollback the Statement objects associated with that Connection.

A Statement is very much what its name implies—a SQL statement. Once you get a Statement object from a Connection, you have what amounts to a blank check that you can write against the transaction represented by that Connection. You do not actually assign SQL to the Statement until you are ready to send the SQL to the database.

This is when it becomes important to know what type of SQL you are sending to the database, because JDBC uses a different method for sending queries than for sending updates. The key difference is the fact that the method for queries returns an instance of java.sql.ResultSet, while the method for nonqueries returns an integer. A ResultSet provides you with access to the data retrieved by a query.

17.3.1 Basic JDBC Database Access Classes

JDBC's most fundamental classes are the Connection, the Statement, and the ResultSet. You will use them everytime you write JDBC code.

> JDBC 最基础的类是 Connection、Statement 和 ResultSet。当编写 JDBC 代码时，需要用到它们。

java.sql.Statement

The Statement class is the most basic of three JDBC classes representing SQL statements. It performs all of the basic SQL statements the book has discussed so far. In general, a simple database transaction uses only one of the three statement execution methods in the Statement class. The first such method, executeQuery( ), takes a SQL String as an argument and returns a ResultSet object. This method should be used for any SQL calls that expect to return data from the database. Update statements, on the other hand, are executed using the executeUpdate( ) method. This method returns the number of affected rows.

> Statement 类是 JDBC 的三个最基础的类中的一个，用来表示 SQL 语句。它可以执行到目前为止本书上所讨论过的所有基础的 SQL 语句。一般，一个简单的数据库交易通常仅用 Statement 类中的三种语句执行方法中的一个。首先，比如方法 executeQuery( )，需要一个 SQL 字符串作为参数并且返回一个 ResultSet 对象。该方法适用于那些需要从数据库返回数据的任何的 SQL 调用。

Finally, the Statement class provides an execute( ) method for situations in which you do not know whether the SQL being executed is a query or update. This usually happens when the application is executing dynamically created SQL statements. If the statement returns a row from the database, the method returns true. Otherwise it returns false. The application can then use the getResultSet( ) method to get the returned row.

> Statement 类提供 execute( )方法来执行 SQL 语句，这些 SQL 语句你事先不知道是查询还是更新。这些 SQL 语句是程序在运行过程中动态生成的。如果 statement 从数据库返回一条记录，execute( )方法返回 true，否则返回 false。应用程序可以通过使用 getResultSet( )方法获取返回的记录。

java.sql.ResultSet

A ResultSet is one or more rows of data returned by a database query. The class simply provides a series of methods for retrieving columns from the results of a database query. The methods for getting a column all take the form:

    type gettype(int | String)

in which the argument represents either the column number or column name desired. A nice side effect of this design is that you can store values in the database as one type and retrieve them as a completely different type. For example, if you need a Date from the database as a String, you can get it as a String by calling result_set.getString(1) instead of result_set.getDate(1).

Because the ResultSet class handles only a single row from the database at any given time, the class provides the next( ) method for making it reference the next row of a result set. If next() returns true, you have another row to process and any subsequent calls you make to the ResultSet object will be in reference to that next row. If there are no rows left, it returns false.

17.3.2　SQL NULL versus Java null

SQL and Java have a serious mismatch in handling null values. Specifically, using methods like getInt(), a Java ResultSet has no way of representing a SQL NULL value for any numeric SQL column. After retrieving a value from a ResultSet, it is therefore necessary to ask the ResultSet if the retrieved value represents a SQL NULL. For Java object types, a SQL NULL will often map to Java null. To avoid running into database oddities, however, it is recommended that you always check for SQL NULL.

> SQL 和 Java 在处理空值时严重不匹配。具体来说，使用 getInt( )方法，一个 Java 的结果集中没有任何数值 SQL 列来代表一个 SQL NULL 值的方式。从 ResultSet 中检索一个值后，有必要问结果集检索到的值是否代表一个 SQL NULL。对于 Java 对象类型，SQL NULL 往往会映射到 Java 的 null 值。为了避免数据库运行时发生一些奇怪的异常，建议在数据库编程时检查 SQL 的结果集是否有 NULL 值。

Checking for SQL NULL involves a single call to the wasNull( ) method in your ResultSet after you retrieve a value. The wasNull( ) method will return true if the last value read by a call to a getXXX( ) method was a SQL NULL. If, for example, your database allowed NULL values for PET_COUNT column because you do not know the number of pets of all your customers, a call to getInt( ) could return some driver attempt at representing NULL, most likely 0. So how do you know in Java who has pets and who has an unknown number of pets? A call to wasNull( ) will tell you if represents an actual in the database or a NULL value in the database.

17.3.3　Clean Up

In the examples provided so far, you may have noticed many objects being closed through a close( ) method. The Connection, Statement, and ResultSet classes all have close( ). A given JDBC implementation may or may not require you to close these objects before reuse. But some might require it, since they are likely to hold precious database resources. It is therefore always a good idea to close any instance of these objects when you are done with them. It is useful to remember that closing a Connection implicitly closes all Statement instances associated with the Connection. Similarly, closing a Statement implicitly closes ResultSet instances associated with it. If you do manage to close a Connection before committing with auto-commit off, any uncommitted transactions will be lost.

17.3.4　Modifying the Database

Example 17.2 shows the simple Update class supplied with the mSQL-JDBC driver for mSQL.

> Example 17.2 给出了一个简单 Update 类，使用 mSQL-JDBC 驱动程序。

Example 17.2
```
import java.sql.*;
public class Update {
  public static void main(String args[]) {
    Connection con = null;
    if( args.length != 2 ) {
```

```
      System.out.println("Syntax: <java Update [number] [string]>");
      return;
    }
    try {
      String driver = "com.imaginary.sql.msql.MsqlDriver";
      Class.forName(driver).newInstance( );
      String url = "jdbc:msql://carthage.imaginary.com/ora";
      con = DriverManager.getConnection(url, "borg", "");
      Statement s = con.createStatement( );
      String test_id = args[0];
      String test_val = args[1];
      int update_count =
      s.executeUpdate("INSERT INTO test (test_id, test_val) " +
      "VALUES(" + test_id + ", '" + test_val + "')");
      System.out.println(update_count + " rows inserted.");
      s.close( );
    }
    catch( Exception e ) {
      e.printStackTrace( );
    }
    finally {
      if( con != null ) {
        try { con.close( ); }
        catch( SQLException e ) { e.printStackTrace( ); }
      }
    }
  }
}
```

Again, making a database call is nothing more than creating a Statement and passing it SQL via one of it's executing methods. Unlike executeQuery(), however, executeUpdate( ) does not return a ResultSet (you should not be expecting any results). Instead, it returns the number of rows affected by the UPDATE, INSERT, or DELETE.

By default, JDBC commits each SQL statement as it is sent to the database; this is called autocommit. However, for more robust error handling, you can set up a Connection object to issue a series of changes that have no effect on the database until you expressly send a commit. Each Connection is separate, and a commit on one has no effect on the statements on another. The Connection class provides the setAutoCommit( ) method so you can turn auto-commit off.

17.4 SQL Data Types and Java Data Types

Support for different data types in SQL2 is poor. Since Java is an object-oriented language, however, data types support is extremely rich. Therefore a huge disconnect exists between what sits in the database and the way you want it represented in your Java application. The SQL concept of a variable width, single-byte character array, for example, is the VARCHAR data type. Java actually has no concept of a variable width, single-byte character array; Java doesn't even have a single-byte character type. The closest thing is the String class.

> 在SQL2标准中对不同数据类型的支持较差。然而，由于Java是一种面向对象的语言，数据类型的支持极为丰富。因此在数据库和应用程序之间，在数据类型上存在巨大的不一致性。例如，变量宽度、单字节字符数组，在SQL中是varchar类型。Java实际上已经没有变量宽度、单字节字符数组的概念；Java甚至没有一个单字节的字符类型。最接近这种类型的是String类。

To make matters worse, many database engines internally support their own data types and loosely translate them to a SQL2 type. All Oracle numeric types, for example, map to the SQL NUMERIC type. JDBC, fortunately, lets you retrieve data in their Java forms defined by a JDBC-specified data type mapping. You do not need to worry that a SQL LONG has a different representation in Sybase than it does in Oracle. You just call the ResultSet getLong( ) method to retrieve numbers you wish to treat as Java longs.

You do need to be somewhat concerned when designing the database, however. If you pull a 64-bit number into a Java application via getInt( ), you risk getting bad data. Similarly, if you save a Java float into a numeric field with a scale of 0, you will lose data. The important rule of thumb for Java programming, however, is that think and work in Java and use the database to support the Java application. Do not let the database drive Java. Table 17.1 shows the JDBC prescribed SQL to Java data type mappings.

These mappings are simply the JDBC specification for direct type mappings and not a law prescribing the format you must use in Java for your SQL data. In other words, you can retrieve an INTEGER column into Java as a long or put a Java Date object in a TIMESTAMP field. Some conversions are, nevertheless, nonsensical. You cannot save a Java boolean into a database DATE field.

Table 17.1   JDBC Specification SQL to Java Datatype Mappings

SQL Type (from java.sql.Types)	Java Type
BIT	boolean
TINYINT	byte
SMALLINT	short
INTEGER	int
BIGINT	long
REAL	float
FLOAT	double
DOUBLE	double
DECIMAL	java.math.BigDecimal
NUMERIC	java.math.BigDecimal
CHAR	java.lang.String
VARCHAR	java.lang.String
LONGVARCHAR	java.lang.String
DATE	java.sql.Date
TIME	java.sql.Time
TIMESTAMP	java.sql.Timestamp
BINARY	byte[ ]
VARBINARY	byte[ ]
LONGVARBINARY	byte[ ]
BLOB	java.sql.Blob
CLOB	java.sql.Clob
ARRAY	java.sql.Array
REF	java.sql.Ref
STRUCT	java.sql.Struct

17.5 Scrollable Result Sets

The single most visible addition to the JDBC API in its 2.0 specification is support for scrollable result sets.

> JDBC API 2.0 规范中增加的一个最重要的特性是对滚动结果集的支持。

17.5.1 Result Set Types

Using scrollable result sets starts with the way in which you create statements. Earlier in the chapter, you learned to create a statement using the createStatement( ) method. The Connection class actually has two versions of createStatement( )—the zero parameter version you have used so far and a two parameter version that supports the creation of Statement instances that generate scrollable ResultSet objects. The default call translates to the following call:

conn.createStatement(ResultSet.TYPE_FORWARD_ONLY, ResultSet.CONCUR_READ_ONLY);

The first argument is the result set type. The value ResultSet.TYPE_FORWARD_ONLY indicates that any ResultSet generated by the Statement returned from createStatement( ) only moves forward (the JDBC 1.x behavior). The second argument is the result set concurrency. The value ResultSet. CONCUR_READ_ONLY specifies that each row from a ResultSet is read-only. As you will see in the next paragraph, rows from a ResultSet can be modified in place if the concurrency specified in the createStatement( ) call allows it.

JDBC defines three types of result sets: TYPE_FORWARD_ONLY, PE_SCROLL_ SENSITIVE, and TYPE_SCROLL_INSENSITIVE. TYPE_FORWARD_ONLY is the only type that is not scrollable. The other two types are distinguished by how they reflect changes made to them. A TYPE_SCROLL_INSENSITIVE ResultSet is unaware of in-place edits made to modifiable instances. TYPE_SCROLL_SENSITIVE, on the other hand, means that you can see changes made to the results if you scroll back to the modified row at a later time. You should keep in mind that this distinction remains only while you leave the result set open. If you close a TYPE_SCROLL_ INSENSITIVE ResultSet and then requery, your new ResultSet reflects any changes made to the original.

17.5.2 Result Set Navigation

When ResultSet is first created, it is considered to be positioned before the first row. Positioning methods such as next() point a ResultSet to actual rows. Your first call to next( ), for example, positions the cursor on the first row. Subsequent calls to next( ) move the ResultSet ahead one row at a time. With a scrollable ResultSet, however, a call to next( ) is not the only way to position a result set.

> 当 ResultSet 对象第一次被创建，它的游标指向第一条记录之前的位置。使用 next( )方法可以使游标指向一条实际的记录。例如，当第一次调用 next( )方法，游标就指向了结果集的第一条记录。然后再使用 next()方法，游标将指向下一条记录。如果是可滚动结果集，除了 next( )方法外，还有其他访问结果集的方法。

The method previous( ) works in an almost identical fashion to next( ). While next( ) moves one row forward, previous( ) moves one row backward. If it moves back beyond the first row, it returns false. Otherwise, it returns true. Because a ResultSet is initially positioned before the first row, you need to move the ResultSet using some other method before you can call previous( ).

17.6 Sample Examples

Example 17.3

```
    //一个简单的数据库查询应用程序
import java.sql.*;
public class Select {
  public static void main(String args[]) {
    String url = "jdbc:msql://carthage.imaginary.com/ora";
    Connection con = null;
    try {
      String driver = "com.imaginary.sql.msql.MsqlDriver";
      Class.forName(driver).newInstance( );
    }
    catch( Exception e ) {
      System.out.println("Failed to load mSQL driver.");
      return;
    }
    try {
      con = DriverManager.getConnection(url, "borg", "");
      Statement select = con.createStatement( );
      ResultSet result = select.executeQuery
        ("SELECT test_id, test_val FROM test");
      System.out.println("Got results:");
      while(result.next( )) {
        //逐条处理结果集中记录
        int key;
        String val;
        key = result.getInt(1);
        if( result.wasNull( ) ) {
          key = -1;
        }
        val = result.getString(2);
        if( result.wasNull( ) ) {
          val = null;
        }
        System.out.println("key = " + key);
        System.out.println("val = " + val);
      }
    }
    catch( Exception e ) {
      e.printStackTrace( );
    }
    finally {
      if( con != null ) {
        try { con.close( ); }
```

```java
      catch( Exception e ) { e.printStackTrace( ); }
    }
  }
 }
}
```

Example 17.4

```java
//数据库数据更新示例
import java.sql.*;
public class UpdateLogic {
  public static void main(String args[]) {
    Connection con = null;
    if( args.length != 2 ) {
      System.out.println("Syntax: <java UpdateLogic [number] [string]>");
      return;
    }
    try {
      String driver = "com.imaginary.sql.msql.MsqlDriver";
      Class.forName(driver).newInstance( );
      String url = "jdbc:msql://carthage.imaginary.com/ora";
      Statement s;
      con = DriverManager.getConnection(url, "borg", "");
      con.setAutoCommit(false);
      s = con.createStatement( );
      s.executeUpdate("INSERT INTO test (test_id, test_val) " +
         "VALUES(" + args[0] + ", '" + args[1] + "')");
      s.close( );
      s = con.createStatement( );
      s.executeUpdate("INSERT into test_desc (test_id, test_desc) " +
         "VALUES(" + args[0] +", '" + args[1] + "')");
      con.commit( );  //提交数据库事务
      System.out.println("Insert succeeded.");
      s.close( );
    }
    catch( SQLException e ) {
      if( con != null ) {
        try { con.rollback( ); } // rollback on error
        catch( SQLException e2 ) { }
      }
      e.printStackTrace( );
    }
    catch (InstantiationException e) {
        e.printStackTrace();
    }
    catch (IllegalAccessException e) {
        e.printStackTrace();
    }
    catch (ClassNotFoundException e) {
        e.printStackTrace();
    }
```

```java
            finally {
                if( con != null ) {
                    try { con.close( ); }
                    catch( SQLException e ) { e.printStackTrace( ); }
                }
            }
        }
    }
```

Example 17.5

```java
//可滚动结果集应用示例
import java.sql.*;
import java.util.*;
public class ReverseSelect {
    public static void main(String argv[]) {
        Connection con = null;
        try {
            String url = "jdbc:msql://carthage.imaginary.com/ora";
            String driver = "com.imaginary.sql.msql.MsqlDriver";
            Properties p = new Properties( );
            Statement stmt;
            ResultSet rs;
            p.put("user", "borg");
            Class.forName(driver).newInstance( );
            con = DriverManager.getConnection(url, "borg", "");
            stmt =
            con.createStatement(ResultSet.TYPE_SCROLL_INSENSITIVE,
            ResultSet.CONCUR_READ_ONLY);
            rs = stmt.executeQuery("SELECT * from test ORDER BY test_id");
            //游标指向rs第一条记录之前
            System.out.println("Got results:");
            //游标指向rs最后一条记录之后
            rs.afterLast( );
            while(rs.previous( )) {
                int a;
                String str;
                a = rs.getInt("test_id");
                if( rs.wasNull( ) ) {
                    a = -1;
                }
                str = rs.getString("test_val");
                if( rs.wasNull( ) ) {
                    str = null;
                }
                System.out.print("\ttest_id= " + a);
                System.out.println("/str= '" + str + "'");
            }
            System.out.println("Done.");
        }
        catch( Exception e ) {
            e.printStackTrace( );
```

```
      }
      finally {
        if( con != null ) {
          try { con.close( ); }
          catch( SQLException e ) { e.printStackTrace( ); }
        }
      }
    }
  }
```

17.7 Exercise for you

1. How to connect to the Oracle DB、MS SQL Server、MySQL?
2. Write a program to insert a record into the actual DB which you create in Oracle DB Manager.
3. Write a program to get the record which you want from actual DB which you create.

反侵权盗版声明

电子工业出版社依法对本作品享有专有出版权。任何未经权利人书面许可，复制、销售或通过信息网络传播本作品的行为；歪曲、篡改、剽窃本作品的行为，均违反《中华人民共和国著作权法》，其行为人应承担相应的民事责任和行政责任，构成犯罪的，将被依法追究刑事责任。

为了维护市场秩序，保护权利人的合法权益，我社将依法查处和打击侵权盗版的单位和个人。欢迎社会各界人士积极举报侵权盗版行为，本社将奖励举报有功人员，并保证举报人的信息不被泄露。

举报电话：（010）88254396；（010）88258888
传　　真：（010）88254397
E-mail：　dbqq@phei.com.cn
通信地址：北京市万寿路 173 信箱
　　　　　电子工业出版社总编办公室
邮　　编：100036